Nuclear Physics in Retrospect

Proceedings of a Symposium on the 1930s

Nuclear Physics in Retrospect

Proceedings of a Symposium on the 1930s

Edited by
ROGER H. STUEWER

School of Physics and Astronomy
University of Minnesota

University of Minnesota Press, Minneapolis

Copyright © 1979 by the University of Minnesota.
All rights reserved.
Printed in the United States of America.
Published by the University of Minnesota Press,
2037 University Avenue Southeast, Minneapolis, Minnesota 55455,
and published in Canada by Burns & MacEachern
Limited, Don Mills, Ontario

Library of Congress Cataloging in Publication Data

Symposium on the History of Nuclear Physics, University
   of Minnesota, 1977.
   Nuclear physics in retrospect.

   Includes indexes.
   1. Nuclear physics — History — Congresses.
I. Stuewer, Roger H.   II. Title
QC773.S95 1977   539.7'09   78-8192
ISBN 0-8166-0869-5

The University of Minnesota
is an equal opportunity
educator and employer.

Dedicated to the memories
of Eugene Feenberg and Laura Fermi
whose recent deaths
brought home to everyone
the importance of
having held this Symposium.

Preface

This volume constitutes the proceedings of a Symposium on the History of Nuclear Physics held at the University of Minnesota in May 1977. The idea for the symposium originated in the spring of 1976 during discussions with my colleagues in the history of science and technology, Alan E. Shapiro, Edwin T. Layton, and Malcolm J. Kottler. It took shape through further extensive discussions with Morton Hamermesh, who offered invaluable advice and help, and who agreed to serve with me on its Organizing Committee. By the beginning of the summer we had assembled the balance of the Organizing Committee, Maurice Goldhaber (Brookhaven National Laboratory), Edwin M. McMillan (Berkeley), Alfred O. Nier (Minnesota), and Spencer R. Weart (American Institute of Physics Center for History of Physics). Professors Goldhaber and McMillan also agreed to deliver two of the principal lectures at the symposium, and Professor Nier agreed to serve as chairman of the opening session.

The primary objective of the symposium, as we formulated it, was to chart key developments in the history of nuclear physics, principally in the 1930s, by inviting eight distinguished nuclear physicists to lecture on and discuss these developments. Significant aspects of this history had been touched upon at two A.I.P. conferences in May 1967 and May 1969, but these conferences were exploratory in nature and entirely discussion oriented. Our goal, rather, was to structure the symposium around the eight principal lectures, asking each lecturer to base his treatment on his personal knowledge and experience, and to include, if possible, information from his personal correspondence or other unpublished materials. In this way we hoped to obtain genuine and revealing historical insights which might otherwise be lost to future generations.

With this plan in mind, invitations were sent out in the name of the Organizing Committee, and during the course of the summer and early fall Hans A. Bethe (Cornell), Otto R. Frisch (Cambridge), Sir Rudolf Peierls (Oxford and Washington), Emilio G. Segrè (Berkeley), John A. Wheeler (Princeton and Texas), and Eugene P. Wigner (Princeton) agreed to deliver the six additional principal lectures at the symposium. In succeeding months we also asked H. H. Barschall (Wisconsin), Herman Feshbach (M.I.T.), William A. Fowler (Caltech), Robert Serber (Columbia), and R. R. Wilson (Fermilab) to serve as session chairmen, and we extended invitations to over twenty-five other distinguished nuclear physicists, historians of physics, and graduate students in the history of physics to participate as discussants in the symposium. A copy of the 1967 and 1969 A.I.P. conference proceedings, *Exploring the History of Nuclear Physics* (New York: 1972), and a copy of a draft chronology prepared by the A.I.P. Center for History of Physics dated October 21, 1968, was distributed to help prepare everyone for the lectures and discussions.

The symposium was held May 18-21, 1977, in Mayo Memorial Auditorium on the campus of the University of Minnesota. All the lectures and discussions were videotaped using high-quality color equipment, and with financial support from the University of Minnesota, including direct support from the School of Physics and Astronomy, the Institute of Technology, and the central administration of the University of Minnesota. Special thanks are due to Peter Roll for his extensive efforts in arranging and coordinating this aspect of the symposium, as well as to several key individuals in the university's Department of Media Resources and at KTCA-TV. Travel support for the speakers and other participants, which is gratefully acknowledged, was provided by the University of Minnesota's Program in the History of Science and Technology, supported in part by a grant from the Northwest Area Foundation, the University of Minnesota's Small Grants Program, the 3M Company of St. Paul, the Sloan Foundation, and the National Science Foundation. Local arrangements were handled responsibly by Joe Kroll and Diane Campbell of the Nolte Center for Continuing Education. The symposium, finally, owes a large debt of gratitude to Henry Koffler, Vice President for Academic Affairs of the University of Minnesota, for his constant support and for committing part of his busy schedule to the symposium in opening it, welcoming everyone to it, and serving as master of ceremonies at the banquet on May 20.

Following the close of the symposium, the lectures and discussions were transcribed and edited for publication. Professor McMillan added an appendix to his paper consisting of personal correspondence he had in connection with his study of the early history of accelerators, and Professor Wheeler prepared an expanded version of his lecture for publication. Various discussants added remarks which appear in square brackets in the appropriate places.

No one present at the symposium will forget its stimulating lectures and discussions; the general atmosphere of high enthusiasm, friendliness, and goodwill; and the lively conversations outside the lecture hall, which opened up numerous new lines of communication among the physicists, senior historians, and graduate students in the history of physics present. On Thursday, May 19, a reception was held at the splendid Bakken Museum in Minneapolis, amid the music of a string trio and surrounded by part of the fine collection of electrical instruments assembled by Earl Bakken, chief executive officer at Medtronic, and supervised by Dennis Stillings, director of the Museum. Mr. Bakken and Mr. Stillings have everyone's sincere thanks for hosting this reception. On Friday, May 20, a banquet was held at the Marquette Inn in Minneapolis, where the symposium guests were staying. A program of after-dinner music was provided by the University of Minnesota Woodwind Ensemble, John E. Anderson, director. Otto R. Frisch entertained the group by reading his account of the opportunities and perils of a coal reactor. Henry Koffler, as master of ceremonies, thanked Maurine Bielawski for her dedicated secretarial work and others for their contributions to the symposium. At the close of the evening, Dr. Koffler warmed everyone's heart by presenting a bouquet of roses to Laura Fermi.

Roger H. Stuewer
December 1978

Contents

Conference Participants xiii

Welcome, by Alfred O. Nier, *University of Minnesota* 3

Introduction, by Henry Koffler, *University of Minnesota* 5

The Happy Thirties, by Hans A. Bethe, *Cornell University* 9
    Introduction, by Alfred O. Nier, *University of Minnesota*

Nuclear Physics in Rome, by Emilio G. Segrè, *University of California, Berkeley* 33
    Introduction, by William A. Fowler, *California Institute of Technology*

Experimental Work with Nuclei: Hamburg, London, Copenhagen, by Otto R. Frisch, *University of Cambridge* 63
    Introduction, by William A. Fowler, *California Institute of Technology*

xii   Contents

The Nuclear Photoelectric Effect and Remarks on Higher Multipole Transitions: A Personal History, by Maurice Goldhaber, *Brookhaven National Laboratory* 81
Introduction, by Herman Feshbach, *Massachusetts Institute of Technology*

Early History of Particle Accelerators, by Edwin M. McMillan, *University of California, Berkeley* 111
Introduction, by H. H. Barschall, *University of Wisconsin*

The Neutron: The Impact of Its Discovery and Its Uses, by Eugene P. Wigner, *Princeton University* 157
Introduction, by H. H. Barschall, *University of Wisconsin*

The Development of Our Ideas on the Nuclear Forces, by Rudolf Peierls, *University of Oxford and University of Washington* 179
Introduction, by Robert Serber, *Columbia University*

Some Men and Moments in the History of Nuclear Physics: The Interplay of Colleagues and Motivations, by John A. Wheeler, *Princeton University and University of Texas at Austin* 213
Introduction, by R. R. Wilson, *Fermi National Accelerator Laboratory*

Name Index 325
Subject Index 333

Conference Participants

Contributors to This Volume

Hans A. Bethe was born in Strasbourg, Germany, and received his Ph.D. degree from the University of Munich in 1928. He came to the United States and joined the faculty of Cornell University in 1935. He was awarded the Nobel Prize in physics in 1967.

Otto R. Frisch, O.B.E., F.R.S., is a native of Vienna, Austria. He received his Ph.D. degree from the University of Vienna in 1926. A Fellow of Trinity College, Cambridge, he was appointed Jacksonian Professor of Natural Philosophy at the University of Cambridge in 1947.

Maurice Goldhaber, who was born in Lemberg, Austria, obtained his Ph.D degree from the University of Cambridge in 1936. He immigrated to the United States and joined the faculty of the University of Illinois in 1938. Since 1950 he has been on the staff of Brookhaven National Laboratory, where he served as director from 1961 to 1973.

Edwin M. McMillan is from Redondo Beach, California, and holds a Ph.D. degree from Princeton University (awarded in 1932). After two years as a National Research Council Fellow at the University of California, Berkeley, he began work at the Radiation Laboratory, and the following year he became a member of the

Berkeley faculty. He served as director of the Lawrence Radiation Laboratory from 1958 to 1971 and of the Lawrence Berkeley Laboratory from 1971 to 1973. He received the Nobel Prize in chemistry in 1951.

Sir Rudolf Peierls, Kt., C.B.E., F.R.S., was born in Berlin, Germany, and obtained his Ph.D. degree from the University of Leipzig in 1929. He is a Fellow of New College, Oxford, and was appointed Wykeham Professor of Physics at the University of Oxford in 1963. In recent years he has also been a visiting professor at the University of Washington.

Emilio G. Segrè, a native of Tivoli, Italy, was the first of Enrico Fermi's students to receive the Ph.D. degree, in 1928, from the University of Rome. He came to the United States and joined the staff of the University of California, Berkeley, Radiation Laboratory in 1938, and became a member of the Berkeley faculty in 1946. He was awarded the Nobel Prize in physics in 1959.

John A. Wheeler, born in Jacksonville, Florida, received his Ph.D. degree from Johns Hopkins University in 1933. He joined the faculty of Princeton University in 1938 and became Joseph Henry Professor of Physics in 1966. At present he is also Professor of Physics at the University of Texas at Austin.

Eugene P. Wigner was born in Budapest, Hungary, and obtained his doctorate in engineering from the Technische Hochschule in Berlin in 1925. He became a full-time member of the faculty of Princeton University in 1933. He spent one year, 1937-38, at the University of Wisconsin and then returned to Princeton as Thomas D. Jones Professor of Mathematical Physics. He received the Nobel Prize in physics in 1963.

Session Chairmen

H. H. Barschall was born in Berlin, Germany, received his Ph.D. degree from Princeton University in 1940, and is currently Professor of Physics and Nuclear Engineering at the University of Wisconsin.

Herman Feshbach, a native of New York, received his Ph.D. degree from the Massachusetts Institute of Technology in 1942. He is presently head of the Department of Physics at M.I.T.

Conference Participants xv

William A. Fowler was born in Pittsburgh, obtained his Ph.D. degree from the California Institute of Technology in 1936, and is currently Institute Professor of Physics at Caltech.

Alfred O. Nier, who was born in St. Paul, received his Ph.D. degree from the University of Minnesota in 1936. He is Regents' Professor of Physics at the University of Minnesota.

Robert Serber was born in Philadelphia, obtained his Ph.D. degree from the University of Wisconsin in 1934, and is presently chairman of the Department of Physics at Columbia University.

R. R. Wilson is from Frontier, Wyoming, and holds a Ph.D. degree from the University of California (awarded in 1940). He is past director of the Fermi National Accelerator Laboratory.

Other Participants

Finn Aaserud, Johns Hopkins University
Lawrence Badash, University of California, Santa Barbara
Franco Belloni, University of California, Berkeley
Joan Bromberg, Smithsonian Institution
David C. Cassidy, University of California, Berkeley
Charles L. Critchfield, Los Alamos Scientific Laboratory
Eugene Feenberg, Washington University
Laura Fermi, Chicago, Illinois
Peter Galison, Harvard University
Gertrude S. Goldhaber, Brookhaven National Laboratory
Morton Hamermesh, University of Minnesota

Paul A. Hanle, Smithsonian Institution
Erwin N. Hiebert, Harvard University
Martin J. Klein, Yale University
Henry Koffler, University of Minnesota
M. Stanley Livingston, Los Alamos Scientific Laboratory
Marjorie C. Malley, Blue Springs, Missouri
John May, University of California, Berkeley
John L. Michel, University of Wisconsin
Don Moyer, Northwestern University
Peter Roll, University of Minnesota
Spyros S. Sakellariadis, University of Pittsburgh

Robert Seidel, University of California, Berkeley
Daniel Siegel, University of Wisconsin
Katherine R. Sopka, Harvard University
John Stachel, Institute for Advanced Study
Roger H. Stuewer, University of Minnesota
Hans E. Suess, University of California, San Diego
Loyd S. Swenson, Jr., University of Houston
Spencer R. Weart, A.I.P. Center for History of Physics

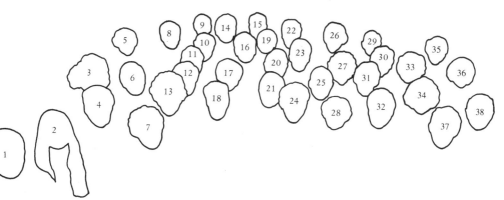

1. Lawrence Badash
2. Marjorie C. Malley
3. Finn Aaserud
4. Emilio G. Segrè
5. Paul A. Hanle
6. John A. Wheeler
7. Katherine R. Sopka
8. Spencer R. Weart
9. Roger H. Stuewer
10. William A. Fowler
11. Herman Feshbach
12. Hans A. Bethe
13. Spyros S. Sakellariadis
14. Charles L. Critchfield
15. H. H. Barschall
16. Morton Hamermesh
17. Franco Belloni
18. Otto R. Frisch
19. Hans E. Suess
20. Edwin M. McMillan
21. Rudolf Peierls
22. Loyd S. Swenson, Jr.
23. Eugene Feenberg
24. Maurice Goldhaber
25. Alfred O. Nier
26. Don Moyer
27. Peter Galison
28. Laura Fermi
29. Erwin N. Hiebert
30. John May
31. M. Stanley Livingston
32. R. R. Wilson
33. Robert Seidel
34. John L. Michel
35. David C. Cassidy
36. Daniel Siegel
37. Robert Serber
38. Eugene P. Wigner

Missing from photo: Joan Bromberg, Gertrude S. Goldhaber, Martin J. Klein, Henry Koffler, Peter Roll, John Stachel.

Welcome
Alfred O. Nier

Introduction
Henry Koffler

Welcome

Alfred O. Nier

On behalf of the Organizing Committee I should like to welcome all of you to this Symposium on the History of Nuclear Physics. No attempt has been made to cover the entire history. Instead, emphasis has been placed on the 1930s, a period when great progress was made, both in experimental discovery, using techniques previously out of reach, and in theoretical understanding, using the newly developed quantum mechanics.

It was an exhilarating time to live. I was a graduate student in physics here at Minnesota in the early 1930s, and the late Professor John T. Tate was the research adviser of many of us. He was also editor of the *Physical Review*, which at that time was emerging as the world's most important physics journal. Manuscripts covering new discoveries crossed his desk continuously. He taught a course entitled "Contemporary Experimental Physics" in which the newest discoveries were presented and discussed practically as they came in. Students took the course once for credit and then sat in the class for the rest of their graduate days, as it was the place where the exciting new concepts and results were revealed. We are fortunate in having with us here as participants in this symposium some of the most important of those who made the history to which I refer. They are joined by scholars of the history of science. We are thankful to all these people and to those who made this symposium possible.

The TV cameras, which are quite noticeable, will be used to make a videotape of the entire symposium for archival purposes as well as for distribution for educational purposes. One of my colleagues remarked that in employing this recording technique we run into a kind of uncertainty principle. On the one hand, a too-diligent attempt to stage the performance would lead to a deterioration of the product. On the other hand, if no recordings were made, there would be no interference but from a historical point of view we would miss an important aspect of the symposium. Our objective, of course, is to achieve a compromise between these extremes.

This is a busy time at the university. Our legislature has been meeting and deciding our fate for the next two years. It is an especially busy time for our top administrators. We are therefore very grateful to Dr. Henry Koffler, our Vice President for Academic Affairs, and himself a scientist, who has taken time from his busy schedule to be with us tonight to open the symposium.

Introduction

Henry Koffler

The purpose of this symposium is to explore key developments in the history of nuclear physics in the 1930s as seen and recalled by eight of the major contributors of that history, and by a group of their distinguished colleagues who will serve as session chairmen and discussants at the symposium. Included among the latter are a number of historians of physics and graduate students in the history of physics who will no doubt be directing questions to the speakers during the discussion period following each of the lectures.

The 1930s opened, as we may recall, with several fundamental discoveries in experimental and theoretical nuclear physics, as well as with key developments in the instrumentation. The discovery of the neutron by James Chadwick, the discovery of deuterium by Harold Urey, the invention and construction of the cyclotron by Ernest Lawrence and Stanley Livingston, and the disintegration of lithium by John Cockcroft and E. T. S. Walton using their high voltage proton accelerator are just four of the outstanding developments in nuclear physics which opened up numerous avenues of research in England, Germany, Italy, France, Denmark, the United States, and other countries.

Of course, as every scientist and historian realizes, these developments themselves did not and could not have sprung out of infertile scientific soil. Rather, each constituted the culmination of years of basic research and development, not only in the particular laboratories in which these discoveries and inventions were made, but in others as well, though it was not an accident that they were made where they were. I think it is obvious to all of us—it is almost a cliché to a working scientist—that although one can plan research in science, one cannot plan discovery. We need only recall, for example, that the *potential* for discovering the neutron certainly existed in Bothe's laboratory in Berlin, and in Curie and Joliot's in Paris, but it was *actually* discovered in Chadwick's in Cambridge. Thus, though discovery in science certainly cannot be planned, the potential for it can be enhanced by assembling the best scientists in a given laboratory and then allowing them the freedom to pursue their ideas and research plans. This assembly of talent and freedom of research led to the rapid discovery and development of the basic experimental knowledge and theoretical concepts of nuclear physics during the 1930s, in much the same way that the assembly and freedom of the best scientists in other locations in the 1920s led to the development of our basic physical theory of the microstructure of matter, or quantum mechanics.

But the 1930s constituted a watershed in science in another respect as well—and this has, I might say, personal meaning to me. One need only glance at the program for this symposium to see that five of our distinguished speakers and chairmen came to this country in the 1930s; two made their homes in England; and seven were American-born. This even distribution is significant in two respects. First, few would deny that by the mid-1930s there was a strong indigenous scientific base in England and the United States. We at the University of Minnesota are proud to have taken a part in this coming-of-age in physics in this country through, for example, the contributions of J. H. Van Vleck in the mid-1920s and the extensive efforts of John T. Tate as editor of the *Physical Review*, as Professor Nier mentioned, and the founder of the *Reviews of Modern Physics*.

The second point which is impossible to overlook, however, is that half of our speakers and chairmen were "illustrious immigrants," to use Mrs. Fermi's characterization, and made enormous contributions to the basic scientific efforts of their adopted countries. Many things have been said about the great impact of this intellectual migration precipitated by the rise of dictators in Germany and Italy. I will make only perhaps the most general point here: the intrusion of politics and racial prejudice on science in Germany and Italy had serious detrimental effects on scientific research and development in those countries in the 1930s, at the same time elevating to a new high point the scientific stature of the adopted lands of those who emigrated. When the war came and the consequent state of emergency channeled and directed research in nuclear physics to a specific wartime goal, this country and England were prepared to meet the emergency through the joint efforts of native-born citizens and gifted immigrant scientists.

The same factors have been important in the modern development of my own field of molecular biology. The same synergism resulting from bringing together the best scientists and giving them the freedom to pursue their ideas led to our current understanding of DNA and the molecular basis of living organisms. The early work in my field was especially stimulated by the interest and immigration of a few physicists in the late 1940s who had contributed earlier to quantum mechanics and nuclear physics—Erwin Schrödinger, Max Delbrück, and Leo Szilard, to name three. My own field and all of science have benefited immeasurably not only from discoveries in nuclear physics of the men here today and their associates in the 1930s and 1940s, but also from their vigorous and productive approach to scientific research, which has served as a model for the rest of us since then.

These, then, are some of the more general characteristics I see associated with nuclear physics in the 1930s and 1940s, and some of the lessons to be learned from the developments of that period. These characteristics and lessons are not all that profound and obscure, but it is not too often that we stop to think deeply about them and to understand clearly the most important factors contributing to them. That is why, at the University of Minnesota, we value the study of the history of scientific ideas—the intellectual and social context from which they arose and their scientific, intellectual, and social impacts. And it is why we are especially proud and happy to serve as the host for this symposium on an aspect of physical science to which we have made some significant contributions and which has had such a profound influence on our science and our society.

As I am sure all of you know, the chairman of this opening session, one of our most distinguished and productive faculty members, and a colleague of whom I am very proud, Professor Alfred O. Nier, is an outstanding example of the University of Minnesota's contributors to both the basic and practical aspects of nuclear physics. His development of the mass spectrometer and use of it to separate the isotopes of uranium was built upon some of the work we will hear discussed during the symposium, and it led directly to major breakthroughs in both the basic understanding of the nucleus and in practical applications of that understanding. It also helped establish the climate in which nuclear physics research has flourished at Minnesota, under his leadership and that of the late Professor John Williams and his colleagues and successors. This is a local example of the principle that, although research can be planned, discovery depends not on planning but on bringing the best scientists together and supporting their work and their freedom to follow their ideas wherever they may lead.

The Happy Thirties
Hans A. Bethe

INTRODUCTION: *Alfred O. Nier*

It is most appropriate that we officially start our symposium with a speaker who might be called "Mr. Nuclear Physics." He grew up at a time when quantum mechanics was just coming into being, and quickly saw its applicability to problems in both atomic and nuclear physics. His education was obtained in Germany, so he was close to those who were making the new physics. Thus, as he started his career, he knew and was close to Sommerfeld, Fermi, Bohr, and the others who were making history at that time.

After holding several teaching positions in Germany, he left for England in 1933 and after spending two years there came to Cornell University, where, except for temporary absences, he has been ever since. By the middle of the 1930s his name was virtually a household word among physicists. In 1936 and 1937 he was responsible for three articles totaling 487 pages in the *Reviews of Modern Physics*, which amounted essentially to a compendium of everything known in nuclear physics at that time. The first of these was written with Robert Bacher, the third with Stanley Livingston, who is also with us at this symposium. If you go to the library and look at the 1936 and 1937 volumes of the *Reviews of Modern Physics* you can quickly recognize the pages, as they either are frayed and dirty or have been replaced by shiny new pages after the original ones wore out.

When World War II began, he spent a year at the MIT Radiation Laboratory, and when the Los Alamos Scientific Laboratory was started, he was appointed head of the theoretical division, a position he held until after the war when he returned to Cornell.

In the years following the war he held numerous visiting positions in other institutions and was active in educating the public on the benefits and the perils of the atomic age. He has been a vigorous advocate of the control of nuclear weapons and an active participant in many conferences on the subject. His many contributions to science and his contributions to fields concerned with questions of public policy have won him innumerable honorary degrees, prizes, and awards—a list too long to ennumerate here. In 1967 he was awarded the Nobel Prize in physics for his contributions to the understanding of the energy production in stars. We are privileged to have Professor Hans Bethe talk to us on "The Happy Thirties."

The Happy Thirties
Hans A. Bethe

It is a pleasure to be here and to talk about old times. I've called my talk "The Happy Thirties," which is perhaps a little bit unusual because politically the thirties were anything but happy. Many of us in this room, as has been mentioned, emigrated from Germany and Italy because of the dictatorships prevailing in these countries, and it was only by great good fortune and by the wonderful hospitality of this country that we were then able to lead happy and productive lives.

However, in nuclear physics certainly the thirties were a very happy period. As I was reading the proceedings of the first conference on nuclear physics,[1] I saw that people were wondering when nuclear physics started. Did it begin in 1932 or much earlier? The starting date became earlier and earlier as people at the conference continued talking, but whatever may be said about the origin of nuclear physics, it is certain that 1932 marks a completely new start for both experiment and theory. The neutron was discovered, and as far as theory was concerned, it was now possible for the first time to think of a rational quantum mechanics of the nucleus. Regardless of the particular difficulties with spin and statistics, it would have been impos-

Nuclear Physics in Retrospect, edited by Roger H. Stuewer, copyright © 1979 by the University of Minnesota.

sible for anybody to do a quantum mechanics of a nucleus composed of protons and electrons. So 1932 was the beginning of the theory of nuclear physics as we know it, and this is therefore where I will start.

Nuclear Forces

One of the remarkable things then was the quick succession of papers. Papers are no longer published as rapidly as they were in those days. Chadwick's first paper in *Nature* was published in February 1932; Heisenberg's paper on the theory of the nucleus was submitted in June and published in July 1932. Now, Heisenberg's paper was much discussed at the second conference,[2] and I have my own ideas about it. First, Heisenberg definitely made the point that only by assuming neutrons and protons to be the constituents of the nucleus is it possible to have a quantum mechanics of the nucleus. Therefore, in this essential point he started our modern ideas. In addition, he had the complete isotopic spin notation: the neutron was +1 and the proton was −1 for the z-component of the isotopic spin, and the x- and y-components would transform a proton into a neutron and *vice versa*. That's all there is in this paper.[3]

But Heisenberg still thought of the neutron in some way as composed of a proton and an electron. I think only in some way; as was said in one of the previous conferences, he pushed the difficulties of spin and statistics back into the neutron and away from the nucleus. He said, in effect, "let's describe the nucleus as if the neutron were an elementary particle, and let's worry later about the neutron." He had the picture that there might be an exchange of an electron between two interacting neutrons or between an interacting neutron and a proton. In fact, he stated explicitly that the hydrogen molecular ion H_2^+ and the hydrogen molecule H_2 are analogous to the neutron-proton interaction and the neutron-neutron interaction. So he had these particles interact and attract each other, but he had no force between two protons. Needless to say, the force was different from any forces which we had assumed before in physics. That no force existed between protons was disproved explicitly in 1936 by Tuve and his collaborators in the Department of Terrestrial Magnetism in Washington, who showed that there was a strong attractive force between two protons apart from the Coulomb force.

Heisenberg's theory had an asymmetry between neutron and proton. He thought, in fact, that this asymmetry might explain the fact that heavier nuclei contain more neutrons than protons. We know now, and in fact we knew very soon after Heisenberg's paper appeared, that this isn't necessary to explain that asymmetry — the Coulomb force does it all.

The second most important paper was that by Majorana in 1933.[4] He worked at Heisenberg's institute, but Segrè testifies that Majorana had many of these ideas long before he joined Heisenberg. Indeed, his association with Heisenberg mainly resulted in his publishing his ideas instead of keeping them in his drawer. Majorana's paper is completely modern. He got away from the idea that the neutron is a composite, and that exchange forces are due to exchanging of electrons. He didn't work from the analogy with atoms and molecules, but instead from the observed properties of nuclei, particularly the saturation of nuclear forces and the particular stability of the α-particle. He realized that saturation could come about by an exchange force, so he also postulated an exchange force, but one different from Heisenberg's. Heisenberg had the idea that a neutron and a proton located at two positions might exchange their charges between these two positions, but the spins would remain at the original position. Majorana said no; the spin goes with the particle and does not stay at the same position. Consequently, Majorana's forces saturated at He^4, while Heisenberg's forces would have saturated at the deuteron, which of course is not a particularly stable nucleus. The deuteron singlet s-state, in fact, in Heisenberg's theory would have had a repulsion rather than an attraction.

The next important paper was Wigner's.[5] He too was interested in the great stability of the α-particle compared with the deuteron. I believe he did not know Majorana's paper. He did know Heisenberg's, and for some reason he disliked exchange forces. In my opinion, he blamed Heisenberg unjustly for assuming that there are still electrons in the nucleus. Anyway, using ordinary forces between nuclear particles, Wigner showed that the potential had to be made very narrow to explain why the deuteron had only a small binding energy and the α-particle had a very big one. Why he disliked exchange forces so much I don't know. I hope when he comes he will tell us.

One piece of work was done by Rudi Peierls and myself.[6] We were both in Manchester at the time. We discussed the deuteron, which was being investigated experimentally in Cambridge by Chadwick and Goldhaber who had been able to disintegrate the deuteron by γ-rays. So we first investigated this very process—what is the cross section for photo-disintegration—and then the neutron-proton scattering. We found a relation between the binding energy of the deuteron and the scattering of neutrons by protons. There was a slight dependence on the range of the forces, and essentially what we had was a precursor of the effective-range theory, which Rudi will presumably talk about later on. Unfortunately, the results we got for the cross section for neutron-proton scattering were in total disagreement with the experiment. Whereas we found about 4 barns for the cross section in the limit of low energy, the measured values were somewhere between 30 and 80 barns according to geographic position, namely, New York (Columbia University) or Rome. Fermi explained soon afterward that this was partly due to the chemical effect of protons bound in molecules, and using Fermi's theory one could then determine from the experiments that for free protons the scattering cross section would be about 20 barns. But that was still five times bigger than what we had calculated.

The discrepancy was explained to me by Wigner in a noisy subway between Columbia University and Penn Station. He said: "Well, why do you assume that the singlet state has the same binding energy as the triplet state? Let's assume a different binding energy and all will be well." This very simple remark was never published by Wigner, but it was published in the article by Bethe and Bacher in the *Reviews of Modern Physics*,[7] which Professor Nier mentioned previously and whose date is cited incorrectly in the chronology prepared for the earlier conferences.[8] (I have found this chronology very useful, but at various points I will point out a few small mistakes in it.) Our paper was published in 1936. Having the singlet state of the deuteron higher than the triplet state, of course, would reintroduce a little bit of Heisenberg forces.

There the matter stood and nobody could tell how to detect the singlet state until a paper by Teller was published in 1936.[9] He proposed a test by scattering neutrons from para- and ortho-hydrogen molecules. This was then elaborated in a further paper by Teller and Schwinger,[10] and the experiment was later done, with the result that Wigner's hypothesis was of course confirmed—the singlet state was found. It was a virtual state determined by the interference between singlet and triplet scattering.

An important point in determining nuclear forces was the saturation of nuclear forces. Nuclear physicists knew of course long before 1932 that the binding energy of a nucleus is pretty nearly proportional to the number of particles in the nucleus, quite in contrast to the binding energy of atomic electrons, which goes up at some high power of the nuclear charge. Likewise, the volume of the nucleus, which was only vaguely known then but is now very accurately known, is proportional to the number of particles. So here is a force that saturates, and saturates obviously at the α-particle. Thus one had to find a theory that gave saturation. Exchange forces do have this property; ordinary forces, Wigner forces, do not unless they have other special properties. Many proofs were given in many papers that indeed ordinary forces are unacceptable from the point of view of saturation but that either exchange forces or forces with a strong repulsive potential at the center are needed. Such a repulsive potential was introduced by Morse, Fisk, and Schiff,[11] albeit they used a special analytical form, which of course was the Morse potential which they could calculate. I must confess that in my *Reviews of Modern Physics* article I rejected the repulsion at short distances, because I said it was too complicated; forces between elementary particles must be simpler. Other people had exactly the opposite idea and said exchange forces are not acceptable; they are too complicated. This was a matter of taste, and in the end we learned that there are both exchange forces and a very husky repulsive component in the nuclear force.

The Shell Model

In the second conference it was mentioned that the shell model was not discussed much in the 1930s, in spite of the fact that Heisenberg's theory introduced a potential acting on a nucleon, and with a potential one should expect a shell model. I think this remark is wrong in every respect. First, Heisenberg-Majorana forces do not provide the same kind of potential as that found in an atom, because it turns out, by a very quick calculation, that the force between an entire nucleus and one particle is no bigger than the force between two nucleons, or let's say four nucleons. So a true general potential acting on the particle in the nucleus really wasn't postulated by having a nucleon-nucleon potential.

It is also wrong to say that the shell model wasn't being talked about. It was talked about a great deal. Bartlett is given credit for the idea of neutron and proton shells in his 1932 paper.[12] I think he should be given very little credit, because he had only a tiny bit of evidence. He claimed, for instance, that only shells with principal quantum number 1 exist, s, p, d, f, etc., which gives shell closures at entirely the wrong places. For instance, the third shell closure would be at 18 neutrons or protons instead of 20; the next at 32, where there is nothing at all; the next is correct at 50—but that is an accident.

The two people who do deserve credit are Elsasser and Guggenheimer, each then working in France and publishing in the *Journal de physique* in 1933 and 1934.[13] Each published several papers, searching for nuclei of special stability and looking at this problem from the point of view of isotope statistics and the abundance of isotopes. This was a reasonable way to look for what we now call magic nuclei, and indeed they found the magic numbers 2, 8, 20, 50, and 82. They couldn't have done much better. They tried to give a theoretical basis in terms of the wave functions in an infinite potential well, and that worked very well for the first three numbers, 2, 8, and 20, but not at all for 50 and 82. They then tried to omit particle states of low orbital momentum ℓ and high principal quantum number n to get agreement, but this was entirely *ad hoc*. Margenau in 1934[14] did the finite square well, hoping that this would improve results, but it didn't help much; the order of levels remained the same. There was just a slight displacement, but the higher magic numbers couldn't be explained.

In my *Reviews of Modern Physics* article with Bacher I didn't try to explain these higher magic numbers, but I did try to get some additional evidence for shells. In fact, I got very good experimental evidence from the binding energy for the existence of the shell containing 8 particles. To do this one has to eliminate the well-known periodicity with a-particles, namely, that nuclei composed of an integral number of a-particles are much more tightly bound than their neighbors. I had to eliminate this effect, so I compared nuclei that differ by an a-particle and found that adding one a-particle to C^{12} increases the mass by 3.996 atomic units, but adding an a-particle to O^{16} increases the mass by 4.999; so there's 3 MeV less binding energy for an a-particle bound to O^{16} than to C^{12}, contrary to the trend calling for more binding with higher atomic number. And I found similar irregularities when adding two neutrons or a neutron and a proton, and so on, to C^{12} and O^{16}, respectively, generally indicating that O^{16} is a closed shell, a doubly closed shell, for both neutrons and protons.

For Ca^{40} there did not then exist any good value for the isotopic weight, but there were well-known irregularities in the isotope pattern, a fact known already to Mendeleeff. K^{39} has less mass than A^{39}, potassium having a nuclear charge one greater. In this neighborhood in the periodic system, the general rule is just the opposite: For nuclei with the same mass number (isobars), the one that has the lower atomic number Z generally has the lower mass. For instance, Cl^{37} has less mass than A^{37}, K^{41} has less mass than Ca^{41}, but with K^{39} and A^{39} it is the other way around. So it was clear that exceeding the neutron number N = 20, or Z = 20, led to less binding of the particle. Then I looked at the nuclear stability line, in particular at the isotopic numbers and the difference in the number of neutrons and protons versus the atomic weight. It again showed that Z = 50 and N = 82 gave higher stability, as had already been found by Elsasser and Guggenheimer.

But this was not the only way in which the shell model was used. There was a paper by Feenberg and Wigner in 1937[15] which I consider very important. They used Russell-Saunders coupling to discuss the binding of nuclei in the p-shell. And they said, very sensibly: We can't reproduce the total binding energy, the Hartree-Fock method is much too clumsy for that. But we can at least get the differences in energy between different levels of a given nucleus and of nuclei of the same mass number. In this paper they found that the most important thing really is the symmetry of the wave function; namely, if lots of particles are put into a symmetric wave function, making something like an a-particle without calling it an a-particle, extra binding occurs, and this goes through the entire range from

He^4 to O^{16}. They found the lowest levels in the p-shell region, and they got quite nice agreement with the energies and spins of the nuclei.

The emphasis on symmetry and on the group theoretical partition of the number of particles was then greatly expanded by Wigner in the paper following directly thereafter.[16] In this paper he investigated the consequences of charge independence, and on the basis of charge independence and a lot of group theory, which is hard for me and for many other people to understand, he predicted the existence of some isotopes such as S^{36} and Ca^{46}, which had not previously been known and which were subsequently found. This emphasized a different feature of nuclear forces, namely, the symmetry property, but it started out from an investigation of shell model levels.

One paper incorrectly listed in the chronology as a shell model paper is Schmidt's 1937 publication.[17] It did not use the nuclear shell model at all. Schmidt investigated the nuclear magnetic moments, and he said that they can be explained by assuming that neutrons as well as protons occur in pairs. When there are only pairs, the total angular momentum of the nucleus is zero; one then adds one more particle with a certain orbital momentum and a certain spin and total angular momentum. He then related the measured total angular momenta of the nuclei to their magnetic moments. This has nothing to do with the shell model, but as you will see in a minute, it was later used to derive a good shell model.

This was about where the shell model rested from about 1938 until after the war, when there was feverish activity in 1948-49. As far as I could find out, at least in the American literature (and I hope that in the discussion Professor Suess will complement that with the German investigations), the first important publication was Maria Mayer's 1948 paper,[18] in which she did a much better statistics of nuclei than was done ever before.[19] She particularly looked at isotones, that is, nuclei having the same number of neutrons, and she found that nuclei having 50 neutrons were far more numerous than any nuclei around them. And this was even more true for 82. She also found that 28 neutrons was a preferred number. She didn't quite dare emphasize the number 28, but as we know now this is a good magic number too. Evidence for proton shells at Z = 50 and 82 had been known before. In addition, she found the neutron shell at 126.

The shells were established beyond doubt, but there was no theory. Two people tried their hands at a theory. One was Feenberg in the next volume of *Physical Review*,[20] and the other was Nordheim, in the same volume.[21] They did indeed look at the Schmidt theory of magnetic moments, and they deduced from the Schmidt theory the orbital momentum of the nuclei right before 50 and before 82, and after 50 and after 82, and this gave a clue to the shells actually involved. Then they again tried to find some theoretical arrangement of shells by considering taking just orbital motion, which would give the right magic numbers, but what they found was quite unsatisfactory. They went in opposite directions, to smaller and larger principal quantum numbers, but neither of these attempts was convincing.

They were given credit by Maria Mayer in her final and most important paper of 1949,[22] very soon thereafter. She got the correct spin-orbit coupling arrangement. Of course, she did not know the mechanism of how spin and orbital momentum could be coupled so well. And this caused a number of people to object to her scheme quite strongly—I think one of them was Wigner. It was not until much, much later, when we analyzed high energy scattering, and in particular the polarization in high energy scattering of protons and neutrons, that we finally found out that there is indeed a very large spin-orbit coupling already in the elementary force between the nucleons.[23] But she didn't know that. She gives credit to Fermi, who knew of her first paper about the statistics and about the magic numbers 50 and 82, and apparently in some conversation he said to her, just as an offhand remark: "Is there any indication of spin-orbit coupling?" That was clearly a very wise remark and may have led her to the right answer.

The Compound Nucleus

Finally, I want to discuss the compound nucleus. In the chronology, credit for the compound nucleus is given to Harkins. I want to say categorically that Harkins deserves no credit whatsoever for this idea. All that he did was to use the compound nucleus as a bookkeeping device, and all that he meant was the following: Suppose that a neutron is interacting with N^{14}. Then the neutron is involved directly in the reaction—a reaction like a chemical reaction occurs, neutrons strike N^{14} to give B^{11} plus He^4. So the neutron participates in the reaction. This contrasts with what people knew from atomic physics about electrons colliding with atoms. Electrons, commonly, when interacting with atoms will produce an excited state of the atom, and the incident electron does not participate except by giving energy to the atom; the electron does not enter the reaction. This difference between the electron interacting with an atom and the neutron interacting with the nucleus—this is what Harkins emphasized in the first extensive paper he wrote on the subject, and then he wrote the equation $N^{14} + n \to N^{15} \to B^{11} + He^4$. But this feature was well known to nuclear physicists before Harkins, and it has nothing to do with the idea of a compound nucleus.

Now, the credit for the compound nucleus belongs, in my opinion, equally to Niels Bohr on the one side and to Breit and Wigner on the other. There was some discussion about this question in the second conference arranged by the American Insititue of Physics, and I think people had different ideas, some giving nearly all the credit to Bohr, others to Breit and Wigner. Bohr's idea of the compound nucleus was being formed apparently in 1935, and I am told, but I have found it hard to confirm this, that I am somewhat responsible for his idea—in a negative sense. In Copenhagen in the summer of 1934 I gave a talk in which I used the extreme direct interaction collision model, that is, I said: "If there is a proton colliding with the nucleus, the collision goes much the same as in the case of atoms; one writes down the proton wave function and that of the nucleus, and that gives the answer." I am told that Bohr went away from this talk shaking his head and being unhappy, as he used to be when thinking hard, and saying: "Well, I am quite sure this is wrong. It cannot go this way." And I am told that this was the start of the thought which then led to the compound nucleus. Bohr's idea was published in *Nature* in 1936,[24] and he did a lot of oral propaganda on this subject. He explained to everybody that nuclear reactions must go on the compound picture. "Perhaps," he said, "if a proton interacts with a deuteron, the collision might be described by a single particle model, but already when there are four particles it won't go that way."

Breit and Wigner's paper[25] was received February 15, 1936, which I think is before Bohr's paper was published (I don't want to talk about priority; I think they didn't know much about each other). In their paper Breit and Wigner first referred to the one-particle theory of the large capture cross section for slow neutrons in certain elements which had been observed in Rome. One such theory had been done by Fermi himself, one was published by me,[26] and there were at least two by two other authors. Breit and Wigner pointed out that on the one-particle theory large scattering also had to be expected whenever there was large capture. Experiments had shown by that time that this was not so, but that in many cases neutrons had a large capture cross section but a very small scattering cross section. This was true, for instance, for neutrons on cadmium, which was investigated in great detail by the Columbia group. They found that the scattering cross section could not be as large as 1 percent of the capture cross section. So Breit and Wigner said the one-particle theory cannot be right.

In this respect I disagree with some of the statements made during the second conference, statements which claimed that Breit and Wigner really didn't have the many-particle picture in mind. As far as I can tell, they did so very definitely, and they said so. Also, they were accused in that second meeting of using perturbation theory, and again I would say they did not. They used analogies to some atomic and molecular processes — molecular processes like predissociation — and the simplest atomic case that I could think of (which is not in the Breit-Wigner paper) is the spectrum of helium. For instance, a state of the helium atom exists in which two electrons are in the 2p state. They interact strongly and a virtual state of the atom results, in fact several with different L and S. So strong interaction is implied if a true virtual state exists, which means the opposite of perturbation theory. It is even possible to have a virtual state with a very long lifetime: With helium, one need only put this pair of p-electrons into a triplet p-state. Then there is no possible radiationless transition to any state in which one electron is in the 1s state and the other electron is free.

The $(p^2)^3P$ state can only decay by an optical transition to a one-particle excited state, a weak optical transition at that, and perturbation theory can be used to calculate the transition probability. This is completely analogous to what Breit and Wigner did in their paper, and in fact they were very conscious of the analogy. Even though they did not mention the p^2 state of helium, they mentioned something very similar, the inverse Auger effect, which had previously been treated by Polanyi and Wigner.[27] So they assumed a quasi-stationary, virtual energy level of the system nucleus-plus-neutron. They further assumed that the radiation width of such a state is 12 eV — how they got that number I don't know. They estimated the neutron width as 0.3 eV, which is also much too big, and they pointed out correctly that the incident neutron should be in an s-state. They knew that there would be interference of the elastic scattering by a compound state, with the shape elastic scattering, that is, with the scattering due to the one-body potential acting on the neutron. All this was already in the paper. They estimated that the resonances of such nuclei as silver, cadmium, etc., should be below 400 eV — why I don't know — and they said that there must be one level below 900 eV in at least 5 percent of all nuclei. Then they tried to estimate the level density taking the ThC″ γ-ray levels, and that part was wrong. They did not realize that the density of nuclear energy levels is much greater for a heavy nucleus than for a lighter one and is much greater for high excitation energy than for low excitation energy. In other words, they did not solve the problem of level density.

On the other hand, I think Breit and Wigner were not given enough credit in the second meeting, nor in my *Reviews of Modern Physics* article, in which I credited them with their formula but not with the basic idea. Bohr's idea was very qualitative. He said that once a nucleon enters a heavy nucleus it will very soon collide with some nucleons already in the nucleus. It will thereby give up a great part of its energy. Then two nucleons will continue with diminished energy. They will again collide, and so the energy will be distributed among many nucleons in the nucleus. He had a nice model of that: a shallow dish with little balls in it.[28] He let one extra ball come into the dish with considerable energy; then very soon the energy of that one would be distributed among many nucleons and they would go merrily around in random motion. Thereupon, he said that once the energy is distributed over many nucleons, it is very difficult to concentrate it back on one nucleon, so there is a very small probability of reemission of the neutron after it is once in the nucleus. This is the basic idea of the compound nucleus, and this indeed was most clearly stated by Bohr in his paper.

Somewhere in the chronology again (I'm sorry that I am pointing out small mistakes in the chronology instead of saying that it has given me a lot of help in tracking down some papers) it says that the compound nucleus was immediately connected to the liquid drop model. I don't think this is correct. You can have one without the other, and the other without the one. It is true that Bohr used both. He used the liquid drop model to calculate the density of nuclear levels. But that could as easily be done in other ways. In fact, it was done by me[29] in 1936, and independently by Oppenheimer and Serber.[30] We assumed free particles in a Fermi sea moving around in the volume of the nucleus and calculated how many levels of the whole nucleus would come about by distributing the energy in various ways among these free particles. Indeed, we got a formula which agrees pretty well with experiments if the right parameters are put in. Bardeen[31] in 1937 took into account the correlations between the particles according to Fermi statistics. With interacting particles, the energy level density was calculated by Bohr and Kalckar in 1937 in the famous paper[32] in the Danish journal which no library seems to have (this paper should have been listed in the chronology but isn't).

Unfortunately, when I wrote my *Reviews of Modern Physics* article I was too much convinced by the authority of Bohr, so I said that the compound nucleus theory applies to everything, including the evaluation of the a-particle emission from natural radioactive nuclei. I persuaded myself, somehow, that the intrinsic width for emission of a-particles should be 1 eV, about the same as the neutron width at about 1 MeV kinetic energy. This was totally wrong, because the a-radioactive nuclei are near their ground state, and therefore the intrinsic level width is much larger. On the basis of this stupid idea, I calculated that the nuclear radius of natural radioactive nuclei is 12 fermis, and this gave me for r_0 in the formula $R = r_0 A^{1/3}$ the value 2.0 fermis, a totally erroneous number. If one uses Gamow's one-particle theory of a-decay, which is much more sensible, one gets $r_0 = 1.5$, which is still too big, but it is closer to the real value of, say, 1.2. Now if I correct this and take the radius of the nucleus to be $1.25 A^{1/3}$, and then use my old formula, I get for the spacing of the neutron resonance levels in U^{238} the value 50 eV, which is about right. With my wrong radius, I found much too small a number. A better discussion of this is given by Bohr and Mottelson.[33] So the free particle model really is quite good, and it was just my stupidity which made me get the wrong answer.

The compound nucleus model dominated the theory of nuclear reactions at least from 1936 to 1954. The level density was observed accurately in the 1950s and 1960s, particularly in neutron resonance measurements, of which we got a large number. The neutron width was observed; the relation between spacing and width was calculated and was in reasonable agreement with observation; the level distribution was obtained; and a lot of phenomenological predictions could be made from the compound nucleus model. As Weisskopf pointed out in the second conference, at Los Alamos when we tried to get cross sections we used the compound nucleus model and usually our predictions were quite reasonable. The compound model could explain many phenomena, for instance, that when any particle is put in a heavy nucleus, what comes out is mostly neutrons. These neutrons come out at low energy, of the order of 1 MeV, even if the energy of the incoming particle is now 10 MeV or more. It explained very well—in fact, it predicted and then it was verified—that very frequently if there is enough energy, two, three, or more neutrons are emitted. Wigner was the only voice in the wilderness who warned repeatedly that people were too fascinated by this one facet of nuclear theory, by which he did not want to detract from the power of the compound nucleus model. He said people should pay more attention to symmetry properties in the nucleus, perhaps to the shell model, and so on.

This one-sided approach was only remedied after the war, especially by the cloudy crystal ball model of Weisskopf, Porter, and Feshbach.[34] This model was very strongly stimulated by experiments, which I would advocate should be added to the "chronology," particularly the experiments by Barschall[35] on the neutron scattering resonances of nuclei in the range of 100 keV to 1MeV as a function of atomic number. When one looks at these, one finds beautiful mountain landscapes with hills and valleys, which were then explained by the Weisskopf-Porter-Feshbach theory in terms of one-particle resonances. So, whatever one does, one always comes back to the original point: one-particle resonances were used at first to explain the low energy neutron resonances, which was wrong, then they were completely discarded, and then they came back in 1954 superposed upon the compound nucleus resonances.

These are the three subjects I wanted to discuss. All three made us very happy during the 1930s.

REFERENCES

1. Charles Weiner, and Elspeth Hart, eds., *Exploring the History of Nuclear Physics* (New York: American Institute of Physics, 1972). The first conference was held May 18-19, 1967, in Brookline, Mass. at the American Academy of Arts and Sciences; the second was held May 18-19, 1969, also at the American Academy of Arts and Sciences.
2. *Ibid.*
3. W. Heisenberg, "Über den Bau der Atomkerne. I.," *Zeitschrift für Physik*, 77 (1932), 1-11.
4. Ettore Majorana, "Über die Kerntheorie," *Zeitschrift für Physik*, 82 (1933), 137-145.
5. E. Wigner, "On the Mass Defect of Helium," *Physical Review*, 43 (1933), 252-257.
6. H. A. Bethe and R. Peierls, "Quantum Theory of the Diplon," *Proceedings of the Royal Society of London*, 148[A] (1935), 146-156; "The Scattering of Neutrons by Protons," *Ibid.*, 149[A] (1935), 176-183.
7. H. A. Bethe and R. F. Bacher, "Nuclear Physics. A. Stationary States of Nuclei," *Reviews of Modern Physics*, 8 (1936), 82-229; see also H. A. Bethe, "B. Nuclear Dynamics, Theoretical," *Ibid.*, 9 (1937), 69-244; M. Stanley Livingston and H. A. Bethe, "C. Nuclear Dynamics, Theoretical," *Ibid.*, 9 (1937), 245-390.
8. Draft chronology prepared by the A.I.P. Center for History of Physics, October 21, 1968.
9. Edward Teller, "Interference of Neutron Waves in Ortho- and Parahydrogen" [Abstract], *Physical Review*, 49 (1936), 420-421.
10. Julian Schwinger and E. Teller, "The Scattering of Neutrons by Ortho- and Parahydrogen," *Physical Review*, 52 (1937), 286-295.
11. Philip M. Morse, J. B. Fisk, and L. I. Schiff, "Collision of Neutron and Proton," *Physical Review*, 50 (1936), 748-754.
12. James H. Bartlett, Jr., "Structure of Atomic Nuclei," *Physical Review*, 41 (1932), 370-371.
13. W. M. Elsasser, "Sur le principe de Pauli dans les noyaux," *Le journal de physique et le radium*, 4 (1933), 549-556; "II," *Ibid.*, 5 (1934), 389-397; "III," *Ibid.*, 5 (1934), 635-639. K. Guggenheimer, "Remarques sur al constitution des noyaux atomiques. I," *Ibid.*, 5 (1934), 253-256; "II," *Ibid.*, 5 (1934), 475-485.
14. Henry Margenau, "Nuclear Energy Levels and the Model of a Potential Hole," *Physical Review*, 46 (1934), 613-615.
15. E. Feenberg and E. Wigner, "On the Structure of the Nuclei between Helium and Oxygen," *Physical Review*, 51 (1937), 95-106.

16. E. Wigner, "On the Consequences of the Symmetry of the Nuclear Hamiltonian on the Spectroscopy of Nuclei," *Physical Review*, 51 (1937), 106-119.

17. Th. Schmidt, "Über die magnetischen Momente der Atomkerne," *Zeitschrift für Physik*, 106 (1937), 358-361.

18. Maria G. Mayer, "On Closed Shells in Nuclei," *Physical Review*, 74 (1948), 235-239.

19. Segrè pointed out to me privately, during the conference, that at the end of the war, in 1945, he prepared a big chart of all known isotopes, stable and radioactive, which was reprinted repeatedly and was disseminated very widely across the U.S. Maria Mayer studied that table intensively, and it undoubtedly helped her greatly in discovering the regularities mentioned.

20. Eugene Feenberg, "Nuclear Shell Structure and Isomerism," *Physical Review*, 75 (1949), 320-322; Eugene Feenberg and Kenyon C. Hammack, "Nuclear Shell Structure," *Ibid.*, 75 (1949), 1877-1893.

21. L. W. Nordheim, "On Spins, Moments, and Shells in Nuclei," *Physical Review*, 75 (1949), 1894-1901.

22. Maria Goeppert Mayer, "On Closed Shells in Nuclei. II," *Physical Review*, 75 (1949), 1969-1970.

23. John Wheeler and Heinz Barschall point out in the discussion of my paper that Barschall discovered in 1940 a strong anomaly in the scattering of 2.5 MeV neutrons by He, which was interpreted by Wheeler and Barschall in terms of a large splitting between the $P_{1/2}$ and $P_{3/2}$ levels of He^5. (See John A. Wheeler and H. H. Barschall, "The Scattering of 2.5-Mev Neutrons in Helium," *Physical Review*, 58 [1940], 682-687.) This large spin-orbit splitting of He^5 became known to many physicists, including myself, but I at least did not remember it when Maria Mayer's paper was published. Moreover, nobody knew the cause of the He^5 spin-orbit coupling in terms of nucleon-nucleon interactions.

24. Niels Bohr, "Neutron Capture and Nuclear Constitution," *Nature*, 137 (1936), 344-348.

25. G. Breit and E. Wigner, "Capture of Slow Neutrons," *Physical Review*, 49 (1936), 519-531.

26. H. A. Bethe, "Theory of Disintegration of Nuclei by Neutrons," *Physical Review*, 47 (1935), 747-759.

27. M. Polanyi and E. Wigner, "Bildung und Zerfall von Molekülen," *Zeitschrift für Physik*, 33 (1925), 429-434.

28. Apparently Otto R. Frisch actually built this model for Bohr.

29. H. A. Bethe, "An Attempt to Calculate the Number of Energy Levels of a Heavy Nucleus," *Physical Review*, 50 (1936), 332-341.

30. J. R. Oppenheimer and R. Serber, "The Density of Nuclear Levels," *Physical Review*, 50 (1936), 391.

31. John Bardeen, "On the Density of Energy Levels of Heavy Nuclei," *Physical Review*, 51 (1937), 799-803.

32. N. Bohr and F. Kalckar, "On the Transmutation of Atomic Nuclei by Impact of Material Particles," *Det kgl. danske videnskabernes selskab. Mathematisk-fysiske meddelelser*, 14, no. 10 (1937), 1-40.

33. A. Bohr and B. R. Mottelson, *Nuclear Structure* (New York: Benjamin, 1969).

34. H. Feshbach, C. E. Porter, and V. F. Weisskopf, "Model for Nuclear Reactions with Neutrons," *Physical Review*, 96 (1954), 448-464.

35. H. H. Barschall, "Regularities in the Total Cross Sections for Fast Neutrons," *Physical Review*, 86 (1952), 431.

DISCUSSION

Feshbach: The question that I would like to ask you is the following: In reviewing nuclear forces in the 1930s, I think one should also include tensor forces, which, as I recall, in the late 1930s and 1940s were a very exciting thing which stimulated all of us a great deal. When we discussed nuclear forces before World War II, one of the very interesting phenomena that showed up at the end of the 1930s was the quadrupole moment of the deuteron, which was accompanied by a number of theoretical papers on tensor forces. It seems to me that this was also an important part of the history of that period. If I recall, there were papers by Rarita and Schwinger and many others, but that's what I can offhand piece together.

Bethe: This certainly was a very important event. We learned that there are tensor forces in the nucleus and that it was necessary to further enlarge our picture. Maybe Peierls will discuss that in his talk.

Nier: Professor Suess, would you like to say something about the shell model at this time?

Suess: There are a few points I could make. One is that what we did in Hamburg and what Maria [Goeppert-Mayer] did in Chicago were completely separate and unrelated. We did not know a thing about each other. Yet, from a completely different approach, and under completely different circumstances, we were led to precisely the same result. But the big difference can be characterized best by what happened: Jensen once remarked that if he had known more theoretical nuclear physics, he would never have believed a word of what I had told him. It was really a difficult job for a mere chemist, who uses different methods than a theoretical physicist, to convince him. I used what is generally considered to be "circumstantial evidence." Chemists are used to considering simultaneously a number of facts and then deriving a conclusion from them, whereas theoretical physicists usually wish to consider the result of one single experiment, or one phenomenon they wish to interpret. I didn't really know a thing about what I was actually doing. I just had these magic numbers — not from Maria [Goeppert-Mayer] but from the great mineralogist Viktor Moritz Goldschmidt, who is not mentioned in our paper but who published them in the Proceedings of the Norwegian

Academy in 1938. I could see from this information that there was "circumstantial evidence" that without question had to have a physical meaning. So I started playing with these numbers and found that they were indeed magic—I mean all kinds of things could be done with them, such as deriving mathematical progressions of these numbers. Next I simply looked up in a textbook the solutions of the Schrödinger equation for the three-dimensional harmonic oscillator. (I did not go through the mathematics, because it could be looked up in any textbook.) Then I wrote down the levels one gets, sorted them in sequence according to the angular momentum, and used the Pauli principle to see how many particles would fit into each level. This gave the wrong numbers, but all one had to do was to reverse the usual sequence of the spin values, and to start with the highest spin. If one includes the highest value in the previous shell, one gets precisely the magic numbers. Well, I then thought that there might be something to this—maybe it's not just a magic trick. I looked up the paper by Schmidt to see whether the parities of the empirical spin values for the odd mass-number species would fit into such a scheme. I drew this up and showed it to Hans Jensen, because to me it was rather convincing that there was something behind it. Jensen said, "Oh, that I have to draw up myself," and he went through the literature and plotted each spin value in the diagram. I remember that when he came to the spin of Mn^{55}, which is 5/2, he said, "That's all nonsense—it doesn't fit—5/2 with even parity doesn't fit into the scheme." The following day he came to me and said, "Well, if there is something to it—if the scheme you drew up means something—it would mean that there is a strong spin-orbit coupling." I asked why there should be a strong spin-orbit coupling. He replied, "Das hat der liebe Gott so gemacht." He then wrote a note to *Die Naturwissenschaften*, but I said, "Wait a minute"—I had an agreement with Haxel that we would publish whatever came into our minds together. So he agreed that we should put Haxel's name on it also. Haxel too had been worrying so much about magic numbers and had realized that the magic numbers had to have some meaning. Thus, it was actually just an accident that we in Germany had this idea of spin-orbit coupling. I certainly did not realize the deeper meaning it had for basic nuclear physics.

Bethe: Did you know the magic numbers for neutrons as well as for protons? Because from the chemistry, from the geochemistry, you find the Z magic numbers easily.

Suess: The neutron numbers stand out even more impressively. For the isotopes that contain 50 and 82 neutrons, there is a big ridge —there is a maximum in the abundance distribution. I plotted this so that it could be recognized beautifully. By the way, Goldschmidt had all these numbers, including 126, and the mathematical sequence, except that he had a few additional numbers in his schemes that don't exist. He had too many, whereas people before didn't recognize all of them.

Barschall: I should like to mention that the first paper to point out the large spin-orbit coupling was written by John Wheeler and published in 1940.

Wheeler: It was on He^5, so it was really Heinz Barschall who stimulated it.

Goldhaber: Just a small point: When Niels Bohr in 1936 came to Cambridge and told us about his ideas on the compound nucleus, I remarked in the discussion that from his ideas there should be large slow neutron cross sections in heavy elements, for instance for Bi^{209}; but there it was known to be small because people were very interested in making RaE (Bi^{210}) to prove that the old natural radioactive families are in the right spots, and they did not succeed with slow neutrons at first with the sources then in existence. So we knew qualitatively that the cross section was very small. Niels Bohr answered, "Are you sure?" Well, we are now sure that bismuth is magic and therefore does not have a large cross section. [N.B. Bohr was wise to ignore this "premature" fact while he was developing his compound nucleus model.]

Stuewer: I would like to ask a more general question. How did it come about that you wrote the famous Bethe Bible? What were the circumstances that led you to write it?

Bethe: That was very simple. I wrote it because I was asked so many questions that I thought it would be simpler to write everything down once and for all. I came to this country in early 1935, and Livingston was already at Cornell. I was invited to many places to give talks, and I was asked more or less the same questions over and over again, so it was much simpler to write it down so that everybody could read it. I also had the benefit of the card file that Stanley Livingston had set up of all the papers that were published and that were relevant to nuclear physics. There was one omission in that card file, as I since found out. That was Yukawa's paper on nuclear forces. So I never paid any attention to that paper and didn't even read it until much later when the meson was discovered.

Swenson: I, too, was struck by the nature of conferences one and two in 1967 and 1969, and the ways in which the initial persons involved in these three cases were talked about. I wonder if you could say something more about why you chose to begin with Heisenberg and Bartlett and Harkins and criticized their first efforts? Could you say something about the quantity and quality of the nature of their judgments and whether perhaps you chose these particular persons to start with as straw men?

Bethe: I did not at all choose straw men. Quite the contrary, I mentioned those men or papers which in my opinion had been wrongly judged either in the chronology of 1968 or in conference number two. Looking at the papers given in the chronology, I found that some were of high value and others were of practically no value. Now, the Harkins business I knew already at that time. He used to go around and tell everybody that he had discovered everything first, and the situation was already clear at that time. Now, seeing his name on the chronology, I went to the trouble of reading his first paper, and I confirmed what had been my previous impression, namely, that there was no indication whatever in that paper of the phenomena which we later called the phenomena of the compound nucleus, namely, resonances, large cross sections, strong dependence on energy, and so on. There isn't the slightest bit of that in these papers. There isn't the slightest evidence in these papers that the things he called compound nuclei stayed together any longer than the time necessary for a neutron to cross the nucleus. This was totally different in the papers by Breit and Wigner, and by Bohr. Likewise, I found the Bartlett paper quoted here, and I knew about the Elsasser and Guggenheimer papers, so I searched in the Bartlett paper for something that would give me some confidence that he had seen the important points, and I didn't find it, whereas Elsasser and Guggenheimer had the important points. [N.B. By contrast, I thought that Heisenberg's paper had been unfairly treated in conference two. It is after all the paper that marked the beginning of actual nuclear theory. It stated that the nucleus was composed of neutrons and protons and that thereby a rational quantum mechanical theory of the nucleus had become possible. After this, Heisenberg proceeded to develop just such a theory. If it contained some mistakes, and some atavistic pictures like the exchange of an electron between neutron and proton, this does not matter in comparison with the great step Heisenberg took.]

Nuclear Physics in Rome
Emilio G. Segrè

INTRODUCTION: *William A. Fowler*

I spent the first days of this week on the plains of San Augustin due west of Socorro, New Mexico. There, history is being made in one of our sister sciences, radio astronomy, with the construction of the VLA, or Very Large Array, which will eventually consist of 27 radio telescopes 25 meters in diameter laid out on a Y-shaped interferometer with 21-kilometer arms. I was reminded that not far away from the VLA Site, in Alamogordo, history was made in physics and in world affairs almost thirty-two years ago. And now by the miracle of air transport I am carried in a short time more than a thousand miles in space; but more to the point, I am also carried back, along with you, more than half a century in time, by the efforts of our colleagues in the history of science, physics, astronomy, and technology here at Minnesota. We of the old guard are all grateful to them for this opportunity to relive an exciting period in our lives.

Our first speaker this morning reminds me of what I like to think of as the Second Renaissance in Italy — those remarkable days in the late twenties and early thirties when nuclear physics and especially slow neutron physics flowered in such beautiful magnificence in Fermi's laboratory. Emilio Segrè was the first of Fermi's Ph.D. graduates from the University of Rome, in 1928, where he remained on the faculty until 1935. He came to Berkeley in 1938, was a group leader at Los Alamos, and became emeritus professor at Berkeley in 1972. He is also emeritus professor at Rome. He has made many contributions in nuclear, high energy, and elementary particle physics. Most notably, he is the codiscoverer of the anti-proton, for which he received the Nobel Prize in 1959. The Organizing Committee could have chosen no one more appropriate nor more distinguished to talk to us about "Nuclear Physics in Rome" — Emilio Segrè.

Nuclear Physics in Rome
Emilio G. Segrè

In this paper I will describe the development of scientific groups in Italy, and not only in Rome. I have indicated Rome in the title, but I think this is not quite exact; I would like to expand it to Italy. This will be a historical account, of course strongly colored by physics.

Let me start by describing the state of physics in Italy around 1920. Italy had a scientific tradition. If you ask Italians, they will always start the story with Galileo, Galvani, Volta, and so on. The truth is that from the time of Galvani and Volta to almost 1920, there was an almost perfect vacuum in physics in Italy. Not in chemistry. In chemistry there was of course Avogadro, who was really of Volta's time, but then there also was Cannizzaro in the middle of the nineteenth century, who is certainly important. But chemistry was also pretty empty. Physics I would say was almost absolutely empty. Mathematics, on the other hand, was flourishing in Italy, and had been flourishing since 1870. There had been a group of Italian geometers, Cremona, Beltrami, and others, who counted in the world of mathematics as first-rate, and by 1920 there were people like Volterra, Levi-Civita, and Peano who were well known all over the

Nuclear Physics in Retrospect, edited by Roger H. Stuewer, copyright © 1979 by the University of Minnesota.

world. These men, however, were truly mathematicians. Even those who were interested in physics considered it from a mathematical point of view. Volterra, for instance, was a professor of mathematical physics — what we would today perhaps call applied mathematics; and Levi-Civita was a professor of mechanics, working mostly in differential geometry and analytical mechanics.

Mathematics prevailed and the result was that Einstein's theory of relativity was known in Italy — not only known, but actively pursued. For instance, Levi-Civita had made important contributions on the mathematical side. There is a very interesting correspondence between Levi-Civita and Einstein, which I understand was revealed in detail not long ago.[1] People were therefore open to relativity. On the other hand no one would teach quantum theory, and it was essentially ignored. The fact that relativity was known in Italy is illustrated in Figure 1. It shows Einstein in Bologna speaking with Enriques, who was a distinguished geometer.

Fig. 1. A. Einstein and F. Enriques in the courtyard of Bologna University, November 1921. (Lawrence Berkeley Laboratory, University of California)

Now, to say that there were no physicists in Italy before 1920 is not quite true. There were professors of physics in all the universities. However, if I mention their names, you probably would not recognize them. Historians of physics may have heard of Righi, but he was a little older, and in 1920 you would have heard of him only if you were an Italian. Other names, for instance, Pochettino, Cantoni, Quirino, Majorana, Puccianti, etc., you probably would ignore.

It was in this atmosphere that Fermi was born. He was essentially self-taught. And self-taught by an admirable teacher. It is always a subject of wonder to me to look at Fermi's little notebooks made when he was eighteen. In them he describes in great detail all that he studied, and he includes a detailed bibliography of the books he read. No pupil guided by an experienced professor could have made a better notebook. He never lost himself in mathematical details or in elegant mathematics; he always saw and understood what was important. It is like Einstein, who said he had a good nose for finding what was important; Fermi also had a good nose.

In any event, after secondary school, Fermi went to the Scuola Normale Superiore in Pisa, which is shown in Figure 2. It is located

Fig. 2. View of the Scuola Normale Superiore in Pisa, where Fermi studied.

in a Renaissance palace, just a hundred yards away from the Leaning Tower. Physically, the piazza and vicinity preserve the aspect and the atmosphere that Galileo saw as a student. The Scuola Normale is a college for the brightest Italian students and has generated a good part of Italy's scientific and literary talent: Volterra studied there; Enriques studied there; Fermi studied there; in more recent times, Bernardini and Piccioni studied there. If I were to make a list of prominent Italian physicists and mathematicians, perhaps 30 to 50 percent of them passed through the Scuola Normale. What Fermi studied there, on his own initiative, is reported in his very interesting correspondence with Persico, which I published in the biography of Fermi.[2] The book shows Fermi's development in physics from age eighteen to twenty-one or twenty-two.

Fermi would have become an important physicist in almost any circumstances, but he could very well have been lost to Italy and to Italian science if it had not been for a strange constellation of circumstances that occurred at that time. At Rome the professor of physics was Orso Mario Corbino, who was an extraordinary personality. He had a brilliant intelligence, not only in physics but also in world affairs, in economics, in most everything. He was a Sicilian, with very much the qualities of a Sicilian—a certain form of mind which is not easy to describe. Corbino's brother wrote an autobiography which is fascinating reading.[3] It shows the mentality and development of this Sicilian family and how they operated. Corbino came from a poor family and in a brilliant career, rose very rapidly to professor of physics in Rome, Senator of the Kingdom, and Minister of Education. He had also made important contributions to physics in his younger years.

After Fermi received his degree at Pisa, he returned to Rome where his family was living. We then see a strange phenomenon— hardly credible, but true, because Fermi himself describes it. Fermi was at that time reserved and shy, and he presented himself hesitantly to this great man, Corbino. Corbino listened to Fermi talk and, shrewd as he was, it didn't take him long to realize that he had an extraordinary person in front of him. Corbino had an undefined and vague aspiration to try and raise the level of physics in Italy. He had tried to do this in various ways, not very successfully. He now saw in the young man before him a tool for a Renaissance of Italian physics. Fermi stated that they saw each other almost daily for several months. It must have been a rather strange encounter: Fermi, age twenty-two, and Corbino, around forty-five, talking physics. From my knowledge of the two people, and from what Fermi said about this encounter, Fermi tried to bring Corbino up to date on modern physics, as he had learned it from Sommerfeld's *Atombau und Spektrallinien*. Corbino had a very critical, extremely fast mind, was well prepared, and obviously got an education. He also became convinced that Fermi would provide the means to rebuild physics in Italy.

We then see the wily Sicilian operating here and there to create a chair of theoretical physics. This may not seem to be such an extraordinary thing, but at that time there were simply no chairs in theoretical physics in Italy. There were some in mathematical physics, one of which was held by Volterra. Corbino, who was very witty, would say that mathematical physics is simply theoretical physics ca. 1830. He wanted the one of 1920. He tried, had his difficulties, and finally succeeded in having a chair created in Rome in which he installed Fermi.

However—and this is a point I should dwell upon in a little more detail—Rome was not the only center of physics in Italy which emerged in the 1920s. There were others in Pisa, Padua, and, above all, in Florence. In Florence there was, of course, no Fermi. Nevertheless, a flourishing school was founded there by the experimentalists Rossi, Bernardini, Occhialini working in cosmic rays and by the theoretician Persico, a close friend and schoolmate of Fermi's. So somehow the atmosphere was favorable not only in Rome but in other places as well. The senior man in Florence, the director of the institute, was Antonio Garbasso. He was also a Senator of the Kingdom, but not as farsighted or intelligent as Corbino. Garbasso had been, at least intellectually, a pupil of Hertz; he had worked in German laboratories. In Italy he also did what he could to favor the development of physics. However, he worked differently from Corbino. Corbino, in spite of his provincialism and Sicilian tendency to push his own, had decided that everyone who knew physics, and was good, was his protégé. Hence, he supported very strongly the Florentines, Rossi, Bernardini, Occhialini. Garbasso, on the other hand, favored mainly his own local people. Furthermore, Garbasso was an enthusiastic Fascist and nationalist, whereas Corbino was not. Corbino had been a minister for Mussolini for a while but never became a member of the Fascist party, which indicates the type of person he was.

Fermi's appointment in Rome occurred in 1927. Earlier he had been in Florence for a year or two with a temporary teaching appointment, but more important, between the ages of twenty-two and twenty-seven he had also studied abroad in Göttingen with Max Born and in Leyden with Ehrenfest. Fermi's stay in Leyden was particularly important. Fermi knew that he was a good physicist—he had no doubts about that—but he had no standards of comparison, and he said: "I am the best physicist in Italy but what does this mean?" And when he went to Ehrenfest, Ehrenfest told him: "You are not only the best physicist in Italy; you are really one of the very best physicists around." This convinced Fermi. We all knew Fermi at a time when he was extremely sure of himself; we did not see the young Fermi, but I know from Fermi himself, and from people who knew the young Fermi, that his interaction with Ehrenfest was quite important for him in developing his self-confidence and in establishing personal connections with leading theoreticians.

The accompanying photo shows the Physics Institute in Rome (Figure 3). The institute doesn't exist anymore; it has become a police station and has also been rebuilt. But this is the way it was when Corbino arranged for Fermi's appointment. (I don't have the time, but it would be interesting to describe the machinations that were necessary to secure a chair for Fermi.) Corbino also brought Rasetti to Rome. Rasetti had been a schoolmate of Fermi's. He was a brilliant student who came from a rich family with a very strong scientific tradition. Rasetti's uncle was a well-known physiologist named Galeotti. Rasetti told me, moreover, that he had learned partial differential equations and thermodynamics from his uncle. For a physiologist in 1910 to know such material was unusual. Rasetti himself was a naturalist at heart: he was a great collector of insects, a great classifier of fossils, a good embryologist—a sort of universal man. By being so universal and by having a very peculiar character, he dispersed his energies too widely. As important as his contributions to physics are, they do not fully measure his potential and talents in the field.

Fig. 3. The Physics Institute of the University of Rome, Via Panisperna 89a.

Rasetti was supposed to do for experimental physics in Rome, what Fermi was doing for theoretical physics, and he did. Together they started building a school. I was the first person who joined that group. I had a very strong interest in physics since my childhood. By hindsight I now know that I probably understood physics better than many of the professors in Italy at that time, and I had read Reiche's *Quantentheorie*.[4] In particular, I saw that there was no one in Rome who knew or at least taught significant modern physics until I met Fermi and Rasetti—first Rasetti, then, shortly thereafter, Fermi. No, that is not quite correct. I once heard Fermi when he was twenty-two or twenty-three, right after he had received his doctor's degree. He gave a talk on quantum theory in a mathematics seminar in Rome, and this made an impression on me. It was very obvious that this man really knew what he was talking about and that he was speaking of modern physics, unheard of in Rome. But at that time I only knew his name and that he was supposed to be a genius.

Later, in 1927, I started mountain climbing with Rasetti, and we became very well acquainted. Rasetti would climb a mountain, try to grab a butterfly (actually not a butterfly but an almost invisible insect of the genus Bythinus), at the same time explaining (or, more accurately, yelling about) phase space and Boltzmann's theorem to me. Then came his conclusions: you should know too that there are only three or four persons in Italy who know Boltzmann's theorem—Fermi, myself, and now you.

The first obvious proof of Fermi's position in world physics came in 1927, when there was an international conference in Como. Fermi had discovered Fermi statistics in 1926, and many famous men were present at this meeting—Rutherford, Lorentz, Compton, Bohr—everyone except Einstein, who didn't want to come into Italy as long as Mussolini was in power. You see the people who attended this conference in the familiar photograph shown in Figure 4. At this conference it became apparent that Italy was represented in physics by Fermi. There were a few older people, but the only Italian who counted was Fermi. I myself went to this conference and could see it. At the same time, I learned a lot of physics by essentially the following method: Rasetti would tell me: "Oh, look, look, there's Paschen." I would say: "All right, who's Paschen?" Rasetti would reply: "You don't know Paschen? The Paschen series? The Paschen-Back effect?" And he would then start telling me who Paschen was. In the same way, I saw Compton and learned about the Compton effect; I saw Franck, and I learned about the Franck-Hertz experiment; I saw other celebrities and learned what they had done. It was quite an education.

In Rome, in 1927, I changed my field from engineering to physics, and I convinced my schoolmate and friend, Ettore Majorana, to make the same change. I told him: "Look, Ettore, you had better come over to physics; it is more suitable for you." So he came to Fermi. But first Majorana tested him to see whether he knew mathematics. Once Majorana was satisfied, he made the switch. Amaldi came a little later. In this way, we assembled a group of physicists in Rome. As I mentioned earlier, at about the same time similar things happened in Florence. In some respects Rossi was the leader there; he was an excellent physicist, but he did not have Fermi's overwhelming superiority. He was more a first among equals with Bernardini and Occhialini.

In Rome experimental physics meant spectroscopy, because there had been a spectroscopic tradition stemming from Puccianti in Pisa

Fig. 4. Some of the attendants of the Como International Physics Conference in 1927.

and from Corbino himself in Rome. The available instrumentation was also for spectroscopic studies. But above all, these were the last years of the pioneering in atomic physics: One could still find fundamentally new things in atoms. All our experimental work in those years was mainly in atomic spectroscopy. These were the years when I personally studied forbidden lines, quadrupole lines, and had the great pleasure of seeing Sommerfeld including my work in a new edition of his *Atombau*. That was one of my first great triumphs.

It was also clear that atomic spectroscopy was in a maturing phase. Quantum mechanics had been developed, so the next step had to be something else, and that something else was fairly obvious. It had to be the nucleus. We were very conscious of this; it is not something that I see by hindsight. In fact, there is documentary proof of this: there is an extraordinary prophetic speech by Corbino in 1929 delivered in his unmistakable Italian style. He discussed this speech with Fermi, and I remember the arguments between us at that time. Although young, we already had a certain investment, especially an experimental investment, in atomic physics, and to change to nuclear physics was troublesome.

Nevertheless, Fermi convinced everyone that the transition had to be made, and we tried to do it — the "we" included Fermi, Rasetti, and myself, shown together in academic garb in Figure 5, taken around 1932 or 1933. Of course, the first step was in a spectroscopic direction, because what we knew experimentally was spectroscopy. So we tried to apply spectroscopy to nuclear physics. This led to the papers on hyperfine structure by Fermi, and the Raman effect by Rasetti, as shown in the oxygen and nitrogen Raman spectra in Figure 6.

To me, this was a very convincing argument which showed that nitrogen is a boson and hence must have an even number of nuclear particles. If one makes a nuclear model with electrons and protons, it contains an odd number of fermions, so there had to be something fundamentally wrong with that picture. I knew of the quantum

Fig. 5. From left to right: F. Rasetti, E. Fermi, and E. Segrè in Academic garb, Rome *circa* 1932.

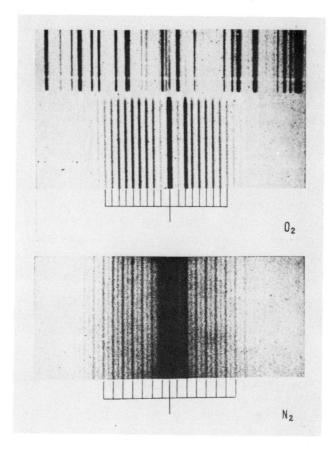

Fig. 6. Raman spectrum of the oxygen and nitrogen molecules. See F. Rasetti, *Zeitschrift für Physik*, 61 (1930), 598. (Lawrence Berkeley Laboratory, University of California)

mechanical arguments against a nucleus made of electrons and protons, but to me the above was much more convincing: this was either yes or no; an integral number can be either even or odd; it cannot be anything in between. The other arguments were all right but not as trenchant. Of course, the statistics of N^{14} was a by-product of our spectroscopic studies and was not originated for nuclear purposes.

To make the conversion to nuclear physics, we also started reading books in the field. In particular, we read very carefully the most important book on nuclear physics at that time, Rutherford's *Radioactive Substances and Their Radiations*. Then, in 1931, we convened a small conference in Rome. The people who came are shown in Figure 7. Others were invited but did not accept the invitation. Marconi, who was then president of the Italian Academy, was the host. He was inactive at that time. Corbino wrote a speech for him, and when he read it he consistently mispronounced in Italian the word for "electron." He would say "eléttrone," and we were all startled. I believe he knew English better than Italian, and he misplaced the accent by saying "eléttrone" in lieu of "elettróne."

Although Rutherford could not come to this conference, we tried to learn about current problems in nuclear physics at the Cavendish Laboratory from those who attended. This was in early 1931, a little before the discovery of the neutron. The time was premature. Never-

Fig. 7. Some of the attendants of the Rome Conference in 1931. In front, first and second lines, left to right: O. Stern, P. Debye, O. Richardson, R. A. Millikan, A. H. Compton, M. Curie, G. Marconi, O. M. Corbino, N. Bohr, F. Aston, N. Parravano, W. Bothe, C. D. Ellis, B. Rossi, A. Sommerfeld, S. Goudsmit.

Nuclear Physics in Rome 47

Fig. 8. Magnet used in Rome in 1934 to determine the sign of the charge of β-rays emitted by artificially radioactive substances. (Lawrence Berkeley Laboratory, University of California)

theless, the conference was very useful. Fermi and Rasetti subsequently made a bismuth-crystal spectrograph; Rasetti went to Meitner to learn experimental techniques; and we made some very simple instruments in Rome, as shown in Figures 8 and 9. The first is a type of trochoidal magnet; the second shows shielded Geiger-Müller counters, which were developed later.

The year 1932 was the year of the neutron. You all know the story of Bothe finding the penetrating radiation, Curie and Joliot finding that it projected protons, and the reaction of Rutherford when Chadwick reported these results to him: "I don't believe it." Rutherford was not the only one who registered disbelief. I was present when Majorana read the paper of Curie and Joliot and said:

Fig. 9. Geiger-Müller counters used in Rome around 1934.

"Oh, look at the idiots; they have discovered the neutral proton, and they don't even recognize it." This is important, because having decided that Curie and Joliot had discovered the neutral proton, which was confirmed by Chadwick very shortly thereafter, Majorana developed the neutron-proton model of the nucleus which Bethe referred to in his paper. This was a rather simple idea, which occurred not only to Heisenberg and Majorana but also to Iwanenko and Tamm, all of whom published it. I am certain that it occurred to many other people as well.

However, Majorana made a rather refined theory, and as I was thinking about Heisenberg and Majorana, a little internal evidence occurred to me which dates Majorana's idea. Majorana explained this theory to us a little before Fermi went to the Congrès International de l'Electricité in Paris in 1932.[5] Before Fermi went to this conference, as an Italian delegate, Majorana had told us about his neutron-proton model, and Fermi, recognizing its importance, encouraged him to publish it. Majorana replied: "No, it's imperfect, and this and that is wrong with it." We all said: "Well, what do you want? This is the first step, you have been thinking about it for a couple of weeks, and you want to have the perfect theory of the nucleus. Let's just start this way." Majorana said: "No." Then Fermi said: "If you want, I will mention it at the conference, give a résumé of it, of course crediting you." And Majorana answered: "Well, if you want to do it, do it. But don't credit me. Credit Professor L. L." Now, Professor L. L. (I won't say the name) was a professor of electrical engineering who knew nothing about the subject. Furthermore, because of his seniority, he was president of the conference. How could anybody do what Majorana asked? And so nothing happened. But this dates Majorana's work. I believe that it must have been either before or simultaneously with, but certainly independent of, Heisenberg's work.

There is also the point that Heisenberg made the neutron out of a proton and an electron. This surprised me when I reread his paper not long ago, because it means that the statistics difficulties are not removed. By contrast, I am quite certain Majorana knew that the neutron was a neutral proton — he called it that. So he knew that it had spin 1/2; he knew that it was the counterpart of the proton.

In the next figure, Figure 10, we see the participants at the 1933 Solvay conference, which was a turning point for physics. At this conference the great discoveries were discussed, compared, and consolidated. All the principal discoverers were there. This was the second Solvay conference, after the one of 1930, which Fermi had attended. Corbino had also been invited to the 1930 conference, but did not attend because he could not speak English or German well, although he could read these languages easily. At the 1933 conference there was a discussion about the neutrino, which had been postulated by Pauli in 1930. Someone asked if the neutrino and neutron were the same. Fermi answered: "No, no, no" and gave a clear explanation of the difference.

After the Solvay conference, Fermi returned to Rome, ruminated on the theory of β-decay, and decided that he had to learn second quantization. He had bypassed creation and annihilation operators in his famous electrodynamics article,[6] because he could not make them out very well. Now, in 1933, he decided he had to understand

Fig. 10. Solvay Conference in Brussels, October 1933. (Lawrence Berkeley Laboratory, University of California)

them. So he sat down and studied them. Then he said: "I think I have understood them. Now I am going to make an exercise to check whether I really understand them, whether I can do something with them." And he went on to set forth his theory of β-decay, which in his own estimation was probably the most important work he did in theory. He told me that he thought that this would be the discovery for which he would be remembered.

This was around Christmas 1933. I remember that we were skiing in the Alps, and one evening he explained his theory to me. By that time Hitler was almost in power in Germany, and we had rather strong feelings against him. Fermi obviously disliked Hitler, but he was more conservative in breaking with the past, and thought we should keep publishing in *Zeitschrift für Physik*.[7] We said: "No, let's publish in an English journal. Let's start publishing in English and forget the Germans." Well, after strong pressure, Fermi was persuaded to send his β-decay article to *Nature* —which promptly rejected it. There was then nothing we could do. It was sent to the *Zeitschrift für Physik*. But that was the last paper sent to that journal from Rome.

Shortly thereafter, Curie and Joliot discovered artificial radioactivity, which really shook me. I remember reading their paper in

Nature and the *Comptes rendus*, and being flabbergasted. Fermi immediately said: "This is a golden opportunity. We have a neutron source; let's make artificial radioactivity with neutrons. It's true that we have only a few neutrons, but the neutrons can die only by nuclear causes, and maybe we can compensate for their paucity by their efficiency." He then tried, and tried, all the light elements in order, without results. When he arrived at fluorine, he obtained success. That was March 25, 1934. Then he very generously asked Amaldi and me to help. At the time Rasetti was in Morocco being decorated by the Sultan—I don't know why—and refused to hasten his return. But eventually he came back, and the great period of neutron work began, the first year of which is summarized in the *Proceedings of the Royal Society*[8] and in Fermi's *Collected Papers*. I will not repeat the results here. Mainly, they showed that everything would become radioactive when bombarded with neutrons, including uranium. This led to the amazing story of the transuranic elements and everything else.

In the summer of 1934, Fermi went to South America and then came back to Rome. Amaldi and I went to Rutherford in Cambridge and were very impressed with him and his entourage. Everyone was very nice to us, helped us, and treated us very well. When we started to work again in Rome in the fall of 1934, we found incomprehensible things, such as samples being activated very strongly if they were irradiated in a certain place in a room, and very weakly if they were irradiated in another place. Miracles. We started to track these things down. At that time, we were joined by Pontecorvo. Finally one day, October 22, we found that if we placed a hydrogenous substance between the neutron and sample, the activity was multiplied by a huge factor. We found this in the morning around 11:00 A.M. We went to lunch, and when we came back at 3:00 P.M. Fermi said: "I know the reason for it. The neutrons are slowed down by elastic collisions with protons, and slow neutrons are more efficient than fast neutrons."

As soon as Fermi said that, everything became clear, because all of us had some skeleton in our closets. For instance, I knew that aluminum had two periods, sometimes one, sometimes another. I couldn't explain this and had arguments and disputes about it. Somebody else had known that the best place to activate a substance was

Fig. 11. The Rome group at the end of 1934. From left to right: O. D'Agostino, E. Segrè, E. Amaldi, F. Rasetti, E. Fermi.

on a wooden table. And so on and so forth. Thus, as soon as we were told that there were slow neutrons, everything fell into place. In the evening of the same day we wrote a letter to *Ricerca scientifica*,[9] in which we said that there were slow neutrons, that they were very efficient in disintegrating nuclei, that they may be thermalized, and we proved these things. It was spectacular. We never had anything like that thereafter. Figure 11 shows a well-known photograph taken in late 1934 of our group in Rome: D'Agostino, myself, Amaldi, Rasetti, and Fermi.

We next began to exploit the slow neutrons. This required a lot of quantitative work, which was extremely well suited to Fermi's temperament. He could calculate, make a classical theory of diffusion, and really work out all aspects of the slow neutron problem. In connection with this work Fermi invented what was later called the Monte Carlo method, but he did not publish anything on it. We also had to do experiments to confirm and guide this work. These experiments were difficult, because all our neutron sources, such as the radon-beryllium source shown in Figure 12, were very weak. We also used the ionization chamber shown in Figure 13. This very simple instrument and very constant radon and beryllium sources

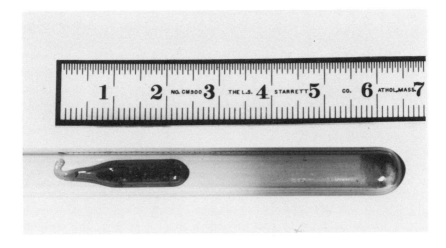

Fig. 12. Neutron source as used in Rome in 1934-36. The ampul contained radon gas and beryllium powder. (Lawrence Berkeley Laboratory, University of California)

Fig. 13. A standard ionization chamber of the type built and used in Rome. This simple instrument was usually connected to an electrometer and was built and used by the members of the Rome group wherever they went. It was jokingly called *Signum Romanum*. The figure shows an American version used by Segrè in Berkeley. The usual electrostatic electrometer is replaced by a DC amplifier and an FP 54 tube, in deference to USA styles. (Lawrence Berkeley Laboratory, University of California)

allowed us to make quite good measurements required by our slow neutron work. This technique became so popular that we made copies of our ionization chamber everywhere we went, and Fermi, joking in Latin, used to call it our *Signum Romanum*. There was a *Signum Romanum* at Columbia, at Berkeley, at Ann Arbor in Michigan, everywhere.

Our work on slow neutrons yielded important results. First, there was the experimental proof of neutron resonances. Then we found the tremendous cross section of cadmium, and the cross section of boron—in the latter case there were all kinds of doubts, because it was an (n,a) instead of an (n,γ) reaction. Fermi even thought for a while that there were two kinds of neutrons, or something very strange, but we finally found out that it was a resonance. Then we found the 1/v law, and the diffusion of neutrons in hydrogenous substances, which was to become so important afterward for the work on the bomb and nuclear reactors. We also made a wheel with a neutron source on its edge, which for the first time allowed the velocity of slow neutrons to be compared with the physical velocity of something moving in the laboratory. Later, we—Rasetti, Dunning, Pegram, myself, and others—made velocity selectors and other types of wheels at Columbia.

By that time, however, the Rome group had come on hard times. There was the Ethiopian war. The political situation in Italy was deteriorating very rapidly, and the group was subject to irrepressible forces, which forced its dissolution. It was a small group. Majorana was sick and became a recluse at home; Rasetti went to the United States, saying he didn't like what was going on in Italy; I was promoted to professor of physics in Palermo, where I discovered the first artificial element technetium; D'Agostino, for some reason, began making insecticides; and Pontecorvo went to Paris. The only people left in Rome were Amaldi and Fermi; they worked like beavers, but in a very depressing atmosphere. I used to commute from Palermo to Rome, only to find them overwhelmed with work, unwilling to talk, and in a bad mood.

In the spring of 1937 Fermi visited me in Palermo and commended the work on technetium. I still remember his complimentary words because I knew that fair evaluation, not exaggerated praise, was his habit. Little did I know then that technetium would become practically important in nuclear medicine. Occhialini couldn't find a job in Italy and soon emigrated to Brazil. In 1938 came the anti-Semitic laws and Mussolini's alignment with Hitler's policies, and everyone left. The only one who remained in Italy was Amaldi, who held the fort until the postwar period.

I now want to say a few things about Fascism, because I have often been asked about the relations of Fascism to science, to physics, and to our group in particular. Simply stated, there were minimal relations. Fermi, through the wily crafts of Corbino had been appointed to the Italian Academy and practically proclaimed the most important physicist in Italy—he was the only physicist in the Italian Academy. That obliged him to assume a certain formal attitude toward Fascism; he had to wear, when necessary, a sword, academic palms, a special hat, a special mantle, and so on, for which he received a pension. But that is about all he did for Fascism. He took no active part for or against the regime.

Fascism on its part showed a total incomprehension of the importance of science. It was a rhetorical regime, a regime devoted to show, to theatricals—not to serious things. The Fascists put Marconi in charge of the National Research Council, but he was an old man and didn't do much. Fermi himself went to Mussolini and tried to get more money for science, to make a modern institute. Mussolini received him, approved the idea, requested a memorandum which was left with him, and said it would be done—and nothing happened. The memorandum was pigeonholed. After the war, just two years ago, I came across, through a historian friend of mine, a secret dossier on Fermi kept by Mussolini's secret police. It contains all kinds of gossipy, stupid things, such as that he was on the payroll of such and such oil company to make synthetic oil. Essentially, however, the secret police had doubts about whether he was a good Fascist —he seemed rather indifferent—which of course was basically a correct appraisal of him. There was in that file, in addition, Fermi's memorandum to Mussolini and a letter of the Minister of Education, Balbino Guliano, who opposed Fermi's idea of a physics institute. He said it would arouse strong jealousies among all the other physicists in Italy. That was the reason the project was pigeonholed.

All this occurred in the mild period of Fascism. When Mussolini made the alliance with Germany, things became much more serious.

People started to quit and to emigrate. You also will notice that in the early 1930s, because of the indifference of Mussolini, we could have quite a few refugees from the Nazi regime in Rome. Some are present here.

The war disrupted all physics work in Italy. Many of the active scientists were in the military, including Amaldi and Bernardini. These two represented a more mature but still active generation, and they were important in the rebirth of physics after the war. Cosmic rays were the natural field of endeavor for a country such as Italy: first, there was a strong tradition going back to prewar years; furthermore, experiments were inexpensive, especially because of the new photographic plate technique, to which Occhialini had made fundamental contributions.

Still during the German occupation of Italy, M. Conversi, E. Pancini, and O. Piccioni at the beginning of their careers made the important observation that negative "mesotrons" from cosmic rays interacted weakly with matter. This fact presented a dilemma which was solved shortly thereafter by C. F. Powell, G. C. Lattes, and Occhialini with the discovery of the pion. The Conversi et al. experiment was performed in dramatic circumstances while the young investigators were hiding from the Germans.

There was also a strong rebirth of theory with new forces such as Wick, Ferretti, and later Regge, Radicati, and many others.

By then Italian physics, which had started as a provincial movement, had merged with the mainstream of physics all over the world.

Fermi revisited Italy in 1949 and gave a series of lectures that helped reestablish connections between Italy and the rest of the world. He saw then his scientific children and grandchildren carrying on his tradition.

REFERENCES

1. J. R. Goodstein, "Levi-Civita, Albert Einstein and Relativity in Italy," *Atti dei Convegni Lincei*, 8 (1975), 43-51.
2. E Segrè, *Enrico Fermi, Physicist* (Chicago: University of Chicago Press, 1970).
3. E. Corbino, *Racconto di una vita* (Napoli: Edizioni Scientifiche nazionali, 1972).
4. Fritz Reiche, *Die Quantentheorie* (Berlin: Springer, 1921).

5. E. Fermi, "La physique du noyau atomique" in *Congrès international d'électricité*, Paris, 1932, comptes rendus, sect. 1., rep. 22, 789-807.
6. Enrico Fermi, "Quantum Theory of Radiation," *Reviews of Modern Physics*, 4 (1932), 87-132.
7. E. Fermi, "Versuch einer Theorie der β-Strahlen. I," *Zeitschrift für Physik*, 88 (1934), 161-177.
8. E. Fermi, E. Amaldi, O. D'Agostino, F. Rasetti, and E. Segrè, "Artificial Radioactivity Produced by Neutron Bombardment," *Proceedings of the Royal Society of London*, 146[A] (1934), 483-500.
9. E. Fermi, E. Amaldi, B. Pontecorvo, F. Rasetti, and E. Segrè, "Azione di sostanze idrogenate sulla radioattività provocata da neutroni," *La ricerca scientifica*, 5 (1934), 282-283.

DISCUSSION

Wilson: There appeared recently a book by Leonardo Sciasia, *La Scomparsa di Majorana* (Einaudi: 1975), a prolific Sicilian author, and I wonder if you could comment on the quite different account of the same period that is portrayed in that book.

Segrè: I would say that it is hard to find much truth in that book. It is based upon a completely gratuitous hypothesis, namely, that Majorana absconded to avoid the possibility of being involved in the making of the atomic bomb. Now, note that Majorana disappeared in early 1937, in the spring, long before the discovery of fission. That in itself tells you that the hypothesis is nonsense. Sciasia's book has political purposes. I hoped it would never be mentioned. Sciasia is a Communist political personality in Palermo. The main purpose of the book is to speak against scientists in general and in particular against American scientists.

Bethe: I think I remember that in 1932 Fermi was already interested in learning the techniques of experimental nuclear physics. Is that correct?

Segrè: That's correct. I said that. I said that we made instruments even earlier than that, a bismuth-crystal spectrograph, a cloud chamber, all built between 1930 and 1934, so that when artificial radioactivity was discovered by I. Curie and F. Joliot we already had a little momentum. This included the making of sources, which was an important thing in which Professor Trabacchi had an important part.

Malley: My question concerns Marie Curie. There are two aspects to my question. I have read a good deal of uncomplimentary things about her personally, that she was a difficult person to deal with. The other aspect of my question concerns her work and that of her school. I can see from my own reading that there was not too much going on until the work of her daughter and son-in-law. I wondered if this was a general impression at that time. I have some correspondence in which people say this privately. If this is true, was she invited to these various conferences as a matter of duty, or was there some other reason? Was there some fear of offending France?

Segrè: I didn't know her well. I only met her occasionally. I had practically no dealings with her. I know that when she asked me to erase the blackboard for her she was never pleased with the result. She said, "The blackboard is dirty. Please erase it again." And I did. That's as far as it went. I tried to visit her laboratory, but the doorman wouldn't let me in. Why, I don't know. Maybe she wasn't there. Her laboratory was important for her own work until the time of the first world war and later for the work of I. Curie and F. Joliot. Her laboratory remained in the forefront of radioactivity. It was not as important as the Cavendish, but it was an important place. After all, what were the places in the world at that time for nuclear studies? There was the Cavendish, Paris, a little activity in Vienna, and a little in Berlin. That's all.

Malley: So it was generally respected. My own impression was that there was nothing after about 1903-4, that it was mainly following up other people's work . . .

Segrè: No, no, Madame Curie had an important part in the disentangling of radio chemistry, and of all the natural radioactive substances for many, many years.

Malley: Are you talking about the period after the first world war?

Segrè: No, until about, say, 1912.

Malley: Related to this, someone recently wrote a biography of her which I felt was unfair to her. It suggested that she was not very productive, and I wonder if you could comment on this. Do you feel that her intellect was up to her reputation?

Segrè: Look, anyone who did what Madame Curie did could be very proud of it. I think she did more than enough.

Peierls: On this question of invitations to conferences, I had the good fortune of being at the 1933 Solvay Conference. That was the first conference to which very young people were invited. Many people were invited to such conferences because they were famous and had done distinguished work in the past but did not have much to say on topical discussions. However, this was evidently not the case for Madame Curie. I do not now recall the details, but Hans Bethe evidently remembers what I said when I returned from that conference, which was that it was impressive to see how much she was abreast with what was going on, and in fact very frequently corrected her daughter and son-in-law on points of detail.

I would also like to comment on your [Segrè's] characterization of Fermi's unusual and extremely impressive way of doing things. I have formed an impression, and I wonder if you would agree, that Fermi deliberately looked only at things that were simple. He hated complications, and that goes with his reluctance to get involved with creation and absorption operators until it was absolutely necessary. Of course, the range of things that were simple to him covered very many things which were complicated to all of us until he explained them. But with certain problems it wasn't possible to get beyond a certain stage without getting involved with complex mathematics, and then he generally left them. He didn't choose to go beyond that. I remember seeing some notes of his on the critical size of U^{235} for fast neutrons, where he did a rough calculationn on the back of an envelope which certainly was very crude and did not get anywhere near the correct number, and he knew that. But he left it there. He didn't think it was worthwhile to do it better, because there would be a certain amount—not very much—of mathematical complexities.

Segrè: I would like to make two comments on this. One is that mathematics was not an obstacle to Fermi. Fermi knew mathematics very, very well and could do anything with mathematics; he just didn't like it. That's different from other people who did not know much mathematics. Fermi was certainly technically highly proficient in mathematics. In fact, he even occasionally wrote papers in pure mathematics. Second, as all who were acquainted with him knew, whenever there was a problem, Fermi reduced it to three or four schemes that he knew quite well, and of course no one but he could see that that problem was a particular case of Golden Rule Number 2, or of the scattering length, or of one of the other five or so things he knew. Everything was reduced to these five things, but it took Fermi to make the reduction, which is what you were saying.

Feenberg: I've been curious for some time about the early history, or possible early history, of the neutron-proton nuclear model. If you look at the experimental information available, say, in 1928, it's perfectly obvious with hindsight that there is a neutron-proton structure of the nucleus. Were there any speculations in the days before the neutron was discovered, or any thought or discussion, of the neutron-proton structure of the nucleus? Was there a neutron-proton theory of the nucleus before the neutron was discovered?

Segrè: Well, of course, nowadays everyone talks of quarks and no one has seen them. No one is upset about making a model of something that no one has seen. Maybe earlier physicists were somewhat more reluctant to make a theory of the nucleus containing neutrons and protons before free neutrons had been seen.

Wheeler: If my memory is not mistaken, I read a paper or a lecture of Lord Rutherford's in which he talked of this long before the neutron was discovered. What does one say? Chance favors the prepared mind. Why Chadwick found the neutron when others who might have been expected to see it did not, I can well believe rests as much on the climate of opinion around Rutherford as on anything else.

Segrè: Rutherford was convinced of the existence of the neutron, and he mentioned it in detail in the Bakerian Lecture of 1920, as everyone knows. The neutron of Rutherford was a collapsed hydrogen atom with the proton inside of the electron, like Heisenberg's picture. But the point of view of course was different. In 1920 no one knew about spins, statistics, or anything like this. Rutherford was really concerned with how neutrons would move, and what they would do, and made some very shrewd guesses on this, and he tried to find the neutron.

Peierls: I think that Rutherford did not think of the neutron as a constituent of the nucleus. He did expect it to exist, but not as an essential ingredient in building up the nucleus. I believe his motivation came from the difficulty of understanding how a-particles can get out of the nucleus when their energy was so much lower than the potential barrier. Of course, that was before quantum mechanics and before Gamow. The thought occurred to him that a-particles somehow were uncharged to get out of the field and acquired their charge later. I think that it was in that context that the concept of the neutron came up, and it had nothing to do with structure.

Badash: I believe that the Bakerian Lecture of 1920, in which Rutherford advanced the idea of the neutron, was based upon mistaken experimental evidence. What happened was that he thought he had a light helium atom, an a-particle with mass three. This turned out to be a different a-particle which had not been known beforehand. But he reasoned that one electron could bind three protons, and this would produce what he had found, or thought he had found. Next he extrapolated and said that one electron might bind two protons to give heavy hydrogen. Then he continued and said that one electron might bind one proton, and this would produce the neutron. In fact these all were found, but the basis of his argument was fallacious.

Fowler: Emilio, you mentioned the vacuum in physics in Rome, but you also mentioned that you had been in engineering before you transferred into physics. What was the general situation in Italy in science, in engineering, in chemistry, in astronomy, and in medicine, for example in X-rays and radiology? What was the whole scientific background at that time?

Segrè: Italy was an underdeveloped country in some fields of science, but it had a tradition of having been the leading country for many centuries in many things. I could assess the various fields of science roughly in this way: strong in mathematics on an international scale—everyone knew of the work of Italian mathematicians in various fields; not much in astronomy; almost nothing in physics; some chemistry, but little. In biology, Italy was fairly good. Golgi had won the Nobel Prize; and there had been well-known studies on, for instance, parasitology and malaria. The parasitological part of medicine and some specialities such as orthopedics had also been strong in Italy. Engineering, in particular electrical engineering, had been good. In fact, there had been Galileo Ferraris, a scientific electrical engineer certainly comparable to the very best engineer in the world. So it was a mixed state of affairs.

Fowler: But then why did you transfer from engineering to physics?

Segrè: First, my true interest, since childhood, had been in physics and not engineering. The first two years in engineering school were very strong scientifically, so I learned mathematics well and physics and chemistry reasonably well. Then when I went on in the engineering school, I found a drop in level, and at the same time the golden opportunity of joining Fermi and Rasetti arose. I recognized it and switched to physics.

May: In somewhat the same vein, I would like to inquire into the nature of support that the various centers of physics in Italy gave to one another, that is, Florence, Padua, Pisa, and Rome. What were the interconnections, the relationships among them? I do not mean the financial support they had, but the interconnections between these centers of physics.

Segrè: Corbino would support anyone who was really good. Garbasso would support anyone who was good and especially from Florence. Other people, important people, would support only stupid people, because they were jealous. This was notably true in chemistry, which was profoundly damaged in Italy by a policy of appointing mediocre people.

May: Were there colloquia that people attended? For example, did the Romans attend the seminars in Florence?

Segrè: Yes. There were exchanges especially between Rome and Florence. I remember going many times to Florence, and Racah, Rossi, and Bernardini coming to Rome to see Fermi. I remember, for instance, Rossi coming to Rome with some good ideas about cosmic ray orbits and the screening of the earth's magnetic field. He could not do some of the difficult mathematics, so he came to Fermi, who immediately did it all for him, and they wrote a very nice paper together on the subject.

Experimental Work with Nuclei: Hamburg, London, Copenhagen
Otto R. Frisch

INTRODUCTION: *William A. Fowler*

We now travel north and west in Europe to hear more about the early days in nuclear physics. Our next speaker, Otto Robert Frisch, is Viennese born and Viennese educated. His contribution to the unraveling of the mystery of nuclear fission is well known to all of you. It was indeed a masterly contribution, but I like to think of Otto Frisch rather for the clever and ingenious and yet simple instrumentation that has marked his life work in experimental physics. Frisch was at Los Alamos during the war and at Harwell in the early days of that establishment. I first really got to know Otto in 1954 when I had a Fulbright in the Cavendish and he was the Jacksonian Professor of Natural Philosophy, a post which he held from 1947 to 1972 — how I envy the designation Natural Philosophy! He was also a Fellow of Trinity College, Cambridge — the College of Newton and Thomson and Rutherford. He will tell us about "Experimental Work with Nuclei: Hamburg, London, Copenhagen." Otto Frisch.

Experimental Work with Nuclei: Hamburg, London, Copenhagen

Otto R. Frisch

This report is largely autobiographic. Born (1904) and educated in Vienna, Austria, I got my degree (Doctor of Philosophy) from the University of Vienna in 1926. About a year later, after a brief employment in industry, I was offered a grant in Berlin where I worked at the Physikalisch-Technische Reichsanstalt (counterpart of the Bureau of Standards). There I first met nuclear physics (Walther Bothe sent a technician to ask me not to whistle in the corridor; it confused him in counting particles) and discussed it with my mother's sister, Lise Meitner, who was working at the Kaiser-Wilhelm Institut für Chemie with Otto Hahn.

At the time her chief worry was the continuous spectrum of the primary β-rays. At first she thought that was a secondary effect, perhaps due to random energy loss by radiation as the β-electron leaves the nucleus; but that explanation was refuted by the well-known experiment of Ellis and Wooster (1927),[1] confirmed by Lise Meitner and one of her students, Wilhelm Orthmann, in 1929.[2] It seemed incredible that the transition between two well-defined long-lived nuclei should lead to the emission of electrons with a continuous energy distribution; Peter Debye is said to have called it "a topic

Nuclear Physics in Retrospect, edited by Roger H. Stuewer, copyright © 1979 by the University of Minnesota.

best not talked about, like new taxes." Pauli's famous letter beginning "Dear radioactive Ladies and Gentlemen" in which he proposed the neutrino was written, as I recall, to Lise Meitner and Hans Geiger in 1930.[3]

My own work, an attempt to design a new unit of luminosity, was abandoned when my grant after three years came to an end, and after a few months of work on spectroscopic problems with Peter Pringsheim at the University of Berlin I got a job at Hamburg, with Otto Stern, head of the Department of Physical Chemistry. He was systematically developing and exploiting the method of molecular beams for studying the diffraction of beams of helium atoms and hydrogen molecules from crystal surfaces, and extending the sensitivity of the Stern-Gerlach method to nuclear moments. When he proposed to measure the magnetic moment of the proton by applying the method to molecular hydrogen, this was widely regarded as an unnecessary tour de force; everybody "knew" that the moment of the proton must be 1800 times smaller than the electron moment. Our measured magnetic moment, about 2.5 times greater, was the first clear proof that the proton did not follow Dirac's theory.

Stern asked Fermi for an estimate of the electric moment of inertia of a hydrogen molecule; Fermi handed that task to Bethe. From this a magnetic moment of about three nuclear magnetons was computed and appeared to agree reasonably with the rather large moment which our experiments indicated. But in the meantime Stern had studied the methods of Farkas for making almost pure parahydrogen by cooling hydrogen with charcoal and decided that we should try that. In parahydrogen the two proton moments must cancel, and indeed at low temperature (liquid air) no magnetic splitting of the beam was found,[4] but even at room temperature the observed broadening of the beam (no real splitting could be seen with those small moments) corresponded to a rotational moment at least three times less than the prediction based on Bethe's estimate. Fermi pointed out that the electrons would slip, and an improved calculation by Gian Carlo Wick agreed with the smaller moment; this showed that the large magnetism we had observed with ordinary hydrogen (i.e., 75 percent orthohydrogen) must be due to a proton moment of two to three nuclear magnetons. The much more accurate value found later by Rabi, by his resonance method, still lay within those limits.[5]

Another experiment, on the spin flip of oriented atoms in rapidly changing magnetic fields, published with Emilio Segrè in 1933, had some bearing on nuclear moments.[6] After we published that paper we got a letter from Rabi who was interested in the fact that we got less spin flip than ideally we should have got; he pointed out that in the weak field where flipping took place the coupling between electron and nuclear spin would be unbroken and that the reduced flip rate gave an indication of the nuclear spin. That letter (addressed to Stern no doubt) might be seen as a step toward Rabi's resonance method.

In 1933 Hitler's racial laws came into force, and the University of Hamburg, rather later than most others, finally had to sack its nonaryan staff members, myself included. The Academic Assistance Council in Great Britian gave me a grant (£250 a year, quite enough in those days), and Blackett offered me a place at Birkbeck College, London, where I arrived in the autumn of 1933. Here I met nuclear physics in earnest.

The first experiment I started, at the suggestion of Blackett, was a search for the annihilation radiation from the positrons which we could make, following the recent discovery (June 1933) by Irène Curie and Frederic Joliot that positrons are emitted from aluminum during bombardment with a-particles. We had just started testing our apparatus (in which the positrons were guided in trochoidal paths along the edge of a strong magnetic field, a method used previously by Jean Thibaud) when positive results were published simultaneously by Curie and Joliot[7] in the Institut du Radium in Paris and Jean Thibaud[8] in the Laboratoire de Physique des Rayons X in Paris, and we abandoned our experiment. But very soon afterward it was noted in Paris that the positron counter went on counting for some minutes after removal of the a-ray source; replacing the counter didn't help, and within hours the existence of the "artificial radioactivity" of the nucleus P^{30} was recognized.[9] Like many others, I jumped on the bandwagon but lost a week by building a spring-operated transfer system for moving the sample in a fraction of a second from the heavily shielded a-ray source to the Geiger counter, trying to stake a claim on short-lived elements. But both of the two nuclei I found, Al^{26} and Cl^{34}, with half-lives of 7 seconds and half an hour, could have been found without that special arrangement. Even so, thanks to Blackett who phoned *Nature* and sent me across London with the manuscript, my note[10] was published inside of a week, probably a record; only Wertenstein in Poland published a paper before me on "induced" radioactivity.[11]

Leo Szilard occasionally turned up and discussed his latest ideas. That was after Fermi had—within a month from the announcement of "induced radioactivity"—published the use of neutrons for that purpose.[12] I remember that Szilard first suggested the use of alternate layers of an element activated by neutrons and another to serve as catcher, the layers to be separated after irradiation and the capture products (isotopic with the activated elements) to be extracted from the catcher layers, thus producing high specific activity. Next he talked about mixing powders rather than using alternate layers of the two materials; this led finally to his experiment with T. A. Chalmers in which the subsequent extraction depended on a change in the chemical bonding of the struck nuclei.[13]

The rest of my time in London I used to construct a cloud chamber in which continued expansion produced a sensitive time of about one second; it was fine for demonstration but never produced scientific results though it was improved in Denmark by a student called Brostrøm, later a professor in Aarhus.

When my grant in England came to an end after a year, I went to Copenhagen where Niels Bohr had arranged for a grant from the Rask-Ørsted foundation; that supported me for nearly five years until I came back to England in the summer of 1939.

In Copenhagen I made myself useful by building counters and amplifiers, and found two more radioactive elements, Na^{22} and Sc^{43}, by α-bombardment.[14] Jackson Laslett came from Ernest Lawrence's lab to help us design a cyclotron, and I also designed some of the pieces; Niels Bohr got the Danish firm of A. B. Thrige in Roskilde to make us a magnet free of charge. George Hevesy raised Kr. 100,000 from the people of Denmark to present Bohr with 0.5 gram of radium for his fiftieth birthday; mixed with beryllium powder it was our source of neutrons until a Cockcroft-Walton accelerator started working early in 1939. We were lucky: none of the dozen people I bullied into grinding beryllium had the allergy, discovered much later.

George Placzek had left for Israel soon after I came to Copenhagen, and then I was the only one with enough knowledge of Italian to give an extempore translation of Fermi's latest results, each time a new issue (or preprint?) of *La ricerca scientifica* arrived. Once we had our neutron source, we got rhodium foil and began to repeat and extend Fermi's work; James Franck took an active part for a while. Hevesy noted that quite a thin gold foil became more active on the side where it was struck by slow neutrons and that this effect became more pronounced behind a sheet of cadmium; here was another example of the "neutron groups" reported from Rome.

It must have been late in 1935 that Bohr conceived his idea of a compound nucleus as a long-lived intermediate state in a nuclear reaction. I vividly remember the occasion: Bohr repeatedly (more than usually) interrupted a colloquium speaker who tried to report on a paper (by Hans Bethe, I believe) on the interaction of neutrons with nuclei; then, having got up once more, Bohr sat down again, his face suddenly quite dead. We watched him for several seconds, getting anxious; but then he stood up again and said, with an apologetic smile, "Now I understand it all"; and he outlined the compound nucleus idea.

George Placzek had come back from Israel (preceded by his famous telegram "Through with Jews forever"—they had insisted that he lecture in Hebrew—and asked me to help him with some experiments, making use of a slow neutron detector, a boron-lined ionization chamber, which I had built. The resulting paper was published in *Nature*,[15] in the same issue in which Bohr's lecture to the Videnskabernes Selskab of January 27, 1936, was published. I hadn't really understood what Placzek was after until we started to draft our joint account, urged on by Bohr. The aim was to measure the absorption in boron of those slow neutrons which were transmitted by cadmium; it was found to be only a few times smaller than the absorption of the thermal neutrons, and Placzek concluded that the width of the cadmium absorption region was 1 eV or less.

In a paper submitted in November 1935,[16] Leo Szilard had described the selective absorption of neutrons and indeed indicated the basic idea of our experiment in the sentence: "An attempt is now being made to determine the energy value of these absorbing regions by studying the absorption in boron and lithium of the 'highly absorbable' components of the residual beam." I certainly hadn't noticed that (probably not read the paper) and probably Placzek hadn't; he was a bit embarrassed when Szilard wrote him a letter pointing it out, quite gently.

In a later paper I measured by the same method, technically improved, the resonance energies of gold, arsenic, and iodine.[17] In gold the self-absorption was measured over a large range of thicknesses and confirmed the shape expected for a single resonance level; in arsenic and iodine there were indications that more than one level contributed. An attempt to observe the Doppler broadening by heating the gold foil was inconclusive.

Hans Halban and Jørgen Koch then joined me, and the main result was three papers, submitted in February and March 1938.[18] The first two concerned the magnetic behavior of neutrons and was based on the discovery that slow neutrons could be partly polarized by passage through magnetized iron. We got a rough value for the magnetic moment, and the correct sign, by observing the effect of transverse and longitudinal magnetic fields of a few gauss, placed between the analyzer and polarizer, flat rings of mild steel, wound with a single toroidal layer of enameled copper wire. The statistical errors were quite large, and doubt was cast (privately) on our results by Felix Bloch, Morton Hamermesh and Hans Staub who measured the polarizing effect very carefully[19] and found that it was negligible with remanent magnetism, which we used. I wish to make no claim, except for having made the first attempt to observe the precession of polarized neutrons.

The other experiment was more significant. We used the method of Fermi and Amaldi, immersing a neutron source in a liquid hydrogenous medium and integrating over its volume. We did that by exploring the distribution of slow neutrons with a small detector, in our case a small boron-lined ionization chamber with pulse amplifier. By comparing the intensity integral in different liquids, we got upper limits for the absorption cross section of deuterium, carbon, and oxygen, and fairly good values for hydrogen and nitrogen, all calibrated by several solutions of boric acid in water.

I believe that was the first experimental evidence that heavy water and graphite were possible moderators when it came to consider nuclear reactors. By the way, Harteck told me a few years ago that soon after the discovery of fission he considered solid carbon dioxide as a moderator for proving the possibility of a low power chain reaction; it would probably have been easily obtained at sufficient purity, but of course useless at higher power.

In the summer of 1938, soon after the take-over of Austria, Lise Meitner was informed of an order (by Himmler, I believe) that scientists were not to be allowed to leave Germany. Her application for permission to travel abroad was indeed refused; her dismissal under the racial laws seemed imminent since all her grandparents had been Jews. Peter Debye communicated her plight to his Dutch colleagues, and she was persuaded to travel to Holland without a visa; a Dutch physicist, Coster, met her train at the frontier, having arranged that the Dutch immigration officer would let her in. From Holland (where there was little nuclear physics going on at the time) she went to Copenhagen, where she stayed for a number of days as Niels Bohr's guest; finally she accepted an offer from Manne Siegbahn in Stockholm, where a cyclotron had just been built (the first on the continent of Europe).

I feel embarrassed to tell once again the story how she asked me to join her at a hotel in Kungälv, a few miles north of Gothenburg, where Swedish friends had invited her to Christmas dinner, and how I found her at breakfast poring over a letter from Otto Hahn. In an earlier letter (which seems to have been lost) she had urged him not to publish the alarming result that radium isotopes resulted from neutron bombardment of uranium, at least not until he had checked that result with the greatest care. That check had now been made, with Fritz Strassmann, and had shown that they had isotopes of barium, not radium.[20] As we talked this over, we came to consider Bohr's analogy between a nucleus and a drop of liquid; I estimated the loss of surface tension due to the nuclear charge and the energy freed in the electrostatic repulsion between the two resulting droplets, overestimated because the fragments were assumed to be of equal size; at the same time Lise Meitner estimated the energy freed according to the packing fraction, also overestimated because the energy subsequently lost by several β-decays was included. So we got good agreement, about 200 MeV, indicating that the process was classically possible; the true figure is more like 160 MeV.

When I told Bohr a few days later, back in Copenhagen, he instantly agreed with our interpretation; he had been aware of the reduced surface tension of heavy nuclei and had correlated it with the narrower spacing of their lowest levels; he called himself a fool for not having predicted fission! I proceeded to draft a joint paper and discuss it with Lise Meitner (back in Stockholm) over the telephone. But it took Placzek to make me think about experimental proof. He was skeptical; the idea that uranium should be liable to fission as well as a-decay, he said, was like dissecting a man killed by a fallen brick and finding that he would have soon died from cancer—an unlikely coincidence!

Placzek suggested a cloud chamber to look for those fission fragments; that was done much later by others and revealed a wealth of secondaries, making the tracks look like Christmas trees. But I quickly made up a very simple ionization chamber—two parallel metal plates separated by a glass ring about 1 cm high; a foil with a layer of uranium hydroxide could be placed on the bottom plate. One of our neutron sources, 100 mg radium mixed with beryllium, was placed close to the chamber; the bias of the amplifier was adjusted until both the a-ray pulses and the noise caused by the γ-rays from the source were suppressed.

My notes of January 13-14 (found and inspected recently for the first time after thirty-eight years) showed that measurements were started in the afternoon, and pulses at about the predicted amplitude and frequency (one or two per minute) were seen within a few hours; but the measurements were continued until six in the morning to verify that the apparatus was working consistently, that the counting rate was only a few counts per hour in the absence of either the uranium or the neutron source, and that the response to variations of the bias and the insertion of paraffin were as expected.

On Monday, January 16, I observed fission pulses from thorium but not from lead. On the same day I sent two letters to the editor of *Nature*, one by Lise Meitner and myself[21] in which our explanation of the results of Hahn and Strassmann was put forward and the term "nuclear fission" proposed; the other in my name only,[22] reporting the results described above. They appeared on February 11 and 18 respectively.

In March Lise Meitner came to Copenhagen for several days, and together we did a recoil experiment, catching recoil fragments and then measuring their radioactive decay; Lise Meitner also did the usual H_2S precipitation both on those recoil nuclei and on irradiated uranium and showed that the decay curves were the same. In our joint paper[23] the conclusion was that all the "transuranic" products studied in Berlin had been fission fragments; with that the whole implausible edifice of parallel isomeric decay chains collapsed.

Actually that conclusion (though it was later confirmed) was not quite convincing; the decay curve was obviously due to many different bodies and was largely statistical in character. Soon afterward I wrote my only theoretical paper[24] (it contained two integral signs!) by assuming an infinite number of bodies with decay energies uniformly distributed, the decay constant of each being proportional to the nth power of the decay energy. That gave, for n = 5 and 7, two curves close to the experimental results of Bjerge, Brostrøm, and Koch; it showed that my very simplistic assumptions were good enough to predict to ± 10 percent the activity of a sample for the next five hours from the activity measured after three minutes. This good agreement proves only that little can be learned by observing such a mixture of activities.

A few months later I went to England; the war broke out, and for some years I ceased to pursue research for its own sake.

REFERENCES

1. C. D. Ellis and W. A. Wooster, "The Average Energy of Disintegration of Radium E," *Proceedings of the Royal Society of London*, 117[A] (1927), 109-123.

2. Lise Meitner and Wilhelm Orthmann, "Über eine absolute Bestimmung der Energie der primären β-Strahlen von Radium E," *Zeitschrift für Physik*, 60 (1930), 143-155.

3. See "Zur älteren und neueren Geschichte des Neutrinos" in Wolfgang Pauli, *Collected Scientific Papers*, ed. R. Kronig and V. F. Weisskopf (New York: Wiley, 1964), vol. 2, pp. 1313-1337, especially pp. 1316-1317.

4. [O.] R. Frisch and O. Stern, "Über die magnetische Ablenkung von Wasserstoffmolekülen und das magnetische Moment des Protons. I," *Zeitschrift für Physik*, 85 (1933), 4-16; I. Estermann and O. Stern, "Über die magnetische Ablenkung von Wasserstoffmolekülen und das magnetische Moment des Protons. II," *Zeitschrift für Physik*, 85 (1933), 17-24.

5. I. I. Rabi, J. M. B. Kellogg, and J. R. Zacharias, "The Magnetic Moment of the Proton," *Physical Review*, 45 (1934), 761; *Ibid.*, 46 (1934), 157-163; "The Sign of the Magnetic Moment of the Proton," *Ibid.*, 49 (1936), 421-422.

6. [O.] R. Frisch and E. Segrè, "Über die Einstellung der Richtungsquantelung. II," *Zeitschrift für Physik*, 80 (1933), 610-616; "Ricerche sulla quantizzazione spaziale," *Nuovo cimento*, 10 (1933), 78-91.

7. Irène Curie et F. Joliot, "Sur l'origine des électrons positifs," *Comptes rendus*, 196 (1933), 1581-1583; "Électrons de matérialisation et de transformation," *Journal de physique et le Radium*, 4 (1933), 494-500.

8. Jean Thibaud, "Déviation électrostatique et charge spécifique de l'électron positif," *Comptes rendus*, 197 (1933), 447-448; "Electrostatic Deflection of Positive Electrons," *Nature*, 132 (1933), 480-481. "Étude des propriétés physique du positron," *Comptes rendus*, 197 (1933), 915-917.

9. Irène Curie et F. Joliot, "Un nouveau type de radioactivité," *Comptes rendus*, 198 (1934), 254-256; Erratum, *Ibid.*, 408. "I. Production artificielle d'éléments radioactifs. II. Preuve chimique de la transmutation des éléments," *Journal de physique et le radium*, 5 (1934), 153-156.

10. O. R. Frisch, "Induced Radioactivity of Sodium and Phosphorus," *Nature*, 133 (1934), 721-722.

11. L. Wertenstein, "An Artifical Radioelement from Nitrogen," *Nature*, 133 (1934), 564-565.

12. E. Fermi, "Radioactivity Induced by Neutron Bombardment," *Nature*, 133 (1934), 757; E. Fermi, E. Amaldi, O. D'Agostino, F. Rasetti, and E. Segrè, "Artificial Radioactivity Produced by Neutron Bombardment," *Proceedings of the Royal Society of London*, 146 [A] (1934), 483-500.

13. Leo Szilard and T. A. Chalmers, "Chemical Separation of the Radioactive Element from Its Bombarded Isotope in the Fermi Effect," *Nature*, 134 (1934), 462; Bernard T. Feld and Gertrud Weiss Szilard, eds., *The Collected Works of Leo Szilard* (Cambridge: M.I.T. Press, 1972), vol. 1, pp. 139-149.

14. O. R. Frisch, "Induced Radioactivity of Fluorine and Calcium," *Nature*, 136 (1935), 220.

15. O. R. Frisch and G. Placzek, "Capture of Slow Neutrons," *Nature*, 137 (1936), 357.

16. Leo Szilard, "Absorption of Residual Neutrons," *Nature*, 136 (1935), 950-951.

17. O. R. Frisch, G. Hevesy, and H. A. C. McKay, "Selective Absorption of Neutrons by Gold," *Nature*, 137 (1936), 149-150; O. R. Frisch, "On the Selective Capture of Slow Neutrons," *Det kgl. danske videnskabernes selskab. Mathematisk-fysiske meddelelser*, 14, no. 12 (1937), 3-31.

18. O. R. Frisch, H. von Halban, Jun., and Jørgen Koch, "A Method of Measuring the Magnetic Moment of Free Neutrons," *Nature*, 139 (1937), 756-757; "Sign of the Magnetic Moment of Free Neutrons," *Ibid.*, 1021; "On the Slowing Down and Capture of Neutrons in Hydrogenous Substances," *Det kgl. danske videnskabernes selskab. Mathematisk-fysiske meddelelser*, 15, no. 10 (1938), 3-37.

19. F. Bloch, M. Hamermesh, and H. Staub, "Neutron Polarization and Ferromagnetic Saturation," *Physical Review*, 62 (1942), 303; *Ibid.*, 64 (1943), 47-56.

20. O. Hahn and F. Strassmann, "Über die Entstehung von Radiumisotopen aus Uran durch Bestrahlen mit schnellen und verlangsamten Neutronen," *Die Naturwissenschaften*, 26 (1938), 755-756; "Über den Nachweiss und das Verhalten der bei der Bestrahlung des Urans mittels Neutronen entstehenden Erdalkalimetalle," *Ibid.*, 27 (1939), 11-15; "Nachweis der Entstehung aktiver Bariumisotope aus Uran und Thorium durch Neutronenbestrahlung; Nachweis weiterer aktiver Bruchstüke bei der Uranspaltung," *Ibid.*, 27 (1939), 89-95.

21. Lise Meitner and O. R. Frisch, "Disintegration of Uranium by Neutrons. A New Type of Nuclear Reaction," *Nature*, 143 (1939), 239-240.

22. O. R. Frisch, "Physical Evidence for the Division of Heavy Nuclei under Neutron Bombardment," *Nature*, 143 (1939), 276.

23. Lise Meitner and O. R. Frisch, "Products of the Fission of the Uranium Nucleus," *Nature*, 143 (1939), 471-472.

24. O. R. Frisch, "Statistical Calculation of Composite Decay Curves," *Nature*, 143 (1939), 852-853.

DISCUSSION

Malley: I have a couple of questions about Lise Meitner's work before about 1920. I know that she was influenced by quantum theory in her β-ray ideas after that date. I have read her papers before that; I know she studied with Planck; I know that her β-ray ideas were supposedly because of the exponential absorption. I am wondering if you have any idea or knowledge of whether she was influenced by quantum theory in her ideas of the homogeneous β-ray groups before 1920, or whether it was just the exponential absorption?

Frisch: I am quite sure that she was very much impressed with the successes of quantum theory and that this was one of her main reasons to pursue physics, that she felt physics was in a great new development and that we now would know the nature of matter and so on much better. She had apparently been fascinated by physics from her young days and was very much convinced that the laws of Nature can be found, if one only works hard enough. At the same time, she sometimes was very convinced of what would be the outcome of an experiment and was for that reason very reluctant to admit, for instance, that the β-ray spectrum should be genuinely continuous, because it was incomprehensible how the transition from one well-defined state of a nucleus to another well-defined state could give anything but a line spectrum, in fact a spectrum consisting of a single line (the many observed lines were known to be secondary to γ-transitions). I have often been asked whether her conviction of what in Nature would turn out to be right, whether her conviction sometimes held her back. I am sure that is the case.

She occasionally got trapped in a blind alley because she was so convinced that what somebody else, or even what she herself, had found could not be true, because it didn't agree with her view of the laws of Nature. On the other hand, I think it was precisely that strong belief in the rationality of Nature which made physics so important and so interesting to her, and she might not have had nearly the persistence she always had if it was just a matter of picking up clues and plowing along as best one could. She was always trying to prove something.

Malley: She was correct in her idea about the γ-rays though, and I am wondering if there was a relation to her theory of the β-rays? The idea that the γ-rays came from the nucleus was pretty unusual before the 1920s, and I am wondering if you know whether she held such an idea privately before that time? I have only found it published by a few people before the 1920s, and not by her.

Frisch: The idea that the γ-ray was emitted after the β-ray was in fact defended by her against, I think, the British School, who for a time thought that the emission of a γ-ray was what then caused the subsequent emission of an α- or β-particle. And Lise Meitner on one occasion was able to prove that at least for an α-emitter which also showed secondary electrons, which in those days were called β-rays, that these electrons could only be correctly accounted for by assuming that the α-emission preceded the emission of those secondary electrons, and that these secondary electrons in turn were, as she thought, created by internal absorption of the γ-rays. This came to be called internal conversion, because it is a much more intense transformation than if you simply assume that the γ-ray, so to speak, walks away from the nucleus and all of a sudden is then captured by an electron. It is rather that the nucleus, excited and ready to emit a γ-ray, can as an alternative transmit its energy to one of the shell electrons. I think she helped clarify these matters, though there were others like Auger in France who also contributed to this. Altogether, the models used in the early days of β-ray studies before the First World War, before Chadwick in 1914 or soon afterward found evidence that the electrons that were in lines had nothing to do with the β-ray itself and that the β-decay manifested itself in a continuous spectrum, were considerably confused, and it is quite difficult to reconstruct what people really were thinking. It doesn't seem to make sense nowadays.

Experimental Work with Nuclei 77

Barschall: I'd like to follow up on your description of what happened when Bohr left Copenhagen. He arrived in this country on January 16, 1939, and, according to the account in Mrs. Fermi's book, the Fermis met Bohr in New York, but Bohr didn't tell Fermi anything about fission. I remember that on the same day (I was at Princeton at the time) we were informed at Princeton of the discovery of fission. Apparently, Fermi heard about the discovery of fission from Lamb, who was at Princeton at the time. My question is: Is the story, which has gone around, correct that Bohr promised you not to talk about it until you had a chance to verify that the fission process produced the large pulses? Or do you have any explanation why Bohr didn't tell Fermi about fisssion, but told us at Princeton about it later the same day?

Frisch: Well, it's certainly true that Bohr said he wouldn't talk to other people until we got our result out. At that time we merely meant to publish the result as fast as possible. Actually, *Nature* took a little over three weeks to publish it, which is not the best, but not a particularly long time for those days I think. On the other hand, Bohr discussed fission on the boat with Rosenfeld, who was traveling with him, and apparently did not tell Rosenfeld that the thing ought to be kept under a hat. So Rosenfeld seems to have talked freely, and Bohr soon realized that there would be no point in keeping it secret at Princeton. I think he discussed it at Princeton with John Wheeler, who was there and who had worked with him and with whom it was natural to discuss it. My impression is that Bohr was not on such easy, natural terms with Fermi to immediately tell him the news. But that's something that John Wheeler may explain much better.

Laura Fermi: I was there too at the pier in New York when Bohr arrived, and there were quite a few people. All the questions revolved around the political situation in Europe, and he may just not have had a chance to talk.

Wheeler: I had my regular morning class on Monday the 16th and then went in on the train to meet the Drottningholm coming in that afternoon; I of course shook hands with those waiting for Bohr and with him and Rosenfeld when they came off. Bohr was staying in New York for a little while, but Rosenfeld went down with me on the train to Princeton. In those days I was in charge of the Monday evening journal club. It was the custom to get three things reported there, and here was something hot, as I had learned from Rosenfeld on the train. Therefore, we got this report that evening in Princeton.

Fowler: My feeling is that Bohr may very well have said it, but it was in a whisper. Are there other questions from the audience? Yes?

Bromberg: I would like to ask whether there was much speculation about energy from some kind of nuclear process in the 1930s, and if there was, what the circumstances were, whether you were talking among yourselves or with engineers or with particular industrial or political problems in mind?

Frisch: This one I can answer very easily. Until the discovery of fission there was no talk whatsoever of that among the people I knew. Rutherford completely discouraged all talk and said it was moonshine to think of power on an industrial scale from nuclear processes. Szilard thought a little further I think. He felt that the only hope would be a chain reaction that would spread, and he made several attempts to find one. I believe he even took out a patent for an unspecified branching process, which would be the source of a chain reaction and hence of power from nuclei. But apart from that, I don't believe anybody took that possibility seriously until the discovery of fission, and then it was not at once seen. I remember that when Møller came a few days later and said these two fragments ought to have enough energy left in them to emit a neutron or two, my first reaction was to say in a very smart-alecky way that it can't be, because if that were so, there would be no uranium deposits left. They would have blown up long ago. But of course uranium contains impurities, although at the time I didn't realize that the heavy isotope is the most serious one and that the layers are quite possibly thin, so that neutrons can easily escape.

Fowler: There was one rather unrelated aspect. You must remember that Millikan believed that the cosmic rays were γ-rays, and he had very early suggested essentially nuclear energy by saying that it was silicon nuclei which we are annihilating.

Bethe: I remember a conversation with Szilard, who always was ahead of his time. I believe the conversation must have been sometime around 1937, when he talked about fusion, in particular the deuteron-deuteron reaction. He said that he knew perfectly well that in a target the energy is lost by the incident particle's collisions; so one has to find some way of keeping the deuterium hot. He said that before the discovery of fission, I am pretty sure, and I know where it was said. It was said in Washington in Potomac Park during the cherry blossom time.

Peierls: I think I can contribute an anecdote relating to this point. I was talking in 1934 with Landau and one of his friends who was an engineer. We were on a walking tour, and the engineer said one day: "What is this one hears about nuclear energy? Is that all science fiction, or is this a real possibility?" Landau said: "Well, it is difficult, you see, because there are reactions in which one can release energy from nuclei, but with charged particles it is very inefficient; they lose their energy on the way. Neutrons are different; they don't lose energy. But at the present the only way we know of making neutrons is by bombardment with charged particles, so we are back to the same problem. But if one day somebody finds a reaction initiated by neutrons that releases secondary neutrons, one is all set."

Segrè: I would like to just quote from Corbino's speech in 1929: "Only technical and financial difficulties, not insurmountable in principle, oppose the realization of this great project [nuclear disintegration]. The object is not only the artificial transmutation of elements in appreciable quantities, but the study of the tremendously energetic phenomena that would occur in some cases of disintegration or recombination of atomic nuclei." Then he evaluates the amount of energy mc^2. "Thus we can conclude that although it is improbable that experimental physics will make great progress in its ordinary domain [he had spoken before of atomic physics], there are many possibilities in the attack of the atomic nucleus. This is a true field for the physics of tomorrow." And then he keeps on going. As you see, he quite explicitly talks about nuclear energy. He doesn't know how to release it—it's 1929, before the discovery of the neutron, but he doesn't consider it to be an impossibility.

The Nuclear Photoelectric Effect and Remarks on Higher Multipole Transitions: A Personal History

Maurice Goldhaber

INTRODUCTION: *Herman Feshbach*

It is my privilege to introduce the next speaker, a pioneer in the development of nuclear physics, Dr. Maurice Goldhaber. You have in your hands a biography so I won't go through it in any detail except to add one item. On the way from Germany to England he stopped off for a number of years in Berlin, where he took courses and even worked with Schrödinger and incidentally also met Gertrude Scharff. I always thought Maurice was a cryptotheorist, and now I know where he got his theoretical background from.

In this introduction it is of course impossible to detail in a complete way his contributions to the field, but I would like to mention a few things besides the topic to which he will devote himself, namely, the photodisintegration of the deuteron. Following that endeavor, Dr. Goldhaber has spent, as far as nuclear physics is concerned, a great deal of time on the electromagnetic properties of nuclei, on the properties of weak interactions, and on the properties of nuclei undergoing weak decays. There are three items that stick in my mind without having made a thorough investigation of his bibliography, namely, the study of isomers which was so important for the shell model, the Goldhaber-Teller model of the giant dipole resonance in which the neutrons oscillate against the protons, and finally the very lovely experiment done with Andy Sunyar and Lee Grodzins in which he measured the helicity of the neutrino. These are all important and exquisite examples of the experimental art. I am very happy to bring Dr. Goldhaber to this podium now.

The Nuclear Photoelectric Effect and Remarks on Higher Multipole Transitions: A Personal History

Maurice Goldhaber

This Symposium on the History of Nuclear Physics concentrates on a time when physicists were switching their major attention from the then recently conquered field of atomic physics to nuclear physics. Our gathering is probably unique in having a large fraction of speakers who had already made important contributions in atomic physics before turning their attention to the nucleus. In this transition the nuclear photoelectric effect has played a role which might be worth recounting.

For some personal reminiscences connected with its discovery and investigation I have to go back more than forty years. However, as we all know, memory is selective. It could hardly be otherwise. "Total recall" probably can play a role only when the "resolving power," both spatial and temporal, is unimportant, e.g., when we are concerned solely with numbers. Different observers remember the same event differently—this might be called, after the Japanese story known to many of us from the famous movie, the "Rashomon effect." This is especially true in the history of science, because here different "observers" are often separated in both space and time. While reproducibility plays an important role in science, history of

Nuclear Physics in Retrospect, edited by Roger H. Stuewer, copyright © 1979 by the University of Minnesota.

science has to emphasize consistency. Thus, by using many different sources, the historian of science can develop a more or less balanced picture, though hardly ever a complete one. The memories of one scientist, subjective as they necessarily are, sometimes telescoping different events into one, and favoring "vivid" ones which he or she has often "replayed" in memory or explicit recollections, can form only a part of the raw material used by the historian. In preparing these remarks, I did not attempt to make more than sporadic use of the published literature and of my own notes, unlike the systematic use a historian of science would consider necessary for a fairly well-rounded picture.

With these provisos, let me come to my story. In the early thirties, while I was a student at the University of Berlin, I got interested in nuclear physics, partly through a stimulating course given by Lise Meitner. She was very enthusiastic about her field, and I still remember how excited she was when reporting the discovery of the neutron in a colloquium. She got so carried away that she spoke of the collision of a neutron with a "brass nucleus." This is not so bad after all because the "brass nucleus" has fewer "isotopes," namely seven, than some other well-known nuclei like tin, which has ten.

From time to time the Nazi students rioted, and one day in March 1933 they had brought all lectures to a halt. This unexpected free time permitted me to go to the library, where I came across a short note in a popular scientific journal which reported from the United States that G. N. Lewis had separated 1 cc of heavy water. The news that a rare isotope—discovered only a year earlier by Urey, Brickwedde, and Murphy—had been isolated in such a large amount impressed me very much. I immediately asked myself: To what use could heavy hydrogen be put? I jotted down a few ideas, one of them being: look for photodisintegration of the heavy hydrogen nucleus (later called the deuteron). Such an effect might explicitly check whether the deuteron is "made" of a proton and a neutron. Remember that only a few years earlier the possibility that nuclei were made of protons and electrons was still taken seriously. The best meaning that can be given to such a naive phrase as "something is made of certain constituents" is that a theory of a process in which these constituents are postulated to play a role leads to agreement with experiment. Photodisintegration of the deuteron can be con-

Fig. 1. Lord Rutherford, April 1934. Picture taken by Professor P. Harteck and kindly given to me.

sidered the nuclear equivalent of the spectroscopy of the hydrogen atom, though nothing like as rich in detail and precision, in fact a rather poor relative as far as its "spectrum" is concerned, which in this case has no discrete lines.

In May 1933 I left Berlin for Cambridge, England, where Rutherford (see Figure 1) had accepted me as a research student at the Cavendish Laboratory. As Herman Feshbach told you in his introduction, I had intended to start a thesis with Schrödinger, but during, or soon after, our first serious discussion we both decided to leave, and it was probably he who recommended me to Rutherford. I started theoretical work with R. H. Fowler in October 1933. In those days one didn't yet feel the hot breath of competition, certainly not at the Cavendish. It was possible to have an idea that one thought important, to sit on it for a year or so without being afraid of losing it, and if it was lost nevertheless, to say to oneself, "Now I can go on to the next idea."

In April 1934 I was writing a paper on the role of spin in nuclear reactions.[1] I had to know the energy released in some of the simple reactions with lithium isotopes which had been studied by Cockcroft and Walton. Somebody at the Cavendish—I believe it was Mott, though I am not quite sure—told me that Chadwick (Figure 2) would know the best values for the masses of these isotopes.

I went out to see Chadwick to ask him about the masses, and after he had enlightened me I suddenly found the courage to tell him my idea that the photodisintegration of the "diplon"—as the heavy hydrogen nucleus was then still called by Rutherford—might be worth trying, using the highest energy γ-rays then available. These were the homogeneous γ-rays of radiothorium, which in equilibrium contains a daughter product, thorium C'', that emits a 2.62 MeV γ-ray. This γ-ray had been discovered by Ellis, and a source of a few millicuries of radiothorium was available at the Cavendish. Thus, the photonuclear experiment would not have been easy to perform in many places but it happened to be relatively easy at the Cavendish. Chadwick, gentleman that he was, listened politely, but seemed to catch fire only when the point was brought out that the mass of the neutron could be determined rather accurately from a measurement of the "photoproton" energy. Perhaps his interest was aroused because the question of the correct neutron mass was hotly debated at that time. It was then not even certain whether the neutron was heavier or lighter than the proton. Our conversation ended with no explicit commitment from Chadwick.

About six weeks later, when I showed him a "Letter" concerned with a then premature subject, later known as "Delayed Neutrons," which I wanted to send to *Nature*,[2] he asked, "Were you the one who suggested the photodisintegration of the diplon to me?" When

Fig. 2. Sir James Chadwick. (From H. Massey and N. Feather, *Biographical Memoirs of Fellows of the Royal Society*, 22 [Nov. 1976]. Reproduced by kind permission of the Royal Society of London.)

I answered yes, he said, "Well, it works — for the first time last night. Would you like to work on it with me?" Since I realized by then that I enjoyed pursuing questions that could be answered by experiment, I was glad to accept this chance to change from theory to experiment. I went to see Fowler to ask his permission, and he immediately agreed. Things were very informal at the Cavendish in those days. Experiment and theory were not yet as separated as they are today in many places, and, anyhow, the overall official supervisor of the research students was Rutherford, who, I suppose, was consulted. I joined Chadwick in his research room. Since he was pretty busy as Assistant Director of Research concerned with the progress of all experimental research students, he allowed me quite a bit of leeway in continuing the further experiments which we undertook. In these experiments I was helped by Chadwick's very capable assistant, H. Nutt, who is known from a line in the "New Cavendish alphabet," composed and recited by N. Feather at the Cavendish dinner, 1933: "N for Nutt who discovered the neutron." This help was, of course, invaluable to a greenhorn in experimentation. Ever since, though, I have tried to remain a cryptotheorist, as Herman Feshbach realized.

After about six weeks of intensive work, we started to write a "Note" to *Nature*.[3] By observing the photoprotons from the reaction

$$D + \gamma \to H + n \tag{1}$$

we could measure the approximate energy release as well as the cross section. We used a sensitive ionization chamber filled with D_2 gas. Since the photoneutron carries away approximately the same energy as the photoproton, we could write, in suitable units, using Einstein's famous two equations for the photoelectric effect and for the equivalence of mass and energy:

$$m_n = m_D - m_H + E_\gamma - 2 E_p. \tag{2}$$

We calibrated the ionization chamber roughly with α-particles and found $E_p \simeq 250$ keV. This led to a neutron mass $m_n = 1.008$, definitely larger than that of the proton. Chadwick, Feather, and Bretscher[4] measured the range of the photoprotons in a cloud chamber containing deuterium and obtained a more accurate Q-value. The best values on the present-day mass scale ($C^{12} \equiv 12$ mass units) are $m_n = 1.008665$ and $m_H = 1.007825$. I remember being quite shocked when it dawned on me that the neutron, an "elementary particle," as I had by that time already learned to speak of it, might decay by β-emission with a half-life that I could roughly estimate from the existing systematics for the β-decay of complex nuclei to be about half an hour or shorter, since the neutron was a simpler system. Of course, we are by now quite used to the idea that elementary particles can decay. I believe the first one for which the lifetime was explicitly measured was the μ-meson, and the half-life of the neutron is now known to be 10.6 min. Since the development of reactors, the spectrum has been measured by modern techniques with high precision, and the neutron does indeed exhibit a nice continuous β-spectrum with an end point of 780 keV, in agreement with the mass difference.

Bethe and Peierls followed the γ-D experiments by a detailed and very successful theory of the deuteron and the photoelectric effect.[5] It is perhaps fortunate that quantum mechanics was developed at a time when electromagnetic interactions were considered to be the only forces on a microscopic scale. One now could assume that quantum mechanics would be valid in the new situation found in nuclei, and thus one had to guess only about the new forces, not about quantum mechanics at the same time. I believe it would have been very hard for theory to make a double step simultaneously, and it was fortunate that we didn't know any particles other than protons and electrons at the time quantum mechanics was created. It is sometimes better if a discovery is postponed: a strange lesson to give anybody! The discovery of the neutron, as we heard from Hans Bethe, removed a major difficulty, permitting Heisenberg, Majorana, and Wigner to extend the use of quantum mechanics to the behavior of protons and neutrons in the nucleus. Bethe and Peierls made the simple assumption that the forces between protons and neutrons were of very short range and pretty strong. They did not need to know the details—this was the beauty of their theory—but were able to calculate the cross section with only one parameter, the binding energy of the deuteron. Refinements of this theory have been made by many physicists, and we especially miss here today one who has perservered in this field over many years but was unable to attend this symposium: Gregory Breit.

To test further the electric dipole character of the photodisintegration of the deuteron, we made a rough measurement of the angular distribution of the photoneutrons (Figure 3). In our simple arrangement we used a cylinder (C) filled with heavy water, and we had two positions for our γ-ray source, A or B, at equal distances from

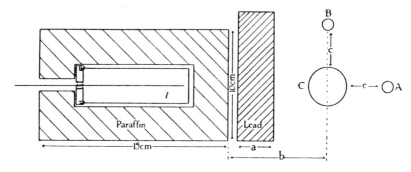

Fig. 3. Arrangement for measuring angular distribution of photoneutrons from D + γ → p + n (J. Chadwick and M. Goldhaber, *Proceedings of the Royal Society of London*, 151 [A] [1935], 479).

C, a lead block to absorb soft γ-rays, a paraffin block to slow down the neutrons, and a boron counter inside the paraffin to detect the slow neutrons. This was one of the early angular distributions studied in nuclear reaction physics, but since the angles were very poorly defined we could not expect any very large effect. We found twice the intensity at position B as at position A. This means that most photoneutrons are emitted at right angles to the γ-ray, which agrees with expectations for an electric dipole transition.

To get the heavy water was quite a business. I went to Rutherford and asked his permission to use heavy water, and he wrote a note to Oliphant, which I kept as a memento (Figure 4). This was a good fraction of the world's supply of heavy water at that time, and it was carefully handled. Since our experimental cross section and angular distribution were in rough agreement with the theory, we could conclude that the deuteron was really "made" of protons and neutrons. Today we know this is only a good approximation, since mesons also play a role. The cross section for photodisintegration of the deuteron has by now been pursued up into billions of electron volts of γ-ray energy, and it is an interesting curve. Figure 5 shows a printout from the photonuclear data bank being maintained at the Bureau of Standards under the direction of Everett Fuller, one of the editors of the Benchmark book on *Photonuclear Reactions*. The curve is theoretical, based on effective range theory with a slight correction for the photomagnetic effect near threshold, which I shall

The Nuclear Photoelectric Effect 91

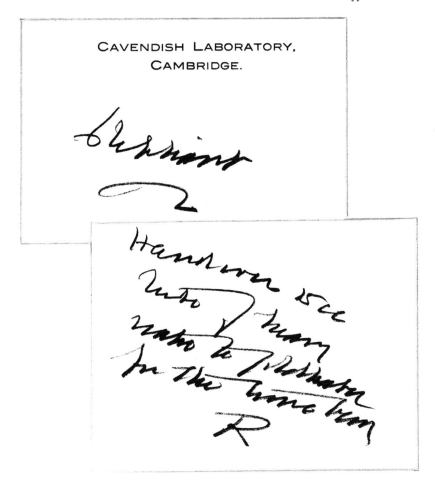

Fig. 4. Reproduction of a penciled note from Rutherford to Oliphant, undated, early 1935: "Hand over 25 cc tube of heavy water to Goldhaber for the time being."

come to shortly. Note the logarithmic energy scale. Anyone with a one-track mind, pursuing the cross section to higher and higher energies, would have found an interesting bump, which was actually found only after the much bigger meson resonance had been discovered. The photomeson production leads indirectly to a resonance also in the "direct" photodisintegration cross section: sometimes it would pay to have a one-track mind!

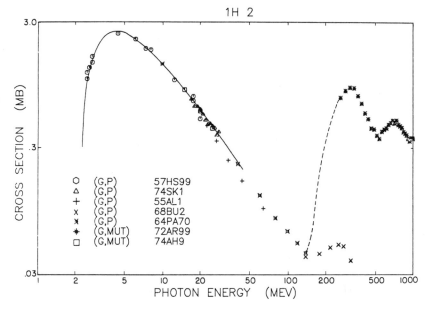

Fig. 5. A summary of theoretical and experimental cross section data on the photodisintegration of the deuteron from threshold to 1 GeV (from Bureau of Standards, photonuclear data bank). The solid curve is based on a Bethe-Peierls effective range theory. The magnetic dipole cross section of L. Hulthén and B. C. Nagle (*Physical Review*, 90 [1953], 62) has been smoothly joined to it. The calculation of F. Partovi (*Annals of Physics*, 27 [1964], 79), represented by the symbol X, extends to meson threshold and includes multipoles up to L = 3. The experimental data at low energies come from the review article by L. Hulthén and M. Sugawara (*Encyclopedia of Physics* [Berlin: Springer, 1957], vol. 39, p. 1). Near 20 MeV the (γ,p) cross section of D. M. Skopik, Y. M. Shin, M. C. Phenneger, and J. J. Murphy, II (*Physical Review*, C9 [1974], 531) coincides nicely with the attenuation measurement of J. Ahrens, H. B. Eppler, H. Gimm, M. Kröning, P. Riehn, H. Wäffler, A. Zieger, and B. Ziegler (*Physical Review Letters*, 52B [1974], 49). The (γ,p) measurements extending to 65 MeV are from L. Allen, Jr. (*Physical Review*, 98 [1955], 98). The (γ,p) cross section at the meson threshold region is from the work of J. Buon, V. Gracco, J. Lefrancois, P. Lehmann, B. Merkel, and Ph. Roy (*Physical Review Letters*, 26B [1968], 595).

An interesting speculation on the possible existence of "slow" neutrons turned up while we were writing our "Note." Some of those present today may remember part of this story.

Our approximate experimental cross section permitted us to estimate the inverse cross section

The Nuclear Photoelectric Effect 93

$$H + n \rightarrow D + \gamma. \qquad (3)$$

In fact, the inverse reaction for ~1-MeV neutrons would just correspond to the same center-of-mass energy as the one we studied in reaction (1), and we could therefore safely make a thermodynamic argument for the cross section expected for reaction (3) for 1-MeV neutrons. A little earlier D. E. Lea at the Cavendish had carried out experiments in which he had observed γ-rays emitted when fast neutrons impinged on paraffin. These were polonium-beryllium neutrons which are on the average even a bit more energetic than 1 MeV, and he interpreted his results as due to reaction (3). But according to our estimate, his observed γ-ray yield was much too large to be explained by this reaction for the fast neutrons he used, by something like a factor of a hundred to a thousand. It was really an impressive discrepancy.

We therefore speculated that in Lea's experiment the fast neutrons were first slowed down and *then* captured as *slow* neutrons. But speculation was somewhat frowned upon at the Cavendish, and Chadwick convinced me not to include it in our final version of the "Letter" to *Nature*; thus the puzzle we had encountered was noted only with the following remark: "A satisfactory explanation is not easy to find and further experiments are desirable" — a typical cautious Cavendish remark. Rutherford allowed himself his famous speculation on the existence of a neutron only in a lecture, his Bakerian Lecture of 1920. In papers he did not speculate readily. By the way, since this question came up in the discussion this morning, Rutherford had made the mistake of thinking that some long-range α-particles were He^3, but when a genius makes a mistake it has its uses, and so he speculated from He^3 down to the neutron. He added a remark which was very important, saying in effect that the neutron seems almost necessary to explain the building up of the nuclei of heavy elements!

I do not know whether Chadwick talked to Rutherford about our speculations, but about three months later, when the news of the discovery of slow neutrons by Fermi and his collaborators in Rome reached the Cavendish, solving the puzzle we had found, Rutherford sought me out and turned out to be well informed about our speculations. In fact, he practically rushed into my lab, and I can only now psychoanalyze this with hindsight: he seems to have had a bad conscience, but I cannot state this as a fact. I wrote down my recollections of these events for a volume honoring Feather's Jubilee at Edinburgh.[6] Peierls, who was aware of our speculations, had pointed out in a letter to me that the cross section for reaction (3) would go to zero at zero neutron energy because electric dipole radiations would then give a zero effect; so to say, a p-wave would be needed. The finite cross section which Otto Frisch and others later measured accurately was explained finally in 1936 by Fermi by using the spin dependence of the n-p forces postulated by Wigner and calculating the spin flip magnetic dipole transition.

Until a few years ago, there was still a small discrepancy between experiment and theory in the spin flip capture of slow neutrons by protons, namely, the experiment gave a cross section about 10 percent larger than the theory. This was explained by Riska and Brown as being due to meson exchange and nucleon resonance effects. Gerry Brown tells me that ~7 percent can be ascribed to meson exchange and ~3 percent to a $\triangle$ resonance, although the latter is somewhat model dependent.

The magnetic dipole transition was taken into account in the cross section for the photodisintegration of the deuteron shown in Figure 5. It is a small effect but important near the threshold. Incidentally, on that curve there are no experimental points near threshold; it would be desirable to check the theoretical cross section curve there, which is based solely on the value of the thermal neutron capture cross section.

In 1934 we knew of only two nuclei, D and Be^9, for which the γ-rays then available had sufficient energy to permit photodisintegration. Szilard and Chalmers were soon able to observe the photoneutrons from Be, using the ingenious Szilard-Chalmers method for separation of radioactive I^{128}. In this method, an organic compound of normal I^{127} is bombarded with neutrons. In the capture of the neutrons, most of the time the I^{128} gets knocked out of the molecule because of the recoil given to it by the neutron that is absorbed and the capture γ-rays that are emitted. The knocked-out I^{128} is then separated, and thus the activity is concentrated. In this way one can discover low intensity neutron activities. Szilard told me a story a little different from what Otto Frisch remembered concerning how he found this separation. He said he was trying to think of a way of incorporating I^{128}, the radioactive iodine, into a molecule that would be of some medical use, for treatment or diagnosis. He was therefore thinking of bombarding a compound with neutrons but realized that owing to the neutron bombardment—remember he was still using fast neutrons—the compound would be destroyed. He turned this apparent defeat into a positive method of concentrating an activity, which was a true Szilardian touch.[7] Chadwick and I studied the angular distribution of the photoneutrons from Be in an arrangement similar to that in Figure 3, using Rn γ-rays, and found the distribution to be symmetrical. We interpreted this as due to a neutron being emitted from an $\ell = 1$ state in Be^9. I believe that is still considered correct. I read it just last night in our old paper; I had forgotten it myself. It was a little bit of "single particle thinking."

In 1937 Bothe and Gentner took an important step forward by using a new γ-ray source of much higher energy, the 17-MeV γ-rays emitted in the $Li^7(p,\gamma)$ reaction, which I believe was found at the Carnegie Institution and at Caltech. This γ-energy is sufficient to photodisintegrate most nuclei. There are very few thresholds above 17 MeV. Bothe and Gentner observed radioactivity in very many nuclei, e.g., $Cu^{63}(\gamma,n) Cu^{62}$ (10 min). It would have been hard to look for the photoneutrons in this case because around an accelerator there is usually some neutron background, and it was much easier to look for the delayed activity that gives an easily detected γ-ray for each photoneutron ejected.

The war interrupted this story, and there was not much progress in the photonuclear field until after the war. Early in 1948 Baldwin and Klaiber in a classic paper[8] reported the use of X-rays up to 100 MeV from the G. E. betatron to induce (γ,n) reactions in a number of nuclei. They found a cross section that increased rapidly from the threshold, reached a maximum, and then decreased rapidly as the energy of the X-rays rose beyond about 20 MeV. They interpreted the decrease as due to competition from (γ,2n) reactions. Since they measured, e.g., in Cu, only the 10-min Cu^{62}, the longer-lived activity of Cu^{61} would have remained unnoticed. I found their explanation unsatisfactory because the drop in intensity seemed much too precipitous, and it seemed worthwhile to try to interpret their data as a resonance process instead. Soon Perlman and Friedlander showed that Cu^{61} was indeed only weakly produced.

In the spring of 1948 I was spending a day at Argonne National Laboratory hunting neutron-induced isomers with Ed ter Mateosian, Mike McKeown, and Carl Muehlhause. I ran into Edward Teller and asked him whether it would not make sense to consider the apparent resonances found by Baldwin and Klaiber as nuclear analogues of the so-called Restrahlen (residual rays) which Rubens first studied long ago in polar crystals. These resonances, which are seen in the infrared, had been explained as due to the motion of a positive ion lattice against the negative, and Teller was well acquainted with this theory. In the nucleus, of course, this would be a vibration of the protons against the neutrons set in motion by the γ-rays, an early example of collective behavior in an electromagnetic transition. With

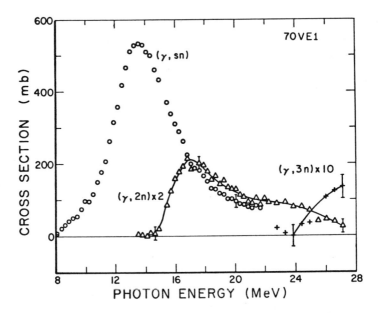

Fig. 6. The photoneutron cross section for gold (from Bureau of Standards, photonuclear data bank). The open circles represent the total photoneutron cross section, $\sigma(\gamma,n) + \sigma(\gamma,2n) + \sigma(\gamma,3n)$. The triangles are twice $\sigma(\gamma,2n)$ and the crosses ten times $\sigma(\gamma,3n)$. The solid lines are merely to guide the eye. The total cross section is fitted by a Lorentz line having the parameters: $E = 13.70$ MeV, $\Gamma = 4.75$ MeV, and $\sigma = 540$ mb. These results were reported by A. Veyssiere, H. Beil, R. Bergère, P. Carlos, and A. Lepretre (*Nuclear Physics*, A159 [1970], 561).

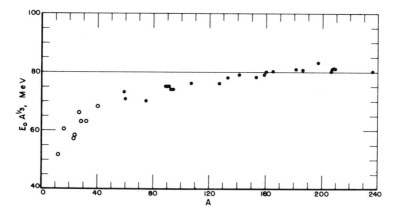

Fig. 7. The giant resonance energy times $A^{1/3}$ (which is proportional to the nuclear radius) as a function of A (from E. Hayward, *Photonuclear Reactions*, NBS Monograph 118). The horizontal line corresponds to the $E_0 = 80A^{-1/3}$ prediction of the hydrodynamic model with up-to-date parameters. The data for the light nuclei are from J. M. Wyckoff, B. Ziegler, H. W. Koch, and R. Uhlig (*Physical Review*, 137B [1965], 576). The data for the heavy nuclei were obtained at Livermore (B. L. Berman, *Atomic and Nuclear Data Tables*, 15 [1975], 319).

various assumptions about restoring forces and the role of the nuclear surface, we were able to calculate the resonance energies E_0, for which we obtained for one model, which turned out to be a happy approximation, a dependence on mass proportional to $A^{-1/3}$, i.e., inversely proportional to the nuclear radius.[9]

Figure 6, a printout (from the photonuclear data bank at the Bureau of Standards) which Fuller kindly sent me, shows the production of photoneutrons from Au^{197}. Note that the (γ,2n) reaction indeed sets in only well after the maximum cross section has been reached.

Figure 7 shows the product of the resonance energy E_0 and $A^{1/3}$ as a function of mass number A. Above A ~130 this product approaches a constant value.

Figure 8 shows an interesting example, Ta^{181}, where the γ-induced nuclear reaction has a double hump appearance, predicted independently by Danos and by Okamoto for nuclei that are strongly deformed and hence have a large static quadrupole moment. The rea-

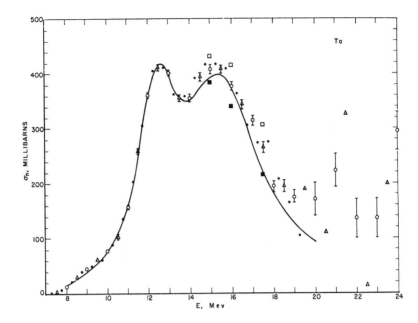

Fig. 8. The photoneutron cross section for tantalum, $\sigma(\gamma,n) + \sigma(\gamma,2n) + \sigma(\gamma,pn)$. Circles, triangles, and alternate dots present four independent determinations of the cross section from the activation curve. The errors indicated are standard deviations based on the counting statistics. The open and closed squares represent various choices for the correction for the emission of two neutrons in the $(\gamma,2n)$ process. The solid curve is the sum of two Lorentz lines having the paramaters: $E_a = 12.45$ MeV, $\sigma_a = 308$ mb, $\Gamma_a = 2.3$ MeV, $E_b = 15.45$ MeV, $\sigma_b = 348$ mb, and $\Gamma_b = 4.4$ MeV. The two resonances correspond to charge oscillations along the one long and two short axes of the nuclear ellipsoid. (From E. G. Fuller and M. S. Weiss, *Physical Review*, 112 [1958], 560. Reproduced with permission.)

son for the double hump is simply that there are different effective radii for the long and the short axes.

Figure 9 shows how this can be explicitly proved by aligning nuclei, in this case Ho[165], which is easier to align than Ta and is also highly deformed. The Ho nucleus is aligned either at right angles or parallel to the photon beam.

For the lightest nuclei, where the photoeffect shows a lot of structure, the single-particle model is much more effective. The reason why our approach works best for heavier nuclei is that a light nucleus is "all surface," while we had postulated a volume effect.

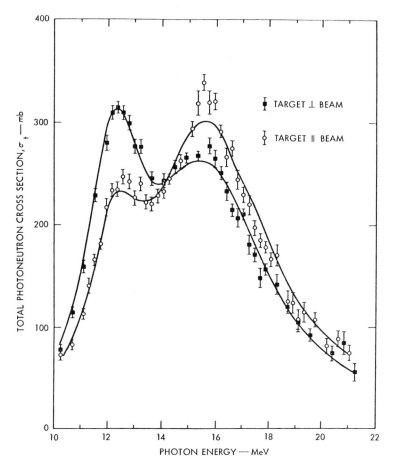

Fig. 9. The total photoneutron cross section, $\sigma(\gamma,n) + \sigma(\gamma,2n) + \sigma(\gamma,pn)$, measured for a holmium target polarized parallel and perpendicular to the incident photon beam. The solid lines are for the situation in which the unpolarized cross section was represented by the sum of two Lorentz lines having the parameters: $E_1 = 12.28$ MeV, $\Gamma_1 = 2.5$ MeV, $\sigma_1 = 215$ mb, $E_2 = 15.78$ MeV, $\Gamma_2 = 5.0$ MeV, and $\sigma_2 = 249$ mb. (From M. A. Kelly, B. L. Berman, R. L. Bramblett, and S. C. Fultz, *Physical Review*, 179 [1969], 1194. Reproduced with permission.)

There have been many variations and improvements of our 1948 paper on the giant dipole resonance, using both the collective and the single-particle approach, starting on the collective side with Steinwedel and Jensen[10] and on the single-particle side with Burkhardt,[11] followed by Wilkinson,[12] Brown and Bolsterli,[13] and others. In a recently published paper[14] Myers, Swiatecki, et al. discuss a macroscopic model in which the motion is treated as a combination of the displacement mode and the acoustic mode.

In 1950 Levinger and Bethe[15] considered the importance of sum rules where no specific models are assumed for the nuclear vibrations. Sum rules permit counting of the number of charges involved in the photoelectric transition, and any excess found is in a way a measure of meson effects that are otherwise hidden. These sum rules are therefore very important. They have now been pursued to high energies. Experiments on total cross sections up to about 100 MeV show how some sum rules saturate by then and others do not. Unknown to all of us, A. B. Migdal[16] had considered nuclear dipole vibrations in a 1944 paper, which reached this country after World War II and to which Levinger drew attention later in his book.[17] Migdal was guided by the observation, already contained in Bethe's 1937 review, that dipole transitions in the low energy regions are comparable in rate only to quadrupole transitions; but remember in those days no absolute rates were known. In a sense, the deuteron photodisintegration rate was the first absolute electromagnetic rate, and because it agreed with theory it gave us courage to go into other ramifications of electromagnetic transition rates.

The nuclear photoelectric effect plays a role in astrophysics, in nuclear technology, in the identification of elements and isotopes, and in other fields. I just heard an interesting talk by J. Peebles where he reminded us that in the first minute after the "Big Bang" the most important reactions going on were $n + p \leftrightarrow d + \gamma$. The rate of these reactions, which can be considered the first and simplest nuclear reactions, determines the ultimate $H:D:He^4$ ratio, which is of such importance for the world we live in. Had we ended up with a fifty-fifty mixture of deuterium and hydrogen, it might have confused some biologically important molecules!

During the late thirties while still at the Cavendish, and more so later at Illinois and Brookhaven, my research interests turned to isomeric transitions in nuclei, very naturally in a way since these were γ-ray transitions but of different multipole order. We were able to classify all the known isomeric transitions into a number of electric and magnetic multipole transitions, which we designated E0–E5, and M1−M4.[18] For most excited nuclear states of relatively low energy, say <1 MeV, the lifetime is now measurable and usually known. Techniques have been developed over the years to measure both very short liftimes, of the order of 10^{-14} sec or less, and very long lifetimes. Some of the longest-lived isomers have lifetimes of several centuries. Thus there is a huge ratio, $\sim 10^{23}$, over which transition rates can be studied. Most people look down their noses at work with something as unfundamental as heavy nuclei. However, if one is interested in high multipole transitions, one is forced to go to heavy nuclei. As we have already seen, for the deuteron there is a great paucity of nuclear levels because of the short range of the forces where most levels get squeezed out and the deuteron has no bound excited state. In fact, going up in mass, the first bound excited state occurs for mass 7, in Li^7 and Be^7; therefore, to study transition rates of different multipoles, one must use heavier nuclei.

Among the results of our classification of isomeric transitions, two are worth mentioning. The M4 transitions were particularly related to the closed shells discussed by Maria Mayer and by Jensen, Suess, and Haxel, because the M4 transitions appeared systematically, e.g., below the 50 shell as $p^{1/2} \leftrightarrow g^{9/2}$ transitions. We found M4 isomers just below 50 but not immediately above until one gets close to the next shell with 82 protons or neutrons. Let us now see how well the transition rates agree with just a simple single particle radiating. The formulas for the expected rates were calculated by Weisskopf and by Moszkowski. Figure 10 shows first of all that the predicted E^{-9} dependence in the γ-ray lifetime for M4 transitions is correct. The line shown has the correct slope. We drew this particular line through the points; the line suggested by Weisskopf's formula would be very close. Weisskopf himself didn't take his absolute values too seriously and was very surprised when he saw this. In fact, we didn't get it into such good shape until one day, on the way by train to M.I.T. to give a colloquium, while going over my slides I noticed that the points looked like two lines. By making a statistical weight correction, whether the transition goes, e.g., from $p^{1/2}$ to $g^{9/2}$ or the other way, equivalent to a factor of 5 in this case, the points were made to coalesce into a single line in time for the colloquium.

Another result worth mentioning (reported in my study with Sunyar) is that in the region of the highly deformed nuclei, which

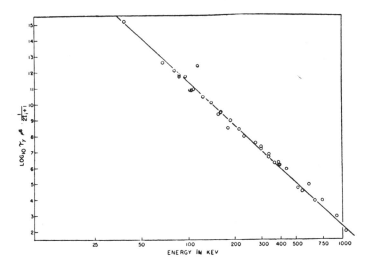

Fig. 10. Normalized lifetime-energy relations for M4 transitions with spin correction. The line shown is fitted to the experimental points and is given by the equation

$$\tau_\gamma \text{ (sec)} = \frac{1.0 \times 10^4 \ (2J_i + 1)}{A^2 E^9} \ (E \text{ in MeV}),$$

where the simplified statistical weight correction $(2J_i + 1)$ permitted us to lump the isomer families from different shells. (From M. Goldhaber and A. W. Sunyar, *Physical Review*, 83 [1951], 906.)

have large static quadrupole moments, we found so-called fast E2 transitions (faster than the single-particle formula predicts by factors up to ~100), and we explained these as a "cooperative" phenomenon whereby the nucleus as a whole was undergoing quadrupole motions. This is, one might say, the dynamic analogue of Rainwater's explanation of the static quadrupole moments, and it helped in the formulation of the unified theory of nuclei by Bohr and Mottelson, who considered these particular fast E2 transitions as related to shape deformation. They had assumed rotational states where the rotor keeps a constant moment of inertia in the various states of spin, 0, 2, 4, 6, and so on. But as more data appeared, it could be shown that ground state bands of all even-even nuclei, except the closed shell nuclei, can be systematically described by one formalism if it is assumed that the moment of inertia increases in a regular way.[19]

In this connection one more investigation may be of interest. It was mentioned this morning that the question was perhaps still unresolved then in the minds of some people (though resolved in most people's minds) whether γ-rays are emitted first and then converted or whether the conversion of γ-rays is really a direct process with the electron removing energy from the nucleus in competition with γ-ray emission. The second was generally believed. Experiments by Segrè and Wiegand, and by Daudel et al. on the lifetime changes in the electron capture decay of Be^7, which manifestly depends on the electron density at the nucleus and which could be changed by changing its chemical binding, indicated a method. We could look for lifetime changes in internal conversion by making different chemical compounds. If the γ-rays were emitted first and later absorbed, it shouldn't matter much what the compound is. However, if the electron can directly induce transitions, the lifetime can of course be changed by pushing the electrons either closer to the nucleus or farther away, which can be done by making different chemical compounds.

In 1951 Bainbridge had a year's leave from Harvard and came to Brookhaven. Together we undertook a search for the effect of chemical structure on lifetime, using the isomer with the lowest transition energy known at the time, Tc^{99}. A 2-keV, highly converted E3 transition of 6-hr half-life is followed by an easily detected M1 γ-ray of 140 keV. Since we were interested in seeing M4 transitions for our collections of points for the M4 line (Figure 10), we had to scrounge M4 transitions wherever we could. Therefore, since we could estimate how long Tc^{99m} would live if the intermediate level at 140 keV were not there, we just calculated what the intensity of the crossover transition should be, and sure enough we found it to have exactly the predicted intensity (see Figure 11). Thus we could put one more point on the M4 line where we were already expecting it. In the course of these experiments, Elizabeth Wilson (now Mrs. Charles Baker) of our Chemistry Department, who was working with us, learned to make very pure Tc^{99} sources. Bainbridge found beautiful lifetime differences between different compounds (metal, oxide, sulfide), and a theory was developed by John Slater, also visiting Brookhaven then, which roughly showed that for those of our compounds in which the electron density was expected to be

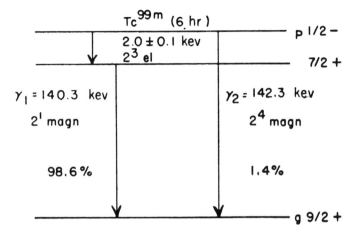

Fig. 11. Tc^{99m} decay scheme. (From J. W. Mihelich, M. Goldhaber, and E. Wilson, *Physical Review*, 82 [1951], 972.)

higher at the nucleus, the lifetime was shorter.[20] Thus we proved explicitly that internal conversion competes with γ-ray emission, a point which, as I said, was by that time already implicitly believed by most people concerned with such questions.

In the last two decades the study of electromagnetic interactions with nuclei and its natural continuation to single nucleons has moved into new fields, particularly since one has learned to use inelastic electron scattering on nuclei and, of course, on nucleons. By treating the electron as a source of virtual photons, one can measure many more giant resonances in nuclei: M1, etc., E2, etc. Excitations that cannot be reached by real photons can be measured with the help of virtual photons, and of course the giant electric dipole resonance also shows up here very nicely. Much work is being done in that way.

The field of electromagnetic interactions of nuclei is now a huge one. To summarize what we have learned: If one wants to "look" at a nucleus in a relatively gentle way, why not use photons; and, if one cannot look with real photons, why not use virtual ones.

I should like to thank Everett Fuller and Evans Hayward of the National Bureau of Standards for their invaluable help in preparing this talk.

The Nuclear Photoelectric Effect 105

REFERENCES

1. M. Goldhaber, "On the Probability of Artificial Nuclear Transformations and Its Connection with the Vector Model of the Nucleus," *Proceedings of the Cambridge Philosophical Society*, 30 (1934), 561-566.
2. M. Goldhaber, "Spontaneous Emission of Neutrons by Artificially Produced Radioactive Bodies," *Nature*, 134 (1934), 25.
3. J. Chadwick and M. Goldhaber, "A 'Nuclear Photo-effect': Disintegration of the Diplon by γ-Rays," *Nature* 134 (1934), 237-238; reprinted in Everett G. Fuller and Evans Hayward, eds., *Photonuclear Reactions [Benchmark Papers in Nuclear Physics*, vol. 2] (Stroudsburg, Pennsylvania: Dowden, Hutchinson & Ross, 1976), 8-9.
4. J. Chadwick, N. Feather, and E. Bretscher, "Measurements of Range and Angle of Projection for the Protons Produced in the Photo-Disintegration of Deuterium," *Proceedings of the Royal Society of London*, 163[A] (1937), 366-375.
5. H. Bethe and R. Peierls, "Quantum Theory of the Diplon," *Proceedings of the Royal Society of London*, 148[A] (1935), 146-156; reprinted in Fuller and Hayward, pp. 10-20.
6. M. Goldhaber, "Remarks on the Prehistory of the Discovery of Slow Neutrons," *Proceedings of the Royal Society of Edinburgh*, 70[A] (1972), 191-197.
7. See M. Goldhaber, Introduction to Szilard's Nuclear Papers in Bernard T. Feld and Gertrud Weiss Szilard, eds., *The Collected Works of Leo Szilard: Scientific Papers*, vol. 1 (Cambridge: MIT Press, 1972), pp. 139-142.
8. G. C. Baldwin and G. S. Klaiber, "X-Ray Yield Curves of γ-n Reactions," *Physical Review*, 73 (1948), 1156-1163; reprinted in Fuller and Hayward, pp. 35-42.
9. M. Goldhaber and E. Teller, "On Nuclear Dipole Vibrations," *Physical Review*, 74 (1948), 1046-1049; reprinted in Fuller and Hayward, pp. 45-48.
10. Helmut Steinwedel and J. H. D. Jensen, "Hydrodynamik von Kerndipolschwingungen," *Zeitschrift für Naturforschung*, 5 (1950), 413-420; translated as "Hydrodynamics of the Nuclear Dipole Oscillation" and reprinted in Fuller and Hayward, pp. 49-64.
11. James L. Burkhardt, "Dipole Photonuclear Reactions and the Independent Particle Model," *Physical Review*, 91 (1953), 420-421; reprinted in Fuller and Hayward, pp. 93-94.
12. D. H. Wilkinson, "Nuclear Photodisintegration," *Physica*, 22 (1956), 1039-1061; reprinted in part in Fuller and Hayward, pp. 95-110.
13. G. E. Brown and M. Bolsterli, "Dipole State in Nuclei," *Physical Review Letters*, 3 (1959), 472-476; reprinted in Fuller and Hayward, pp. 129-133.
14. W. D. Myers, W. J. Swiatecki, T. Kodama, L. J. El-Jaick, and E. R. Hiff, "Droplet Model of the Giant Dipole Resonance," *Physical Review*, C 15 (1977), 2032-2043.
15. J. S. Levinger and H. A. Bethe, "Dipole Transitions in the Nuclear Photo-Effect," *Physical Review*, 78 (1950), 115-129; reprinted in Fuller and Hayward, pp. 65-129.
16. A. Migdal, "Quadrupole and Dipole γ-Radiation of Nuclei," *Journal of Physics (Moscow)*, 8 (1944), 331-336; reprinted in Fuller and Hayward, pp. 27-32.
17. Joseph S. Levinger, *Nuclear Photo-Disintegration* (London: Oxford University Press, 1960).
18. M. Goldhaber and A. W. Sunyar, "Classification of Nuclear Isomers," *Physical Review*, 83 (1951), 906-918; M. Goldhaber and R. D. Hill, "Nuclear Isomerism and Shell Structure," *Reviews of Modern Physics*, 24 (1952), 179-239; and M. Goldhaber and J. Weneser, "Electromagnetic Transitions in Nuclei," *Annual Review of Nuclear Science*, 5 (1955), 1-24.

19. Gertrude Scharff-Goldhaber, Carl B. Dover, and Alan L. Goodman, "The Variable Moment of Inertia (VMI) Model and Theories of Nuclear Collective Motion," *Annual Review of Nuclear Science*, 26 (1976), 239-317.

20. Kenneth T. Bainbridge, M. Goldhaber, and Elizabeth Wilson, "Influence of the Chemical State on the Lifetime of an Isomer," *Physical Review*, 84 (1951), 1260-1261.

DISCUSSION

Hintz: When did you know that you could first disintegrate deuterons with the energy of ThC″ γ-rays? In other words, was the binding energy known from mass measurements on deuterium and the neutron?

Goldhaber: The masses of deuterium and hydrogen were known approximately; the neutron mass was also known approximately. Depending on who was right on the neutron mass, it looked either easy or just touch and go.

Hintz: They were known that well? They were known well enough to give you courage to try the experiment?

Goldhaber: Well, I think you could predict the threshold to 1 MeV at least, and it fell into the right ball park.

Hintz: When did people first start using deuterons as projectiles in accelerators?

Goldhaber: Well, I should perhaps let Willy answer this.

Fowler: It started in 1933 in Berkeley.

Goldhaber: In 1933 in Berkeley, and the same year at Caltech. In the Cavendish, the famous d + d reaction was also found in 1934, I believe, or was it already 1933? You see, the d + d reaction was a very interesting and fundamental reaction, of course, and Rutherford, Oliphant, and Harteck found it. Kempton then investigated it further. Lawrence had used deuterium on copper, and this is how he came to think that the neutron is so light. He found neutrons from heavy elements when by barrier penetration he wouldn't have expected a reaction. Rutherford explained that deuterium had such a large cross section that the little that gets absorbed as you bombard copper, for instance, then gets bombarded by a following deuteron and gives you the (d, d) reaction. This was the reason for the copious yield. If you don't think of the two-step reaction, it makes you think that the deuteron is inherently unstable and just breaks up on collision.

Malley: Do you remember the title of the course that you took in Berlin under Lise Meitner?

Goldhaber: "Kernphysik." It was a one-hour course, one hour a week, which was a lot of nuclear physics for that time.

Kojoian: I have been particularly intrigued by the character of the environment, the atmosphere of physics research, before World War II, as expressed by the participants of that research. My question is probably to all the speakers today: How do you view the advantages and the disadvantages of the character change between the period preceding World War II and the current period, if in fact there has been a character change in the atmosphere of physics research?

Goldhaber: I would like the other speakers of today to answer too, but let me answer first because I happen to have the microphone. I think there has been a character change, and nobody can view it with happiness, but in a way it's forced on us by the circumstances. The first to disintegrate a nucleus was Rutherford, and there is a picture of him holding the apparatus in his lap. I then always remember the later picture when one of the famous cyclotrons was built at Berkeley, and all of the people were sitting in the lap of the cyclotron. Roughly speaking, this gives you an idea of the change. Now, once you have this large scale, which is needed to go to smaller and smaller distances, to higher and higher energies, to understand new phenomena—once this scale is forced on you—you cannot work any longer in the good old-fashioned way, saying, "I have this idea; let me try it." We used to have many ideas which when they excited us sufficiently we tried the same day we had them, or if we wanted to show our patience we waited a day or so. This is now essentially impossible, with the scale forced on fundamental frontier physics.

Feshbach: I think that there is a spectrum of scales, and I think one can still find some small-scale and very interesting work going on, and one shouldn't forget that. I wonder if we can get some comments on this question from the other speakers?

Wilson: I fear that in the field of very high energy physics that is all too true, for to mount a modern experiment is almost as difficult as building a cyclotron used to be. The number of people who form an experiment, then man it three shifts a day, do programming, work up the data, and so on is large—sometimes the group is as large as thirty to fifty. You lament for the good old days. You do notice that on such large experiments usually one or two people are dominant and either are brokers of physics or are leaders of physics. Perhaps they play a role comparable to the role that Fermi played in Italy with his experiments. You might find similarities, if you look for them. It could be that in the minds of these leaders what they do is quite similar to what the "old timers" were doing.

Goldhaber: Yes, I think you have a "Little Italy" in your lab.

Segrè: Tc^{99}, just to inject a practical thing, has by now become a very, very big industry in medicine. It is one of the most powerful diagnostic tools. Just with that isomeric transition that we found many years ago.

Goldhaber: It's good that the lifetime does not change very much —up to fractions of a percent, maybe .1 percent—otherwise instead of a table of radioactive isotopes we would need for each one a sub-table of all the compounds. Maybe one day we will have this if there's a reason. This lifetime change, by the way, has been studied now in many isomers.

Fowler: Maurice, it isn't fair to ask you a question for which you are obviously unprepared, but you did mention the fact that you went to see Chadwick about a question regarding the effect of spin in nuclear reactions. Now, I recall that we had repeated the Cavendish experiments and confirmed this remarkable effect, that when you bombarded lithium with protons, the light isotope gave approximately 10 times as many α's as the heavy isotope which was 12 times as abundant. As I recall, it was you who invoked the conservation of angular momentum and parity to explain this—and that was such a marvelous thing to all of us who had never even heard of parity. [N.B. Some of us had not studied our atomic physics carefully enough.]

Goldhaber: Can I explain it for a second? Cockcroft and Walton had observed this big difference and noted in their paper (anyway,

I was well aware of it somehow in the Cavendish) that the yield of Li^6 + p was much larger going into He^4 + He^3, compared with Li^7 + p going into He^4 + He^4. Now the remarkable point was that the first reaction had a very large energy release, 15.6 MeV, but the second had only about 4 MeV. So it struck me as strange that the one with the higher energy was much weaker, because it had a smaller barrier penetration problem on the way out. Then, on top of that, they had seen Li^6 + d going into He^4 + He^4 with very good yields, unlike the Li^7 + p reaction. I began thinking, what could be the reason for these different yields? [N.B. On rereading my 1934 paper I find that I convinced myself by a complicated spin argument without using the word "parity" explicitly, which others did a little later, that the Li^7 + p reaction was "improbable" because it needed $\ell = 1$ protons, whereas the Li^6 + p reaction which had a good yield could work with $\ell = 0$ protons.] Li^6 was believed by some to have spin 0 because Schüler could not find any hyperfine structure in its atomic spectrum. For spin 0 the highly probable reaction Li^6 + d could not have worked for $\ell = 0$ deuterons which have spin 1, because the two α's had to be in an even orbital angular momentum state. So I stuck out my neck and "determined" the spin of Li^6 as 1, which was later confirmed. The work concerning the Li^6 + p reaction in a sense prepared me quite well when the slow neutrons became known. I said to myself, look, if I replace the proton with a neutron of zero energy and replace the He^3 by H^3, I can predict the Q-value, which is a little higher than for the proton reaction because of the larger neutron mass. Therefore, slow neutrons have "enough energy" to disintegrate Li^6 and the following would be a reaction with a "good yield": Li^6 + n → He^4 + He^3. We took one of Chadwick's ionization chambers, put in lithium fluoride, and obtained a small effect. I then asked his permission to put in lithium metal, but he thought that lithium metal would spoil the chamber—there was a great lack of chemical knowledge in the Cavendish; people feared lithium would oxidize, explode, and so on. So one day when Chadwick was attending a meeting of the Royal Society in London, we quickly put in lithium metal and obtained a big effect; and he then believed it. So this came directly out of the above considerations as a trivial thought, and this was one time when we beat the fast Romans.

Badash: During Rutherford's period at the Cavendish laboratory there was no professorship of theoretical physics, though R. H. Fowler, his son-in-law, was a professor in another department and was around. Would you care to comment on the appreciation of theoretical physics by Rutherford and others in the laboratory?

Goldhaber: The structure was complicated. There was Dirac, who held the Lucasian professorship, once held by Newton. Dirac was not under Rutherford. There was the director of the Cavendish, who was called the "Prof." That was Rutherford. But there were other people who held professorships there. Fowler held an independent professorship located at the Cavendish, but he was otherwise independent and had theoretical students. So there was a theoretical presence. Bethe and Peierls were there for a while and they might tell you more. I think Hartree came later, didn't he?

Peierls: Certainly Fowler was a professor at the time I first came to the Cavendish. Dirac had not yet got his professorship. This was in 1927, I believe. Now, although Fowler belonged to a different faculty, he had room in the Cavendish Laboratory next door to Rutherford or on the same corridor. Certainly, there was extremely close contact. When Dirac later held his professorship, his office was located in the mathematics school, and there was less day-to-day contact because Dirac's interests were more abstract. But there was enough contact, either directly or through Fowler, for Dirac's ideas to be taken very seriously.

Early History of Particle Accelerators
Edwin M. McMillan

INTRODUCTION: *H. H. Barschall*

In his introductory talk on Wednesday, Hans Bethe discussed the question of when nuclear physics started, whether it was 1932 or earlier. There is much less doubt about the geographic area where nuclear physics started. It started in Europe, and yesterday we heard from the pioneers about the activities in Rome, Berlin, Copenhagen, and England. In this country, there was very little nuclear physics during the early period, which was discussed yesterday, except at Berkeley, where Lawrence and Livingston (who is with us today) had developed the cyclotron, and where a vigorous effort in nuclear physics using the cyclotron was under way. Berkeley has remained for a long period of time one of the most active centers of nuclear physics, some of which is called nuclear chemistry at Berkeley.

Our first speaker today is Ed McMillan, who has been connected with the Berkeley laboratory from its beginnings in 1932 and who was director of the Berkeley laboratory, known earlier as the Lawrence Radiation Laboratory and later as the Lawrence Berkeley Laboratory. For fifteen years, Ed McMillan was the director of the laboratory, and his most important contributions have been in the two areas for which the Berkeley nuclear physics laboratory is most reknowned, namely, in nuclear chemistry and in accelerator design. For his work in the transuranic elements, he was awarded the Nobel Prize in chemistry in 1951, but equally important were his contributions to the development of high energy accelerators, especially the invention of the synchrotron, which he made independently and at about the same time as Veksler in the Soviet Union. Since 1932, the developments, first in nuclear physics and then in high energy physics, were dependent on the development of accelerators. We are therefore fortunate that Ed McMillan has agreed to talk to us on the subject of accelerators.

Early History of Particle Accelerators
Edwin M. McMillan

This is the first time I have spoken before a historical group like this, and I am very pleased to see that there is so much interest now in the history of nuclear science, because this particular time is about the latest that it is still possible to get together as good a group of the pioneers as we have here. In a few more years, this will become more and more difficult.

After I agreed to participate in this symposium, the first thing that I did was to go to the library and take out a volume entitled *Exploring the History of Nuclear Physics* (AIP Conference Proceedings No. 7), which is a report of the two earlier conferences on the same subject, in 1967 and 1969. I found that a lively topic in those discussions was "when did nuclear physics start?"—with the year 1932 given as a zeroth order approximation. Segrè argued that it was born in 1911, with Rutherford's interpretation of the α-particle scattering by matter. One could almost as well have carried it back to Becquerel in 1896; radioactivity is a nuclear phenomenon, even though no valid model or theory existed at that time, and the "nucleus" had a biological connotation (as did "plasma" and "fission" before they were taken over by the physicists).

Nuclear Physics in Retrospect, edited by Roger H. Stuewer, copyright © 1979 by the University of Minnesota.

John Wheeler, on p. 56 of *EHNP*, tells how Poincaré suggested to Becquerel that he should try to find a relation between the phenomenon of phosphorescence which he had been studying and Röntgen's newly discovered radiations. Now, the phosphorescence of the uranyl salts is an electronic effect, not a nuclear effect, but by this lucky coincidence the discovery of radioactivity was advanced by who knows how many years. The importance of coincidence in scientific discovery would make an interesting study. Perhaps I don't know too well the field of the history of science. I am just learning a little bit about it, but when I run across these things it occurs to me that some of them are quite prominent, like the phenomenon of coincidence. Science is full of things like that.

Regardless of the arguments about when nuclear physics started, there was general agreement that there was a large increase in the second derivative of interest and effort on or about the year 1932. That was the year of the discovery of the neutron, the positron, and the deuteron, and of the first nuclear disintegration by artificially accelerated particles. What a wealth of new information! The student of today will find it hard to imagine a world in which these things were not known. He or she is taught that nuclei are made of protons and neutrons, that the positron is the anti-particle to the negative electron, and that the deuteron is the "hydrogen atom" of the nuclear world, simple enough for elementary theoretical interpretation. Artificial disintegration by "atom smashers" is familiar to anyone who reads the popular press. Is it possible to single out one of these discoveries as the most influential in the phenomenal growth of nuclear physics as a field of investigation?

I think a good case could be made for the disintegration by artificially accelerated particles. There is a lead time involved here. If, by 1932, no one had developed accelerators, it would have been years before anything other than natural α-particles would have been available for nuclear experimentation. Artificial radioactivity would have been found with the natural sources (as it happened anyhow), but progress would have been slow with the weak sources available, and large areas in the chart of isotopes would have been inaccessible. Many types of nuclear reactions would have gone unobserved. The accelerator opened up a wide territory for investigation, using means simple enough that they were accessible to many institutions, and so the field opened up rapidly. The other discoveries of 1932 are of the kind that would have followed inevitably.

Rutherford's presidential address in 1927[1] is frequently cited as the origin of interest in high voltages for the acceleration of particles, but on reading it I get a somewhat different impression. He mentioned work already under way by the electrical industry in developing sources of high potential, and by Coolidge at the General Electric Company and by Breit and Tuve at the Carnegie Institution in Washington in applying high potentials to vacuum tubes to accelerate particles. The tenor of his address was to point out the importance of what they were doing, giving encouragement for further development as a tool for nuclear research. I had never actually read this address before. It is always quoted. But Rutherford really was talking about what was being done in high voltages, and then he said, that's good stuff—keep it up.

In the early twenties there was a great deal of interest in the production of high voltages. The electrical industry was developing means for long-distance power transmission. I remember the high voltage laboratory at Caltech, with its million volt cascade transformer that had been built for testing transmission line components. One of the features of the annual "exhibit day" was a demonstration of this equipment. A steel weight on a chain would be hung from the lofty ceiling over the high voltage terminal. Then there would be the rythmic clink-clunk-clink-clunk of the tap-changing switch as the voltage was raised, the rising ominous hiss of the corona discharge, and the crash of the spark, which turned into a long sinuous snarling arc as the weight was pulled away from the terminal. It made a great show. I suspect a lot of students got an inspiration from it. (Willy Fowler is nodding his head; he remembers that place. It really was a great show.) Amateur experimenters built Tesla coils, radio transmitters and receivers. The impulse generator (Marx circuit) and the electrostatic machine were well known. Vacuum tubes for accelerating electrons (X-ray tubes) were in wide use. The time was ripe for someone to start the development of accelerators for nuclear research.

In fact, several lines of development were started, which I will outline. Chief attention will be paid to continued efforts that led to the readiness for expansion by the pivotal year 1932. Items such as independent inventions that did not become known until later, and patents not reduced to practice, will be left for the comprehensive accelerator histories, but single contributions that had a known influence through the reading of publications will be mentioned. (I had some arguments with historians on this point. When you are trying to get together a history of accelerators, you want to be complete, so you get all the little things you can dig up, little obscure publications that there is no evidence anyone ever saw, and patents that there is no evidence anybody tried to build, and set these all down in the footnotes. If you are trying to find out, to trace the lines of influence, how something developed and became an important part of science and technology, some of these things have to be left as footnotes that could be included for completeness, but not referred to again. However, the things that are in the direct line deserve special prominence. I think I even won this argument with at least one historian.) In any case, the betatron (which did not come into practice until after the period considered), the linear accelarator, and the cyclotron all had more than one inventor, as did the principle of the belt-type electrostatic generator. In the present context, what is important is their influence on the field of nuclear physics in those early times.

As far as I know, the first person to start working on accelerators for nuclear physics was Merle Tuve at the Department of Terrestrial Magnetism of the Carnegie Institution of Washington. Before I came to this meeting I had a little correspondence with some of the people I have mentioned, one of whom was Merle Tuve (see letter no. 5), who said he was sorry he wasn't able to come, but he sent his best to everyone. I should also say that since this paper is really a study paper, which I originally intended to just hand in to the historians here to give them some ideas to work on, I solicit letters and information from anyone who would like to add something.

I am very much impressed by how important Tuve's part was. He was, as far as I know, the first. He tells about his early work in *EHNP*, pp. 26, 32, 50, and so on. His motivation was very clear and specific, to look for deviations from Maxwell's equations near the nucleus by studying proton scattering. Gregory Breit was his supervisor at that time (1926); I don't know how much of the motivation came from Tuve and how much from Breit, but I have no question that the prime mover in the experimental work was Tuve. (Since I wrote this, I received a reply from Tuve [see letter no. 6], who stated that Breit did have a very important part in this, and perhaps started it. He was the one who got Tuve to come to the Carnegie Institution. Tuve already was interested in high voltages. He said he was interested in all kinds of extremes: high pressures, high temperatures, low pressures, low temperatures, etc.—one of them being high voltages. His voltage source was a Tesla coil immersed in oil, which in his letter to me he speaks of as his "albatross." He feels, I think, that if he had started out with a steady potential source rather than something exotic like a Tesla coil, he would have been the first with artificial disintegration. He doesn't say exactly that. He simply says in parentheses "our albatross." I can see what he means. A Tesla coil gives lots of volts, but it's uncontrollable, unsteady, and hard to do experiments with, and Tuve was stuck with it.)

The Tesla coil is a high-frequency transformer with a loosely coupled resonant secondary, and has the advantage over a line-frequency transformer of being much more compact for a given voltage. In a Tesla coil the potential drop is more or less uniformly distributed along the length, so that the maximum total potential is obtained for a given dimension. The Tesla coil did have that advantage, but that's all. The coil Tuve used was 36 inches long and would produce over 3 million volts. The accelerating tube was immersed in the same oil bath as the coil, so there was no problem of sparking in air. W. D. Coolidge of the General Electric Company in Schenectady had already developed sectionalized high voltage X-ray tubes for medical use; the medical motivation is closely entwined with the physical one in the accelerator field. I think this point must always be noted. It certainly was true at Berkeley, and Willy Fowler just told me yesterday that it was true at Caltech. The cynics will say that this connection occurred because it was the way to raise money; this happens to be true—it's easier to get money for medical purposes than for pure scientific purposes—but there also was, at least in some cases, a sincere medical interest. I know that was the case with Lawrence in Berkeley, because Lawrence had started out in his early life with the idea of being a physician. Medicine attracted him. He had deviated and become a physicist, but he still retained his interest in medicine, which in his case was a sincere interest.

Tuve started his work with sectionalized high voltage X-ray tubes furnished by Coolidge. I remember I saw some when I was a graduate student at Princeton, on a visit to Schenectady. They were typically in two sections. I think he had one that had three, maybe four, but a moderate number of sections. Tuve got them free, which was fortunate because he and Breit broke a lot of them or burned them out by discharges. Tuve soon made his own design, his own tubes, with a larger number of sections, which had many features of modern designs. I think that is the most remarkable feature of the early work of the Carnegie group, that they did develop these highly sectionalized tubes, with sections which do look quite a lot like what people use now in a Van de Graaff or similar machine, and which solved the problem of the electrical breakdown and really made the whole development possible.

Breit and Tuve's earliest publication[2] on this work was in 1928. By 1931 they had demonstrated the acceleration of protons and electrons, but the Tesla coil outfit was never used for nuclear research. It was too unsteady and unreliable. (That was the "albatross," which they didn't say in their publications, by the way. The publications I read are full of praise for it, but looking back now Breit and Tuve may have a slightly different opinion.) After abandoning this method, Tuve and his group combined the accelerating tube they had developed with a Van de Graaff electrostatic generator, and early in 1933 observed the disintegration of lithium and boron by protons. I was a graduate student at Princeton during this period, and on at least one occasion I went down to Washington and saw Tuve's Tesla coil. Tuve also visited Princeton a couple of times. On one visit in 1931, which he describes in his letter to me, he tied a couple of Van de Graaff's early models of the electrostatic machine to the rear bumber of his car and took them down to Washington and used them for his first setup. Thus, although Tuve and Breit didn't demonstrate nuclear disintegration by that means until 1933, they had already taken the first steps toward it in 1931.

Cockcroft and Walton have the honor of achieving the first artificial disintegration. But I believe that Tuve's work had a great influence. It was widely known and discussed, and was mentioned by Rutherford in his 1927 address. Their tube design was excellent and made the Van de Graaff machine a practical accelerator for the first time. In fact, the vacuum tube of the first successful Van de Graaff setup was the highly sectionalized tube which was ideally suited to go with the Van de Graaff machine as a voltage source, while Van de Graaff in Princeton (and I saw a great deal of him) was fussing around with textolite tubes of various exotic designs, which were all great, except that they wouldn't work when there was a vacuum inside. Van de Graaff could get lots of volts and sparks, but every time he put a vacuum in his tube it would break down and discharge. So it was the bringing together of Van de Graaff from Princeton and Tuve from Washington starting in 1931 that led to what we now call the Van de Graaff machine.

The next in time were A. Brasch and F. Lange of the Physical Institute of the University of Berlin, who started their work in 1927. They are best known for their dramatic attempt to capture lightning in the Alps, which was abandoned when one of their men was killed by a lightning stroke; but they also used transformers and impulse generators to apply potentials to sectionalized vacuum tubes. Their first publication was in 1930.[3] In 1933 they repeated the lithium disintegration of Cockcroft and Walton using an impulse generator voltage source. The impulse generator, or Marx circuit, is a device in which a bank of capacitors is charged up in parallel and then discharged in series. (I remember that Willy Fowler had what he called "the world's most powerful impulse generator" at Caltech's "high volts" lab, which he once tried out on R. W. Wood. It actually consisted of a firecracker, a large firecracker, which was set off by a little spark coil. It fooled even R. W. Wood.) As far as I know, Brasch and Lange published only one paper thereafter, in 1934, from Berlin. That paper also has Szilard as one of the authors. It is an interesting paper — the only actual contribution to science I know of to come out of that particular work.

I don't know how to evaluate the influence of this work. It was widely known, and the attempt to use lightning had a dramatic impact; it certainly aroused interest, but I can't trace any influence other than by example. I remember Brasch's coming around and giving some lectures once. I'm not certain whether I heard him at Caltech or at Berkeley. He was describing a sectionalized tube made essentially of sheets of metal and insulator with holes cut in it. It was just stacked up with rubber gaskets between, and for the insulating material he used water. That's as good as oil, because the impulse lasts such a short time that no appreciable current flows down. It was very interesting. I would like to know more about what started them, what their motivations were, what happened to them, and so on. I suspect that the political situation in Germany had something to do with the termination of their effort (we are talking about 1934), but I can't document it. The only documents I have are their published papers. I would like to get further information. After 1932 several people in Germany set up low voltage rectifier-operated accelerators, but none seem to have led to important developments. The whole field in Germany, starting with a lot of fanfare with Brasch and Lange, seems to have simply faded out in the 1930s.

I come next to the work of Lauritsen at Caltech, starting in 1928 with the construction of a high voltage X-ray tube[4] using the cascade transformer I mentioned earlier. (Incidentally, in Stan Livingston's book on accelerators, this is given as a three-section cascade, while it was actually a four-section cascade. It had four transformers in sequence. The fourth was difficult to see, because it was down in a pit, and the other three were on a sort of sectionalized platform, with each layer insulated from the next, going up to the fourth on top.) This apparatus was on a much larger physical scale than Tuve's, even though the limiting voltage was lower (about 1 MV). The power output was also much larger. That is another thing you have to consider. Lauritsen's cascade transformer would give about an ampere of current at about one million volts, which is a lot of power, and if a breakdown in a vacuum tube occurs, things are really destroyed by the discharge; however, with a Tesla coil, there is only a very small power output at 3 million volts, owing to the small current. Lauritsen's installation went through several modifications, and in 1933 a positive ion source was installed and the production of neutrons by a-particles and deuterons on beryllium was observed. I consider the work of Lauritsen and his group to be important not because of decisive contributions to the accelerator art but because it led to the creation of an active center of nuclear research, the Kellogg Laboratory.

I was at Berkeley through much of this period and observed the interaction between the Berkeley group and the Caltech group, with Oppenheimer and Serber (who is with us today) and others oscillating back and forth as sort of roving ambassadors. There was a strong rivalry. I remember the great scorn that certain people had for some of the results claimed by other people. I won't go into details on this. Some of the scorn I think I would share myself, but there was also a lot of mutual respect. I think it was an extremely healthy situation. And there certainly was openness. Nobody was concealing anything. There was the danger of getting your ear talked off by somebody who tried to tell you more than you really wanted to know. It was a great period.

Lauritsen's start with an existing cascade transformer was a matter of opportunity. It had been set up for testing power lines and power line components, and for some reason, whether told by Millikan or not (which I hope Willy Fowler will determine from the archives if he doesn't know already), Lauritsen built these tubes and started something which turned out to be great. Some important work was done with the transformer source, but this was clearly not the best method for a number of reasons. One of the problems was that the operators would get headaches from the ozone made by the ever-present corona discharge. I've been there when it was running. You could hear the sizzling. The whole room was buzzing and sizzling with corona, and you could see the little purple glow around everything. The most important reason for changing, of course, was that to do precise work you need a steady voltage source, and all these pulsing and alternating methods finally had to be given up. So the transformer was dropped, and Van de Graaff machines were installed in the Kellogg Laboratory.

Also in 1928 (I am not sure whether it was earlier or later in the year than Lauritsen) Cockcroft at the Cavendish Laboratory proposed to Rutherford to start developing means for accelerating protons to high energy. This work was motivated by Gamow's calculations of the probability of penetration of nuclear potential barriers, which indicated that protons should penetrate the boron nucleus to a measurable extent at energies as low as 300 keV, and the lithium nucleus at even lower energies. This is one case where theoretical predictions really influenced the accelerator builders, as Cockcroft stated in his Nobel Lecture. Rutherford gave his full support to this program, and Cockcroft was soon joined by Walton, who had been trying without success to make a linear accelerator and an induction accelerator (betatron). Together they considered four voltage sources: the Tesla coil, the impulse generator, low-frequency alternating potentials, and rectified steady potentials. They chose the last as being the most suitable for nuclear research, as giving the most constant bombarding energy. That was a very wise choice. In their first joint publication[5] they described a single-stage rectifier setup giving 280 keV protons. With this setup they bombarded lithium and other targets, looking for γ-rays with a gold leaf electroscope, but did not find any clear indication of an effect.

Just then they were faced with the necessity of vacating the laboratory room they were using, but a larger room was available and they decided to try for a higher energy. Their next apparatus had a four-stage voltage multiplier, giving up to 500 kilovolts. This time, instead of looking for γ-rays, they looked for α-particles with a zinc sulfide scintillation screen and found scintillations, at 125 kilovolts! Remember that they had had 280 kilovolts two years before. So had they looked for α-particles in 1930, the discovery of the first completely artificial disintegration would have been made two years earlier than it actually was. Perhaps this was a case of an unfavorable coincidence, where the necessity of moving created undue haste and precluded a longer search for effects using their first apparatus. I wrote to Walton (see letter no. 9), who is still living in Dublin, and received one reply from him (see letter no. 10), but I want to ask him specifically whether this statement is true. Perhaps if they had not had to move, they would have put a window in to count scintillations and would have found them in 1930. They were still the first to observe artificial disintegration, however, and the Cavendish group, with Oliphant and others, exploited the new method extensively in the next few years.

The last two efforts that I will discuss, by Lawrence and Van de Graaff, had their beginnings in 1929. I first met Van de Graaff when I arrived at Princeton in the fall of 1929 as a graduate student. He was building small models of his belt-type electrostatic generator. I remember one with a tinfoil-covered cylindrical conductor, about two feet high altogether, and a larger model with a copper flagpole ball on top. He told me that he had worked for a while at a power plant in his home state of Alabama (in a town called Bugtussle, by the way, which is now called Warrior — the people thought Bugtussle was not a very dignified name, so they changed it). Van de Graaff had become fascinated with electrical power, and he showed me his designs for a power system using DC transmission with electrostatic generators and motors, which he had calculated would give very high efficiency. Earlier, while a Rhodes Scholar at Oxford, he had learned the value of high voltages for nuclear physics, and when he came to Princeton as a National Research Fellow in 1929, the development of the electrostatic generator became his research project. His friend Joe Morris in the physics department took a great interest in his work and gave much encouragement and help; it was Joe Morris who went to Trenton to buy the flagpole ball. I remember this trip. He went down, bought it, and essentially told Van de Graaff: "Now, you build a generator to charge this up." He wanted something for a good-looking lecture-demonstration model. K. T. Compton, the department chairman, gave his support. The first publication of this work is an abstract of a paper given at the Schenectady meeting of the American Physical Society in September 1931,[6] describing a seven-foot model.

In 1930 Van de Graaff built a cylindrical vacuum chamber and installed in it an electrostatic generator with a segmented glass and metal belt, in an attempt to reach higher voltages with vacuum insulation. Often the belt broke, making a real mess, but even when this problem was solved the system did not work; the tiny sparks produced when the charging wire touched the metal segments were enough to initiate breakdown of the vacuum. Don Mueller, a student of H. A. Barton, came from Cornell to build a pressurized machine. This was successful in the sense that a higher voltage could be reached under a pressure higher than one atmosphere, but it was not useful for nuclear research, because the textolite accelerating tube would not hold voltage when evacuated. It is interesting that Herb, Parkinson, and Kerst went through the same sequence of operating first in vacuum, then under pressure in their later successful development of the pressurized Van de Graaff machine.

The development of the Van de Graaff machine now divides. On the one hand, starting in 1931, Tuve applied the electrostatic generator to his already developed accelerating tube, with great success. Van de Graaff himself went to MIT in 1932 and engaged in the building of the huge electrostatic generator at Round Hill, Massachusetts, which made very impressive sparks but was useless for nuclear research, again because of trouble with the accelerating tube. Still later he became one of the founders of the High Voltage Engineering Corporation, noted for its advanced developments in the field. I might add that I got to know Van de Graaff very well at Princeton, and he was one of the finest people I have known in physics, what I call one of "Nature's Gentlemen"—always friendly, always helpful, never interested in advancing himself or his own ego. He was a great guy.

The other effort, starting in 1929, was that of Lawrence at Berkeley. Ising and Wideröe were "precursors" of Lawrence. This is a case where somebody wrote something, published it, and somebody else read it and was influenced to do something. It was not dropped down a well but somehow came out someplace where it was useful. Ising (not the Ising of the "Ising model") had proposed in 1924[7] a primitive linear accelerator using timed electrical pulses applied to a series of cylindrical electrodes, and Wideröe had tried this out with two stages of acceleration using high-frequency excitation and had shown that it would work. His doctoral thesis describing his investigation was published in 1928,[8] and Lawrence ran across it while looking at journals in the University of California library in the spring of 1929. Ising himself did no further work along this line; when I met him in Stockholm in 1951 he was interested in the navigation of birds. Wideröe became a prolific inventor in the field of particle accelerators and worked for the Brown-Boveri Company in Switzerland.

Lawrence had already been thinking of the problem of accelerating particles before he saw Wideröe's paper, so his mind was receptive to the idea of the linear accelerator. This sentence brought me a lot of comment from one historian, who asked how I knew that? I had not said anything about how I knew that Lawrence had been thinking of the problem of accelerating particles, so that his mind might be receptive. What had he been thinking about? I had to admit I didn't know. I wasn't with Lawrence at that time. In 1929 I was still in Pasadena before going to Princeton. I therefore wrote to various people on this point, and some of the replies I got are rather interesting. I can't go into everything, but I can find no evidence that Lawrence left any notebooks, diaries, or other records which speak of any interest in nuclear physics or in anything other than what he was currently working on: experiments involving the timing of small intervals. He was working with Jesse Beams on spinning tops. There's mention of that. The closest I got was a statement from Tuve (see letter no. 6), which I think is correct. Shortly before leaving Yale to go to Berkeley, Lawrence was talking to Tuve, who was of course a boyhood friend of his and had attended the same high school. Tuve asked Lawrence what he was planning to do at Berkeley, and Lawrence started talking about cutting the tails off quanta to see what would happen to them, which of course is the spinning top work. Tuve says he then told Lawrence that he should stop picking research problems as if he were picking out cookies at a party, and try to think of some field that was rich with things to observe, and so on. He suggested nuclear physics with particle accelerators. I suspect that this story is correct in essence and that Tuve did have an influence on Lawrence. That's as much as I have on this point. Lawrence himself says in his Nobel Lecture that he was already thinking about these things; his mind was receptive to the linear accelerator. He must have been thinking along those lines, otherwise something a little oddball like a linear accelerator wouldn't have excited him.

Lawrence realized the difficulty of applying the linear accelerator to particles as light as protons with the high-frequency techniques available at that time. Then came the "flash of inspiration," the invention of the cyclotron. I think one can only describe it that way. From one of my other letters I got a little more information on this point from Tom Johnson (see letter no. 2), who was with Lawrence when he saw that Wideröe paper and immediately thought of the cyclotron. He hadn't figured out the timing problem yet, and that night he got worried about it. He then worked it out, realized that the timing came out right, and the next morning came back to Tom and said, "It's all right. The timing works out." I call this a flash of inspiration—it is not that easy to get from a linear accelerator to a cyclotron. By applying a magnetic field, the excessively long path in the linear accelerator could be wound up into a compact spiral, while the constant angular velocity in the magnetic field allowed the proper timing of the accelerating impulses produced by an alternating electric field between two electrodes. I consider this to be the single most important invention in the history of accelerators; it brought forth a basic idea of great power, and one capable of later elaborations and variations, such as the use of phase stability and strong focusing. All the big proton synchrotons are really just an extension of the cyclotron principle.

Experimental trial of Lawrence's idea was delayed for a year by his other duties, which is another thing I would like to get a better explanation for. It took about a year before Lawrence did anything about his idea. There is a story of how Otto Stern encouraged him, which apparently is true. Stern was visiting Berkeley and Lawrence told him about his idea, and Stern said: "Why don't you get on with it?" The first apparatus built by Edlefson was encouraging, and a later model built by Livingston (1931) was a clear success. The first publication is an abstract of a report given before a meeting of the National Academy of Sciences in Berkeley in September 1930.[9]

The year 1931 saw not only the first successful operation of a cyclotron but also the founding of the Radiation Laboratory at Berkeley. Lawrence realized that the program of development and research that he was contemplating would need support beyond the capability of the university, and he approached some private foundations, receiving assurance of financial aid. Professor Leonard Fuller, who was chairman of the Department of Electrical Engineering at the university and a vice president of the Federal Telegraph Company, knew of a large magnet the company had built for a Poulsen arc but had never delivered, and he arranged for this magnet to be given to the university for Lawrence's use. Another coincidence! It happened that Leonard Fuller was then splitting his time between Berkeley and Palo Alto and suggested that Lawrence use this magnet. I think that without that big magnet the cyclotron development would have been much slower. Remember, we are talking about 1931; the depression was on, people were broke, nobody had any money, so one was really dependent upon grants and gifts for this kind of work.

It turned out that there was a building available on the campus, an old civil engineering testing laboratory near Le Conte Hall, and in August President Sproul of the university arranged for Lawrence to have the use of this building for his researches. I found in the

Lawrence Collection at the Bancroft Library the letter that Lawrence wrote to someone the day after this meeting with Sproul, which was August 16. Everything pulled together. He had the money, the formal gift of the magnet, and the building. The magnet was installed in January 1932, and the laboratory was under way. (I found it hard to pin that date down. Childs in his biography of Lawrence said the magnet was installed in October 1931. I looked in the files and got what looked to me like evidence that it was in December 1931, which I stated in a little note I wrote for a laboratory publication in Berkeley. That date is incorrect; it is January 1932. I finally tied that down by finding four documents. One of the pitfalls of historical research is finding something that looks like evidence, such as Lawrence's writing to somebody that he expects the magnet to be installed by Christmas, when it actually wasn't installed until January.)

A good account of Lawrence's plans at that time is given in a letter from Joseph Boyce to John Cockcroft dated January 8, 1932, about a week before the big magnet was moved in. In this letter, which was quoted in a 1972 article by Charles Weiner,[10] Boyce said:

> But the place on the coast where things are really going on is Berkeley. Lawrence is just moving into an old wooden building back of the physics building where he hopes to have six different high-speed particle outfits. One is to move over the present device by which he whirls protons in a magnetic field in a very high frequency tuned electric field and so is able to give them velocities a little in excess of a million volts . . . [Livingston's 11-inch cyclotron, which actually stayed in Le Conte Hall]. Then there is the Hg ion outfit. . . . Then a similar device with higher applied voltages and longer electrodes to use with protons [Linear accelerators of the Wideröe type]. The fourth is a whirling device for protons in a magnet with pole pieces 45 inches in diameter . . . [the 27½-inch cyclotron in the Federal magnet]. [The word "cyclotron" had not been coined yet. It was usually called a "whirling device," sometimes a "whirligig," or something like that.] Then a small tesla-coil X-ray outfit is already installed [the Sloan X-ray tube], and the remaining room is reserved for a Van de Graaff electrostatic generator [this letter is the only reference I know of to this plan, which was never carried out]. On paper this sounds like a wild damn fool program, but Lawrence is a very able director, has many graduate students, adequate financial backing, and in his work so far with protons and mercury ions has achieved sufficient success to justify great confidence in his future. . . .

It was clear that Lawrence intended to explore all avenues, not just to develop his own invention, but it soon became apparent that the cyclotron was the important thing. This was a large enough effort that it could not be done without some degree of organization and a great deal of cooperative effort by many people. Thus was born a new style of research in nuclear physics, which may be considered a second invention of Lawrence's. With the addition of efforts in biology, medicine, and chemistry the Radiation Laboratory became the prototype of the modern multipurpose laboratory.

This brief account shows that when Cockcroft and Walton announced their result in 1932 there were others ready and eager to check their findings and to carry them further. Disintegration experiments were soon reported by the groups in Washington (Tuve), Pasadena (Lauritsen), Berkeley (Lawrence), and Berlin (Brasch). The first three became important centers for nuclear research. Thus, counting the Cavendish Laboratory, there were four active centers of accelerator development and use. There was a great deal of communication between these centers; a critical mass existed, and soon other centers were established. This situation, I believe, gave the impetus to the great expansion of nuclear physics that started around 1932. In Weiner's 1972 article he emphasizes the special role of Berkeley in this expansion, not only by its direct contributions but also by its encouragement of the diffusion of the cyclotron art through the furnishing of information to all who were interested, and by its introduction of the style of research necessary for efforts of such magnitude.

Rutherford published a lecture[11] given in Cambridge in November 1936, describing some of the early results. In it he mentions explicitly Van de Graaff; Tuve, Hafstad, and Dahl in connection with the electrostatic generator; and Lawrence in connection with the cyclotron; but he makes no mention of Lauritsen. However, he quotes without attribution some results of the Pasadena group (the discovery of the 17 million-volt γ-ray, which was clearly made in Pasadena). So all four centers I mentioned had already made their influence felt by 1936. There were a few other small efforts, such as those I mentioned in Germany, but these never got off the ground. By 1936, Herb, Parkinson, and Kerst were already working on the pressurized Van de Graaff. The conditions were right for the next big expansion.

The development of accelerators after 1932 became widespread and complex and will not be discussed here, but further historical investigation of the period up to about 1932 seems very worthwhile. This was the formative and exploratory period when some methods were tried and discarded and the basic successful lines of development were started. The technical details are reasonably well covered in publications, but many materials in the form of personal communications (especially between the different groups) and personal papers and notebooks remain to be explored. These should give further insight into the motivations for the work and the cross-fertilization that took place.

This is the end of my talk. I would be very interested in learning of any materials from this formative period. I would be interested in obtaining copies of letters or other documents, because I think these materials should be preserved.

REFERENCES

1. Ernest Rutherford, "Scientific Aspects of Intense Magnetic Fields and High Voltages," *Nature*, 120 (1927), 809-811.
2. G. Breit and M. A. Tuve, "The Production and Application of High Voltages in the Laboratory," *Nature*, 121 (1928), 535-536.
3. A. Brasch and F. Lange, "Ein Vakuum-Entladungsrohr für sehr hohe Spannungen," *Die Naturwissenschaften*, 18 (1930), 16.
4. C. C. Lauritsen and R. D. Bennett, "A New High Potential X-Ray Tube," *Physical Review*, 32 (1928), 850-857.
5. J. D. Cockcroft and E. T. S. Walton, "Experiments with High Velocity Positive Ions," *Proceedings of the Royal Society of London*, 129[A] (1930), 477-489.
6. Robert J. Van de Graaff, "A 1,500,000 volt electrostatic generator" [Abstract], *Physical Review*, 38 (1931), 1919-1920.
7. Gustaf Ising, "Prinzip einer Methode zur Herstellung von Kanalstrahlen hoher Voltzahl," *Arkiv för Mathematik, Astronomi och Fysik*, 18 (1924), 1-4.
8. Rolf Wideröe, "Über ein neues Prinzip zur Herstellung hoher Spannungen," *Archiv für Elektrotechnik*, 21 (1928), 387-406.
9. E. O. Lawrence and N. E. Edlefson, "On the Production of High Speed Protons," *Science*, 72 (1930), 376-377.
10. Charles Weiner, "1932—Moving into the new physics," *Physics Today* 25 (May 1972), 40-49.
11. Ernest Rutherford, *The Newer Alchemy* (Cambridge: Cambridge University Press, 1937).

130 Edwin M. McMillan

APPENDIX

Correspondence between E. M. McMillan and T. H. Johnson, J. J. Brady, M. A. Tuve, G. Breit, and E. T. S. Walton. These letters are published with some editing by E. M. McMillan and with the consent of the authors.

[1]
E. M. McMillan to T. H. Johnson
April 4, 1977

Dear Tom,

I have recently been getting together some historical material on the beginnings of particle accelerators, and one of the questions that I would like to answer is the following:

When Lawrence saw the Wideröe article, he was already thinking about the problem of getting high-speed particles, as he mentions in his Nobel Lecture. The question is, what was his line of thinking? Is there any record in correspondence, or elswhere? You were there, and I thought you might be able to help me in this. If you have any correspondence that you would like to keep, I could make copies and return the originals, or you could send me copies. . . .

                                            Best,
                                            [signature]
                                            Ed [McMillan]

[2]
T. H. Johnson to E. M. McMillan
April 9, 1977

Dear Ed,

I'm afraid I can't give you any very substantial help on your problem. I was with Ernest that night, probably in March or April 1929, when he first saw the Wideröe article and we discussed the ideas it invoked. I believe it was the first time the idea of a multiple acceleration had occurred to Ernest, and the next day we discussed the matter again. The idea of bending the orbits in a magnetic field occurred to Ernest immediately on reading Wideröe, but that evening he was worried about synchronization and it was not until the next day that he realized the periods would be the same for successive orbits.

As to his general line of thinking on the problems of acceleration and nuclear physics, there was no correspondence and probably no notebooks, as far as I am aware. Ernest asked me to stay on with him to develop the cyclotron, an offer that I had to reject for personal reasons, so at that stage there was probably no one else with whom he was corresponding on this subject. My recollection is that his ideas of nuclear physics were rather nebulous and far ranging. It was a large and untouched field where interesting problems would develop. He had observed Swann's strugggles with the "dropping steel ball" electrostatic generator, with guarded scepticism as to the validity of the method, and he was or had been in close touch with the experiments of his old buddy Tuve, who was of course stimulated, if not directed, by Breit in developing a tube for high voltages and a modified Tesla coil generator. As you know, Ernest was not inclined to waste time speculating on problems beyond his own experiments, and he had certainly not developed any sophisticated ideas in the field until he was able to back them up in the laboratory. . . .

Sincerely,
[signature]
Tom [Johnson]

[3]
McMillan to Brady, April 4, 1977, asking the same questions as in the letter to Johnson.

[4]
J. J. Brady to E. M. McMillan
April 21, 1977

Dear Ed:

In reply to your letter, I shall be glad to answer your questions as I am possibly the only one who does know that very early history. When Stan Livingston was about to choose a research project, he asked my advice, and I recommended he work with Lawrence on what became later known as the cyclotron.

I was one of the first to hear of Lawrence's idea of the cyclotron. He came into my research room (room 216, Le Conte, at that time) one morning and asked me to come to the blackboard. He said he had been reading a German article by Wideröe the evening before, which gave him a new idea. He proceeded to write a few simple equations on the board, and he pointed out that the radius of the particle orbit disappeared from the equations relating the magnetic field and the frequency of the oscillator. He pointed our that resonance would be maintained regardless of the radius. From his remarks to me, I think he was greatly influenced by Rutherford's 1919 paper on the disintegration of nitrogen by alpha particles, in which Rutherford predicted the possibility of a variety of disintegrations if one had available laboratory-accelerated particles.

Lawrence remarked to me also at that time that methods such as the one by Lauritsen, at Cal. Tech., would produce limited high voltages all right "but what can they do with them after they get them!" He said they couldn't build a vacuum tube which would stand a million volts. He said the same thing later to me in connection with the Van de Graaff generator.

The above statements indicate some of the thinking that Lawrence did at the time he conceived the idea of the cyclotron.

Sincerely,
[signature]
James J. Brady
Professor Emeritus of Physics

[5]
E. M. McMillan to M. A. Tuve
April 6, 1977

Dear Merle,

I was asked to participate in a Symposium on the History of Nuclear Physics that is to be held in Minneapolis in May 1977, and I chose as a topic the early history of accelerators. I have written a "study paper" which (in the current draft) I enclose. There are some questions that you could help me with.

One question is, who really started the work at DTM — you or Breit? What were the dates of work on the induction accelerator and the Tesla coil? When did you decide to use the Van de Graaff electrostatic generator? Do you have any correspondence or notes for documentation, that you could send me copies of? The DTM effort seems to be the first in this field (Coolidge was not thinking of nuclear applications, as far as I know) and therefore has a special importance.

The second question is in connection with the statement . . . that Lawrence was already thinking of accelerating particles when he saw the Wideröe paper.

The only documentation of this that I have is EOL's Nobel lecture of 1951 and the Childs biography, "An American Genius." I don't remember Ernest ever talking about this period to me after I came to Berkeley at the end of 1932, and it remains a disconnected, almost abstract statement, which has been criticized as such by one person to whom I showed my study paper. I think the criticism is valid. Do you remember any conversations with Ernest in that connection? Have you any correspondence bearing on it? Unfortunately, Ernest did not keep a diary or any sort of consistent notes. Those who make history are often not prone to record it.

With very best regards,
[signature]
Ed McMillan

[6]
M. A. Tuve to E. M. McMillan
April 21, 1977

Dear Ed,

It was good to have your letter. . . . I'll try to answer your questions about the early days. I'm sorry I had to decline the invitation to Minnesota in May. I have cardiac limitations, and can't handle air travel and/or the stimulus of a group and excitement. Please greet them for me.

The high voltage project at the Carnegie DTM was started by Gregory Breit, when he helped to arrange a staff appointment for me at Carnegie, and also helped persuade me to accept it so that we could together start work on high-voltage tubes for nuclear physics. The idea was an old one for both of us, prompted, no doubt, by the classification of α-particle energies in "electron-volts." I had started to apply for an NRC Fellowhsip to attempt high-voltage tubes under Rutherford. There is a bit of history, leading back to Minnesota, connected with the offer to me of a staff position at Carnegie by Dr. Fleming, the Director of the DTM. Perhaps you'd like to hear it.

Ernie Lawrence and I took MA degrees in Physics in Minnesota in June, 1923, and Gregory Breit arrived in early July, just as Ernie left for Chicago with W. F. G. Swann. I was busy with some further experimental work under Jack Tate before reporting to K. T. Compton, at Princeton, as an Instructor in Physics. Breit and I hit it off well, and enjoyed many lively discussions about basic ideas and experiments. I liked *extremes*—very high temperatures (stars), very high pressures, very great distances, very small dimension or quantities ($h = \sqrt[3]{10^{-81}}$ thrilled me), etc., and I disliked complexities, like metals, chemical physics, etc. I preferred "simple" but perhaps hopeless questions, such as how the nuclei (protons and electrons then) might be held together. Breit and I admired Rutherford immensely. Breit then also learned of my "ham radio" experience and my efforts with 80-cm waves on Lecher wires using Army VT-II tubes with bases removed. I had been trying for highest possible frequencies. They were feeble.

In September 1924, when I had moved to Johns Hopkins (my Instructor's post went to Henry Smyth as he returned from England), Breit asked me to handle the very short wave receiver in Baltimore if he could arrange for a parabolic reflector at DTM. He proposed to beam short waves upward on a slant, expecting them to be reflected down to Baltimore by the "electric field layer" in the upper atmosphere. This had for decades been assumed for explaining the daily variations of the compass. I was highly dubious about the reflection coefficient and the attainable receiver sensitivity for such short waves, so I countered with the idea of using dots or pulses and looking for "echoes" at more customary frequencies, as discussed by Swann and J. G. Frayne at Minnesota while I was there. This idea was approved, so I came to the DTM for a summer job with Breit in June 1925. We successfully recorded the pulse "echoes" on 71 meters. It took some weeks to arrange for evening use of the Navy transmitter we used, which then promptly proved that the echoes varied in "height" and hence were *not* echoes from the Blue Ridge mountains 65 miles west of us!

Having this new technique for upper air studies related to Terrestrial Magnetism, the Director, Dr. Fleming, obviously wanted more done with it. I returned to Hopkins, and he asked me to consider returning to the DTM. In late December 1925 (possibly early January 1926), no doubt with Breit's urging, Dr. Fleming offered me a staff appointment to begin July 1, 1926. I replied that I was applying for an NRC Fellowship, so I could go to Cambridge to study with Rutherford. I said I was proposing to try to put a very high voltage on a vacuum tube, to produce a controllable "beam" of high-energy protons for studies of atomic nuclei. Fleming responded, "Can you do that in your one year of a Fellowship?" I said, "Of course not, but I can start it." Fleming then said, "Why don't you do it here; then you don't have to stop at the end of the year." I said, incredibly, "At a Department of Terrestrial Magnetism!?" He said, "If you will spend a part of your time on the radio experiments, the rest can be on high-voltage tubes; Carnegie cannot start a new department for every new problem, so we use the existing departments to do new things."

So I backed away from the NRC Fellowship Board, and agreed to join DTM to work with Breit, primarily on high-voltage particle beams. We started July 1, 1926, after I took my degree at Johns Hopkins, using the radio echo work with Breit as my thesis. From the start our joint aim was nuclear physics.

We wanted to use high-energy protons to repeat Rutherford's α-particle scattering experiments of 1910-1912 and his experiments on nitrogen disintegration with the emission of protons, 1919; also other light elements, 1923. The broad aim was to understand how atomic nuclei, composed (1926) of protons and electrons, could hold so stably together—some modification of the electrical universe square law must be involved. We expected to find that some non-electrical short-range forces were involved.

We had suitable high-voltage tubes by 1928, but the Tesla coil was grossly unsuitable as a voltage source. We used it (1930) to demonstrate gamma-rays and beta-rays above 1.25 million volts, also proton tracks in the cloud chamber. We constructed a 2-meter spherical Van de Graaff electrostatic generator (1200 kilovolts) and used it on a tube outdoors in May 1932; no housing was available. In October 1932 we began operation (indoors) of a 1-meter Van de Graaff unit, and verified (1933) many of the light-element nuclear reactions reported (1932) by Cockcroft and Walton from Cambridge.

Joe Morris brought Van de Graaff over to meet me at DTM in April 1931, after his return from Townsend at Oxford. I went up to Princeton in the autumn of 1931 . . . and hauled back to DTM, on my car's rear bumper, the two "desk model" Van de Graaff generators with copper spheres about 22 inches diameter. These worked OK on our cascaded tubes at 600 kV or so, but there was not time to keep them long enough to fit on an ion source. So I delayed until autumn 1932 any tests with protons on a steady D.C. source!

You asked about Ernie's early interest in high-energy beams. I have a vivid memory of one such discussion, because I "sternly lectured" him on the subject, and he seemed to accept the bawling out and was very sober about it. This was in late spring or early summer of 1928. Ernie spent the afternoon with me at DTM just before leaving for the long trip to Berkeley. You know, of course, that Ernie and I grew up together (doing ham radio together!) in Canton, So. Dakota, and I helped bring him to Minnesota and W. F. G. Swann (later Chicago, Yale and Bartol). So we always talked freely to each other.

I asked Ernie what research he was aiming to do at Berkeley. He responded, rather vaguely, with some small notions about high-speed rotating mirrors, chopping the tails off quanta and other single-shot ideas. I then talked to him like a Dutch uncle. I said it was high time for him to quit selecting research problems like choosing cookies at a party; it was time for him to pick a field of research that was full of fresh questions to be answered, and sure of rich results after techniques were worked out. I said that any undergraduate could see that nuclear physics using artificial beams of high-energy protons and helium ions was such a field, and that he should stake out a territory there to work and to grow in. Our 1½ million volt tubes and the confounded Tesla coils (our albatross) were right there in the room with us; also the assemblage of stuff for our early attempts at what was later called a betatron, using toroidal paths and a changing magnetic field. I told him to consider carefully the possible indirect methods of accelerating particles (linear accelerators, etc.). We were at the time asking for quotations on a cascade of transformers and rectifiers to 1 MV (later $150,000 and out of our reach), but my primary heavy-footed emphasis to Ernie was on the nuclear disintegration of light elements, as first done by Rutherford in 1919, and on billiard-ball scattering of α-particles, as per Rutherford in 1908-1911. We would first use protons, and later helium ions. Ernie was very sober and did not seem to resent the rather harsh way I went after his quantum tails. I asked him to join in the questing, and he seemed pleased. I think he was vaguely searching for an identifiable field full of specific problems, and this discussion clearly impressed him. He had followed our Carnegie efforts from the start, of course.

Please be indulgent with me and these rambling details of life fifty years ago. All of my "dead files" (64 storage file boxes!) went to the Library of Congress 10 years ago. I might have saved some letters from Ernie, but surely not many — neither he nor I was much of a letter-writer. Whether I helped mould his choices or not, he was ready and alert when the cyclotron idea popped up. We were all so naive in those days!.

Best regards,
[signature]
Merle [Tuve]

P.S. You may share this letter with the Minnesota seminar in May.
M.

[7]
E. M. McMillan to G. Breit
May 24, 1977

Dear Professor Breit,
I have just come back from the Symposium on the History of Nuclear Physics in Minneapolis, where I gave a talk on the early history of particle accelerators. . . . It is clear that the work of you and Tuve was very important, and I have been collecting a small archive of statements from those of the pioneers who survive, so I am writing you to see if you would be willing to write me your memories of what happened in the beginning. I am particularly

interested in motivations, origins of ideas, and mutual influences between individuals and groups.

>Sincerely,
>[signature]
>Edwin M. McMillan

[8]
G. Breit to E. M. McMillan
June 5, 1977

Dear Professor McMillan:
    This is a partial reply to your letter of 24 May which you wrote on your return from the Minneapolis Symposium. I hope to write to you again soon. There were differences in background of my former collaborator M. A. Tuve and myself. I am about a year older. Being mainly acquainted with my own, I will mention some of it.
    During World War One, I was assigned to the National Bureau of Standards, where I was mainly concerned with calibrating Kolster Decremeters — essentially wave meters — which through a measurement of sharpness of resonance gave the band width of the spark transmitters. I had the kind advice of the assistant head of the Radio Section at the U.S. Bureau of Standards, L. E. Whittemore. Professor J. S. Ames of the Johns Hopkins University, whose student I was, provided me with a little money, which with the kind advice of Mr. Whittemore of the Radio Section (Dr. Kolster was the Head), made it possible to buy some condensers, Litz wire, and other supplies. I spent a year on experiments in Baltimore, some of which were done jointly with E. O. Hulburt, who had returned from France as an army captain. I wrote my master's thesis for Johns Hopkins on my own time while serving as an apprentice at the NBS. Marconi's observations of long-distance transmission appeared inexplicable unless one invoked the participation of a second sphere surrounding the first. The mathematical side was essentially as in G. N. Watson's papers.*
    My doctor's dissertation at JHU was on the Distributed Capacity Effects of Inductance Coils. It was partly concerned with measurements and partly with calculations. Several "Scientific Papers" (issuance of these has been discontinued) of the Bureau of Standards were written by me on properties of inductance coils.

I spent a year in Leyden (Holland) as a National Research Council Fellow, working partly with Ehrenfest on quantum theory (which was then coming into its own) and partly on applications of radio techniques to low-temperature physics with Kamerlingh Onnes. The former and part of the experimental work with Onnes was published.

The second year of the National Research Council Fellowship was spent at Harvard, where I had the use of a room at the Cruft Laboratory, the director of which was G. W. Pierce, the inventor. He had bats flying close to the ceiling and avoiding obstacles. Among the speculations regarding how the bats did it, there was the simplest one that they had a way of sending out a signal which on reception indicated when they were in danger of a collision. I believe, therefore, that the correct answer to who invented "radar" is, "the bats."

At the completion of my year as a NRC Fellow at Harvard [1923], I went to the University of Minnesota as an assistant professor. The invitation to the Department of Terrestrial Magnetism came while I was in Minneapolis. My teaching duties were light, and since Van Vleck and I went to Minnesota from Harvard, where we were on friendly terms, and since Van enjoyed teaching (and I not especially), I went to the Department of Terrestrial Magnetism. Lawrence was still in Minneapolis and showed me around the town. He was on the point of joining Swann in Chicago, and Tuve was still in Minneapolis but left before the summer's end to go to Princeton.

In Minneapolis I had the use of two rooms (Van preferred to work at home) that were previously occupied by A. H. Compton. In one of these there was an X-ray machine for medical use. This type of machine worked on the principle of an electric commutator on a shaft driven by a synchronous motor. This system is equivalent to an electric rectifier. (The voltage to be rectified has to be supplied, of course, in synchronism with the shaft driving the commutator.) My acquaintance with this type of machine proved useful when I went to the Department of Terrestrial Magnetism of the Carnegie Institution of Washington because it made it financially possible to get started on nuclear disintegration experiments.[1]

I joined the Department of Terrestrial Magnetism at the invitation of Dr. Louis A. Bauer, who had asked Albert Einstein whether his General Relativity could explain an effect which appeared to be present in observations made in Dr. Bauer's Department. According to these observations, the line integral over a closed curve on the earth's surface was $\neq 0$ for the magnetic field taken on the surface. I was told by Bauer that Einstein said that it might. My Hopkins training told me that one should first see whether the observations of Dr. Bauer's staff were accurate enough to justify the conclusion. Some simple arithmetic showed that their observations were not accurate enough for this purpose. I therefore told Dr. Bauer that it appeared that I should look for another job.

Early History of Particle Accelerators   139

Dr. Bauer persuaded me to stay. He offered me the use of the non-magnetic building which was previously occupied by Dr. and Mrs. Barnett, who were engaged in experiments on magnetization by rotation, and by W. F. G. Swann, who soon moved to Chicago taking his student, E. O. Lawrence, along.

In the meantime, Merle Tuve received a fellowship at Princeton which was not renewed for reasons unknown to me. At all events, there was an assistantship at JHU that came in handily.

The importance of nuclear disintegration experiments was beginning to be realized while I was still at the University of Minnesota. Lawrence, Tuve and I talked about it in our occasional meetings, and experiments were done rather early at Cambridge (scintillation method) and also in Vienna (Pettersson and Kirsch, as I remember it). With this in mind, I made some crude experiments with Tesla coils at Minneapolis and learned for myself how sparks in air were related to potential differences between electrodes.

When Tuve joined me at DTM [1926], we went on with the radio work [reflection of radio waves from the ionosphere] and also started on developing equipment for nuclear disintegration. At later stages, Dahl and Hafstad joined us. Berkner took over the radio part later on.

We were much and generously helped by Coolidge, of the General Electric Co., in Schenectady, N.Y. A trip to GE gave me access to the manufacturing procedure. He even gave me one of their largest X-ray tubes to take home. He was using at the time large tubes with a thin metal window (Lenard rays) for the electrons to go through. The maximum safe voltage was about 300 kV. (At DTM we were interested in a polarity opposite to that used by Coolidge.)

The thin foil window was not needed by us if several tubes were to be used in series: we made experiments to determine what mattered most to avoid puncture of the glass. It soon became clear that outgassing of the tube was of help but only up to a point, and most for sealed-off tubes. Experiments were made with the tubes immersed in transformer oil. (We used Wemco C.) (We also worked on what was the progenitor of what in Don Kerst's hands became the betatron.) Van de Graaff's invention of a belt to convey the charge to an electrode was adaptable to accelerating protons down a tube to ground. By the time this came about, I was concentrating on theoretical physics and left DTM for New York University, keeping in touch with my DTM friends as a consultant. A successful design was evolved at DTM employing a belt drive to carry positive charge up to, and protons down from, the electrode in a multi-section tube. The uniformity of charge distribution along the tube was secured by corona points in air at atmospheric pressure. A number of developments were made at the University of Wisconsin by R. G. Herb and collaborators. The charge along the belt also had to be made uniform. The High Voltage Engineering Co. used to recommend the use of old crank case oil for the belt. . . .

It is hard to point to any single idea in the field as the most important. Theory and practice form a closely interwoven fabric.

> Sincerely,
> [signature]
> Gregory Breit

*A troublesome point arose in connection with the ground wave which appeared likely in view of a calculation by Arnold Sommerfeld for a partially conducting earth. Later work by Hermann Weyl, the renowned mathematician, appeared to indicate that Sommerfeld's type of ground wave can hardly be real.

1. The phrase "made it financially possible" is amplified in a reply, dated August 4, 1977, to a question concerning that point:

    "The availability of a voltage source in the form of a commercially available X-ray machine made it possible to get started on the behaviour of accelerating tubes. The financial side of the matter was important because we had little money available and the X-ray machines of the type I became acquainted with in A. H. Compton's two rooms were rapidly going out of style because rectifiers for AC voltage could be and were built at some additional expense which was not large by ordinary standards but was too large for us. Smoothing of the voltage was secured by a home made glass condenser in which sheets of metal were made to stick to sheets of full size grade B window glass with transformer oil. The margin left (about 7 inches wide) showed a corona when the voltage was on and prevented puncture of the glass. The operation took place about as follows:

    "Two metal spheres (about 12 inches diameter) in air were connected to the condenser. When the voltage between the spheres built up sufficiently, a spark passed between the spheres, the glass plate condenser would be discharged and the process could be repeated. The object of the two spheres was to provide a measurement. If the transformer voltage was held just below that necessary to give the spark one had a steady voltage of known magnitude."

[9]
E. M. McMillan to E. T. S. Walton
March 14, 1977

Dear Professor Walton,
    I am collecting historical material relating to the beginnings of particle accelerators, and I would like to ask you a question: Did you get the idea of the voltage multiplier from the paper by Schenkel that is quoted in your paper (with Cockcroft) in Proc. Roy. Sci. 129A, or did you (or Cockcroft) have the idea independently? Also, did you know of the work of Greinacher, who made the same modification to the Schenkel circuit?
    Also, I would appreciate it if you could fill in some dates regarding your work with the induction accelerator and the linear accelerator, when you got

together with Cockcroft, and when the various phases of your work together occurred. When you started, were you aware of the work of others regarding accelerators?

With my best regards,
[signature]
Edwin M. McMillan

[10]
E. T. S. Walton to E. M. McMillan
April 11, 1977

Dear Edwin:

I hope that you don't object to my use of your first name, and I hope you will reciprocate this less formal mode of address.

I was very interested to get your letter and pleased to hear that you are collecting material about early particle accelerators. I hope that what follows will answer satisfactorily the questions you have asked.

Cockcroft had the idea of modifying the Schenkel circuit, and at the time I don't think that he knew of . . . Greinacher's paper. It was certainly unknown to me. My contribution to the circuit theory, as given in the Proc. Roy. Soc. 136A paper, was in showing that the circuit could be thought of as a method of maintaining equal potentials across each of a number of capacitors connected in series. The arrangement could therefore be regarded basically as a transformer for stepping up or stepping down a D.C. voltage. Indeed we took out a patent on this aspect of the circuit as we thought that it might be used for the long distance transmission of electric power by D.C. which would have important advantages.

I will now give you a short account of my early work, which I hope will contain the information you need.

I entered the Cavendish Laboratory as a research student in October 1927 and, as was usual in those days, spent most of the first term gaining some experience in the use of high vacua and in radioactive measurements. Towards the end of the term, Rutherford sent for me in order to discuss a research problem for me to work on. He asked me if I had any suggestions to make, and I said that I would like to try producing fast particles. Knowing that the potentials required to produce particles likely to be energetic enough to disrupt nuclei would be impossible there and then, I suggested a method for accelerating electrons in a circular electric field. Rutherford immediately suggested a more practical method of producing the circular field. It was really the method being used at that time in the laboratory by J. J. Thomson in his study of electrodeless discharges in gases. An account of attempts to get the method to work and and of calculations on radial stability were submitted to the Cambridge Philosophical Society in June 1929 and published in October.

On realising that I would probably not be able to make the induction method work, I turned my mind to devising other indirect methods and suggested to Rutherford the method of the linear accelerator in early December 1928. The use of caesium ions was suggested as they were easy to produce in a good vacuum and their high mass eased the problem of producing high voltages at high frequencies. The idea was new to Rutherford, who after making a few quick simple calculations, agreed that the method was feasible and worth trying. No one in Cambridge had heard of Ising's paper of 1924 until the publication of Wideröe's paper in December 1928. The things which made the experiments difficult for me were mainly, (1) my high-frequency source was a quenched spark generator, and (2) very little was known about the focusing of ion beams. Indeed, I had unwittingly destroyed the focusing action between the cylinders by covering the ends with gauze to secure a field-free space inside them. It is interesting to note that Ising, in his diagram, showed the ends of the cylinders covered with gauze. While I was working on these experiments, Gamow visited the Cavendish Laboratory and gave a lecture on his recent calculations on the emission of α-particles by penetration through the nuclear potential barrier. Cockcroft saw that his formula could be applied in reverse to calculate the probability of the penetration of particles into nuclei. Simple calculations indicated that 100 kV protons should penetrate the nuclei of light elements in sufficient numbers to be easily detectable if disintegrations followed penetration in most cases. This vist of Gamow was early in 1929.

Cockcroft pointed out to Rutherford the significance of these calculations. The decision was taken that I should abandon attempts to produce fast particles by indirect means, and that Cockcroft and I should work together on building an apparatus to produce fast protons using some hundreds of kilovolts. At that time Cockcroft was working at the bench next to me on the deposition of metallic vapours on surfaces cooled to liquid air temperatures. We built two high-voltage accelerators which are described in the published papers. I shall be glad to give further details if you need any.

As regards our knowledge of the work of others on accelerators, we were familiar with the papers published in the Phys. Rev. and by the Carnegie Institution, but I do not recall reading of any work done on the continent of Europe other than that of Ising, Wideröe, and Brasch and Lange. At the moment I can't think of anything that we missed.

I was very fortunate in the timing of my suggestion of an induction accelerator to Rutherford. It was unknown to me then that about a week or so previously [November 30, 1927] Rutherford in his presidential address to the Royal Society had pointed out the need for the development of high-voltage apparatus which could be used in nuclear disintegration experiments. Hence his mind was very receptive to suggestions in this area, and he may have been glad to find someone interested in such experiments!

Please let me know if I can be of any further help,

Best wishes,
Yours sincerely,
[signature]
E. T. S. Walton

[11]
E. M. McMillan to E. T. S. Walton
April, 1977

Dear Ernest,

Your letter is very interesting, and I thank you for it. I am enclosing a copy of a "study paper" I wrote for a Symposium on the History of Nuclear Physics, to be held in Minneapolis in May. Your comments will be greatly appreciated. In particular, the date for the Gamow visit given in your letter does not agree with that in your Nobel Lecture.

I have recently learned that a circuit like Greinacher's was patented by Slepian . . . (U.S. patent 1,666,463, filed February 26, 1921, issued April 17, 1928). His 1922 patent on an induction accelerator is well known and mentioned in the histories of accelerators.

Of course any more details and dates that you could furnish would be useful. I find the subject of how accelerators got started very fascinating, and it is also important because so much grew out of these small beginnings. Do you have any sketches or pictures of your linear accelerator?

Very best regards,
[signature]
Ed McMillan

[12]
E. T. S. Walton to E. M. McMillan
May 14, 1977

Dear Edwin:

Thank you very much for your letter and for the script of your lecture, which I was very interested to read. I am glad to have a copy of it.

Tracking down the discrepancy in dates has taken a considerable amount of time. When writing my previous letter to you, the date for Gamow's visit was taken from the Royal Society memoir on Cockcroft, which was written by Oliphant and Penney. It did not occur to me to refer to the Nobel lecture. My notebook for the period August 1928 to October 1929 is very sketchy, but it does indicate that early in February 1929 I was working both on the linear accelerator and on the production of fast, positive ions accelerated by a 60 kilovolt D.C. set available in the laboratory. I found another bit of information in a report which I sent to the Commissioners for the Exhibition of 1851, who had provided a three-year scholarship for me. The information was to the effect that I had started working with Cockcroft in November 1928. I cannot now remember the source of the date 1928 for the Gamow visit given in my Nobel lecture. Possibily it was from memory, which in my case might be unreliable after a lapse of over 20 years.

A photostatic copy of the relevant part of the Royal Society memoir is enclosed as you may not have it readily available. According to it, a typescript copy of Gamow's paper was received in the Cavendish Lab. in November 1928, and he gave the lecture in Cambridge in January 1929. I cannot recall ever having seen this preview of his paper. I think that these two events may have caused confusion about dates. I am inclined to think that January 1929 is correct for Gamow's visit. The authors of the memoir must have felt certain that it was in January as otherwise they would have given only the year. A typographical error in the year is ruled out as January 1928 or January 1930 would certainly be incorrect.

Thank you for the copy of the Slepian patent of April 1928, which I was very interested to see. The circuit is very similar to the one which we used. A comparison of his single-phase diagram with our circuit shows that he used 5 capacitors and five rectifiers to produce a multiplication of five, while we used 4 capacitors and 4 rectifiers to give a multiplication of four. So there is not much of a difference except that in the Slepian circuit no part of the rectifiers is at earth potential. This would cause difficulties when using continuously evacuated rectifiers, as we were forced to do. However, Slepian's circuit, as used in multiphase versions, seems to save considerably on the number of capacitors required when compared with the obvious way to adapt our circuit to multiphase operation. I must have a look at other possible multiphase versions of our circuit.

I am sorry not to be able to send you any pictures or diagrams of my linear accelerator. No photographs of it were ever taken. About six months ago, I found a sheet of paper containing a diagram explaining the principle. It was dated December 1, 1928, and is likely to be the diagram which I showed to Rutherford when I put the proposition to him. At the moment I have only a very poor photostatic copy. The original is on loan to the Smithsonian Museum for their forthcoming exhibition on particle accelerators. I expect you know all about this. I could not lend my notebooks as they have been deposited in the library of Churchill College, Cambridge, where Cockcroft's notebooks are preserved.

Now for some comments on the script for your lecture. . . . You state that the four-stage multiplier would go up to 500 kV. In theory, it should have given 800 kV, but corona losses limited the output to 700 kV. You also state that we found our first scintillations at 125 kV. I have no record of this, but I would suspect that it was much higher. The details of the first scintillations are as follows. I usually spent 15 to 30 minutes each morning conditioning the apparatus by raising the voltage gradually and switching off when the liberation of gas caused a breakdown in the rectifiers or accelerating tube. Eventually we got to the stage of putting a lithium target and fluorescent screen into the apparatus. Next morning, I carried out the usual procedures while Cockcroft went to the Mond Laboratory to help Kapitza. When the voltage reached a fairly high value, I left the control table and crawled across the room to the little hut under the apparatus. On looking through the microscope, I immediately saw scintillations which seemed very like what I had read about the appearance of a-particle scintillations but which I had never previously seen. After applying a few simple checks, I telephoned Cockcroft, who returned immediately and confirmed my observations. We then got Rutherford to come along and observe them. It was most interesting to watch his reactions.

I have never regarded the move to the larger room (an old lecture theatre) as anything but fortunate. The higher voltage enabled us to investigate immediately elements up to about fluorine. For most work we could produce far more disintegrations than we could cope with. This meant that we were operating the apparatus mostly well below its limits, which meant that breakdowns were not very frequent. If we had remained in the old room with its low ceiling, observations would have been difficult. Indeed we would have had to lie down on the floor to see the scintillations—not a relaxed position for reliable counts.

The main reason for looking for X-ray or γ-ray emission first was that at the time there was considerable interest in penetrating radiation produced by a-particle bombardment of beryllium and boron. Another reason was that it was a very simple thing to do, especially as I had a significant amount of experience in using a gold leaf electroscope. . . .

> Best wishes,
> Yours sincerely,
> [signature]
> Ernest Walton

[13]
E. M. McMillan to E. T. S. Walton
May 23, 1977 [Extract]

Dear Ernest,
    My values for the voltage were taken from Cockcroft's Nobel lecture, of which a partial copy is enclosed. I will change the top voltage from 500 to 700 keV, as you suggest, in my revised script. But the 125 keV is italicized by Cockcroft for emphasis, and I would like to keep this value, as he seems so sure of it. Do you have the date on which you saw the first scintillations?

[14]
E. M. McMillan to E. T. S. Walton
August 4, 1977 [Extract]

    . . . One interesting point: the volume of Arch. f. Elekrotech. with Wideröe's paper is dated December 17, 1928, just a short while after you presented your idea for a linac to Rutherford. What was your reaction when you saw that paper?

[15]
E. T. S. Walton to E. M. McMillan
September 1, 1977

Dear Edwin,
    I have now consulted the microfilm of my notebooks which was made by our library before they were sent to Cambridge. The a-particles from lithium were first observed on 14 April, 1932. I made no attempt to count scintillations in my first observations and so I did not record either the proton current or voltage used. I do not think that I had enough faith to crawl across the room

to look for scintillations until a reasonably high voltage was achieved. On the same day, Cockcroft and I counted scintillations starting at 252 kV and going down in a series of steps to 126 kV. So, in a sense, both of us were right about the voltage used. In the succeeding few weeks we bombarded various elements and the voltages used never exceeded 550 kV. This gave us an adequate number of disintegrations up to about fluorine and the apparatus would work less reliably at higher voltages. . . .

You asked about my reaction on seeing the Wideröe paper. I am not an easily excited person and I have no recollection of feeling very annoyed. I was only mildly disappointed. No injustice had been done to me and so I had no cause to nurse any grievance or ill feeling. As far as I can recall I merely decided to try to be quicker off the mark next time and any slight feelings of disappointment vanished very quickly. There was no doubt that Ising had four years of lead on the basic idea of the linear accelerator but he missed the more practical idea of using alternating voltages between the cylinders.[1]

Best wishes,
Yours sincerely,
[signature]
Ernest Walton

1. Ising mentioned this possibility in a footnote, but rejected it as being too difficult.

DISCUSSION

Livingston: I would like to add a personal story, which I think Ed wasn't present to observe at Berkeley and which illustrates the atmosphere of Lawrence's laboratory in I believe 1933, when artificial radioactivity was discovered in Paris by Curie and Joliot and reported in the *Comptes rendus*. Those of us working in the laboratory with the 27-inch cyclotron had about 2 million-volt protons, and we had recently been given deuterium in the form of heavy water by G. N. Lewis of the Chemistry Department. We were the first people that I know of who used deuterons as projectiles. We had them running at about 2 million volts, and we were busy. We were kept busy because it was a very active, moving laboratory.

Malcolm Henderson, for example, had a linear amplifier system attached to thin vacuum ionization chambers for detection. We had a Geiger point-counter which could be tuned down to detect α-particles only. Various things were all ready. We were taking data at that time, measuring long-range α-particles from disintegrations by deuterons on various targets—a target wheel was placed in the vacuum where we could turn to different ones to observe the radiation coming out through a thin mica window into these detectors, which were inserted close to the target. We were operating this when Lawrence came running into the laboratory early one morning with a copy of the *Comptes rendus* and described to us with great excitement the observation of induced radioactivity. In their paper Curie and Joliot mentioned that similar kinds of radioactivities could certainly be obtained from other interactions, including—and they mentioned it specifically—the possibility of deuterons bombarding other targets. And we were there with that target on the wheel with the instruments available and operating on other experiments. We immediately turned our attention to trying to observe this phenomenon. It just so happened that we had been taking one-minute or two-minute runs by a switch—a big double-pole switch, one of which turned on the oscillator and started the cyclotron, the other turned the controls to the counters on or off. We disconnected the counter switch, twisted the wires together so that it did not turn off, pushed the oscillator switch down for a matter of just a few minutes to activate the target, pulled it up, and we heard the radioactivity: click, click, click, click—the induced radioactivity was there! Within one-half hour from the time Lawrence came into the lab with this news, we were observing induced radioactivity.

Bethe: I would like to ask Stan Livingston especially what Lawrence and he thought as they were constructing the cyclotron. I think one of the most important questions is: How clearly did people have in mind that they were going to produce nuclear reactions? The speaker [McMillan] mentioned that Tuve had in mind a modification of the scattering due to the non-Coulomb field near the nucleus, which was certainly very close to knowing that there was nuclear physics. On the other hand, he also mentioned that Rutherford pushed Cockcroft into actually using his particles to produce nuclear disintegrations. I would like to know this from Stan, and I think it would be interesting to find out at what time in Cambridge it was clear that the Cockcroft-Walton generator would be used for producing nuclear disintegrations.

McMillan: That is an interesting point. I wonder who here would know that? Not me.

Livingston: Well, Hans, I can tell you of my memory of my discussions with Ernest at the time he suggested the cyclotron principle as a doctorate thesis and offered me the chance to use it for a doctoral study. His words were that we can make high energy particles —he called them high voltage particles—without the use of high voltages. It was his intention to make these high voltage particles to use them for studies of the nucleus. I don't believe that the word "nucleus" came into our conversation. Different words were being used then, and I cannot recall exactly what he used . . . it was the "atom"; the "inside," the "core." Anyway, I knew perfectly well that Lawrence's intent was to study somehow the interior of the atom with high voltage particles that were of a higher voltage than could be obtained by other available methods, the direct voltage methods. His intent is given in the title of the first paper: "The Production of High Voltage Particles Without the Use of High Voltages."

Fowler: Since Ed asked about it, the story in Pasadena was of course quite different. The High Voltage Laboratory at Caltech had been built by the Southern California Edison Company to test the towers for the transmission lines from Boulder Dam—or Hoover Dam, depending on your politics—which was being constructed at that time. Lauritsen had been working with Millikan on cold emission. Millikan realized that there were other lucrative applications of the million-volt, one thousand-amp cascade transformer in the High Voltage Laboratory. There was money in medical radiology, and so he asked Lauritsen to build tubes for cancer therapy at high voltage. The highest voltage tubes in those days were the Coolidge tubes of about 100 to 200 kilovolts. Lauritsen built the first million-volt X-ray tube for cancer therapy. But in addition, he had great interest in physics, and a major thing he did, with John Read, a visitor from England, was to show that the Klein-Nishina formula was correct for the cross section of the Compton effect. Some of you will remember also the Dirac-Gordon formula, which did not take into account the spin of the electron, whereas Klein and Nishina included this. Read and Lauritsen showed that up to a million volts the Klein-Nishina formula was the correct one.

The story of Caltech's entrance into nuclear physics began when Cockcroft and Walton's results were announced. It was very simple to do, because these were AC-powered accelerators, and all that Lauritsen and Crane had to do was to take out the electron-producing filament and put in an ion source, and they were in business. There was absolutely no motivation originally, as far as I have ever been able to ascertain, of doing nuclear physics. The early tubes were built to do radiology, and to study the cross section of the Compton effect among other problems in X-ray physics. But when word came that Cockcroft and Walton had disintegrated the nucleus, Lauritsen and Crane, within a few days, were bombarding nuclei, first with α-particles, then with deuterons (Gilbert Lewis had sent down a small amount of deuterium), and finally with protons.

I can only add a footnote to your story, Stan. I suppose you remember that there was a great race to see who would be the first to publish on the production of artificial radioactivity by deuteron bombardment, and Charlie Lauritsen and Dick Crane pulled a fast one on you fellows at Berkeley by submitting, on February 27, 1934, their paper to *Science*, whereas on February 27, 1934, you submitted your paper to the *Physical Review*. So it's the same old story as it is now—their paper came out on March 9 and your paper appeared in the April 1 issue. But Charlie and Dick covered their tracks by also submitting a paper on March 1 to the *Physical Review*, so on page 428 of the April 1, 1934, issue there is a paper by Henderson, Livingston, and Lawrence, and on page 429 there is one by Crane and Lauritsen.

Feshbach: I just wanted to add a few comments on the Van de Graaff story while he was at MIT. That large machine he built at Round Hill was eventually transferred to MIT, and some experiments were done with it, namely, electron scattering by nuclei. At that time there were some violent anomalies in electron scattering by nuclei that had been observed by Barber and Champion in England and Scherrer and Zunti in Switzerland. The MIT experiments finally exposed the error in those experiments and got the correct answer, which was that there was no anomaly. Then, during the war, Van de Graaff's laboratory was asked to build a compact source of

X-rays using the electrostatic generator. It was a pressurized affair which used sulphur hexafluoride as the gas, and one of the great difficulties was keeping it from sparking over. Finally, a system of leakage resistors was used which made it work in a reliable fashion, which it had to do for the purposes of the Navy. If you want more information on this, I would recommend consulting John Trump and Bill Buechner.

Peierls: Could I attempt a partial answer to the question Hans Bethe raised? I am thinking about what motivated the accelerator work of Cockcroft and Walton. You [McMillan] have already mentioned that this was motivated by the Gamow prediction, and therefore very deliberately intended for nuclear bombardment and nuclear transmutations. Now, there is a story, which I only have as a rumor from usually unreliable sources, but maybe in your searches you have come across more definite evidence for it. It is said that Cockcroft and Walton were interested in raising the voltage of their equipment, its reliability, and so on, more and more, as so often happens when you are involved with technical problems, and that eventually Rutherford lost patience and said, "If you don't put a scintillation screen in and look for α-particles by the end of the week, I'll sack the lot of you." And they then went and found them.

McMillan: That sounds in character.

Sopka: You already mentioned that Merle Tuve and Ernest Lawrence were boyhood chums. In one of the Lawrence biographies, there is an indication that there was a coolness between the two in their mature years. Would you have any comments on whether this did take place and what might have caused it?

McMillan: I also have the impression that it took place. I have no idea why. Of course, when I first knew Lawrence, it was at the end of 1932, and hence long after he and Tuve had gone their separate ways, so I never saw them together. This is all second hand, but I think what you say is correct. I think there is perhaps one basis for it, and this could be the principal basis. Around 1933, when Lawrence and some of the other people at Berkeley, particularly Gilbert Lewis, first started using deuterium in the ion source in the cyclotron, they observed that when any target was bombarded by deuterons, protons and neutrons were emitted. The protons had some characteristic energy; the neutrons were always in there also, but I don't think they had any means of measuring their energy at that time. (I should point out that although I was in Berkeley and came at the end of 1932, I didn't actually go to work in the lab until the spring of 1934. I started out, I guess in a foolish way, trying to measure the magnetic moment of the proton, which was already being done in Hamburg. What hope did I have as a young research fellow of beating out these guys? As soon as their result was published, I went into the lab in 1934. So I know what happened in 1933 only from the fact that I was around and talked to people.) Anyway, Lewis and Lawrence—I never knew which one more so—promoted the interpretation, the concept, that the deuteron was unstable, that on impact with anything it would come apart and emit these characteristic protons and neutrons, as a set spring would break apart at some impact. And Lawrence didn't get much support for this idea. I know that Lawrence presented it at the Solvay Congress in 1933—I wasn't there, but I don't think it was accepted very widely. As a National Research Council Fellow, I was formally under Lawrence, so I talked with him. I remember Ernest trying to convince me that he was right in this breakup of the deuteron theory, and I refused to be convinced of it. I think Ernest was quite disappointed in me at that time. The correct explanation was thought out by several people, and one person who had a fair amount to do with it was Tuve. I've looked at some of the old correspondence between Tuve and Lawrence during this time, and the tone of the letters was quite touchy. Tuve was clearly trying to point out that he thought Lawrence was wrong without directly saying so—this may have contributed to their break. I think that any break between Tuve and Lawrence was not personal; it was technical; it was scientific. It could have been just a strong rivalry, and it could have been this dispute. Some of these letters I recall—they're in the Lawrence files, and I looked at them—are pretty sharp. Of course, it turned out that what was happening (Maurice, weren't you in on that one, too?) was that the deuterium was simply sticking in the target and was then being hit by the following deuterons, so it was the (d, d) reaction that was being seen. The Caltech group, everybody, suddenly pounced on Lawrence and showed that he was wrong, and Lawrence kept quiet about it since then.

Goldhaber: As I mentioned yesterday, when I first talked to Chadwick about the possibility of measuring the neutron mass, there was a big debate going on which produced a lot of letters in *Nature*. There was the heavy mass of the Joliots, the very light one of Lawrence, and the intermediate one of Chadwick. I think the debate closed when we measured the neutron mass. There were suddenly no more letters.

McMillan: Also, there was the explanation which Tuve was certain of, and which I believe was published. I know in the letters I read that there was some discussion about whether he should publish it, and I know that Lawrence and the Pasadena group were involved in it.

Goldhaber: That's correct. The explanation was then given by, I believe, Rutherford, Oliphant, and Harteck, that the (d, d) cross section was so large at very low energies. They built a low energy accelerator to observe what happened in Lawrence's case and found that the deuterium was absorbed in the target and then bombarded. I think that settled it. May I say just one more thing? Here is a small elaboration of your remarks on Szilard's collaboration with Brasch and Lange. I didn't realize that that was the only scientific paper Brasch and Lange published out of this . . .

McMillan: It was the only one I could find.

Goldhaber: It could be, but what I remember is the following. As I mentioned, Szilard and Chalmers had photodisintegrated beryllium with radium γ-rays, and at that time there was still this vague notion that Be^9 might be unstable, because the Be^9 mass had been measured incorrectly. So Szilard then thought of a chain reaction for the first time — somehow, you start tickling beryllium and produce a neutron, which will tickle other beryllium nuclei, etc. If the neutron had really not been bound but in a positive state, he might have been right. Anyhow, he was therefore interested in knowing the threshold for beryllium disintegration, and he did it in a way that started a new way of doing physics, namely, long-distance physics. He wouldn't have dreamed of going to Germany at that time, but he sent one of those Szilard-Chalmers kinds of liquids — this time replacing iodine with bromine, which produces a long-lived activity — to Brasch and Lange in Berlin. They bombarded it with their X-rays and sent it back to him. He then separated out the bromine and found the activity. But Brasch and Lange could not maintain their voltages very well, and if I remember correctly all they could say in the end was that less than 1.8 MeV X-ray energy was needed. Today the threshhold is of course very well known; it is 1.66 MeV — as a mnemonic device you can just tell your students $1^{2}/_{3}$ MeV.

Fowler: I want to address a question probably to Stan, and perhaps Ed, but first of all here is a postscript about Brasch. Brasch was brought to Pasadena in either 1937 or 1938 by Lewis Strauss, who later on, through his interest in Brasch's work with Lauritsen in Pasadena, became chairman, the first chairman, of the Atomic Energy Commission. The question I wanted to ask Stan . . .

Bethe: The fourth chairman, or something like that . . .

Fowler: Thank you. But he was chairman at one time.

Bethe: He was chairman under the Eisenhower administration. The first chairman was Lilienthal.

Fowler: You are quite right, Hans. Thank you. Is there any place, Stan, where there is written down the successive sizes of the cyclotrons that were built? What was the first one? The second one? The third one? In radii and maybe in voltage, if you remember.

Livingston: Yes. I gave a series of lectures at Harvard in 1969 or 1970. I wrote up these four lectures. One of them was on the cyclotron and gave a detailed study, from my point of view, of the sequential development of the cyclotron. It is published by Harvard Press in a small monograph. For the historians I might say that another of the articles is on the early race for high voltages, which gives my studies and observations of this race between the different groups that were working in the days before 1932 to develop high energy particles for disintegration. Another is on strong focusing.

Fowler: Could you give us just a few remarks on the first cyclotron?

Livingston: The first one, which was not an operating machine but proved the principle, was my doctorate thesis. It was a 4-inch diameter pole-face magnet, and with that I observed resonance and calculated an energy of 80 kilovolts, with a maximum applied voltage on the dees of 1,000 volts. The second one was the first million-volt cyclotron and was a 10-inch magnet, which was built with windings, not cooled, so it could only be operated for a relatively few hours before one had to stop for the day. This one reached 1.2 million volts. The vacuum chamber of that second one is about 10½ inches across and is on exhibit in the Kensington Museum of Science in London. Lawrence sent it to Cockcroft, who had it mounted in the museum there. I have seen exhibits of some of the earlier 4-inch chambers in the Lawrence Museum of Science.

McMillan: Stan, could I point out that there is also the zeroth-order cyclotron, the things that Edlefson built, the glass and wax things.

Livingston: I didn't mean to leave this out, but it was not a working cyclotron.

McMillan: No, it wasn't a working cyclotron, but . . .

Livingston: It was a glass vial made with an internal coating of silver to make the dees, and that was used by Edlefson to try to observe resonance. I think Lawrence's words were, "It was a failure, but it did show promise." And then there was the final one, with which I was concerned, the 27-inch. I was there while we erected that magnet in the big laboratory, the old lab now, and we built this one and I helped develop it through several million volts; with deuterons we got up to five million volts by the time I left about the summer of 1934.

Feenberg: I just want to contribute one more to the many footnotes on the history of the cyclotron. In 1930 I was working a few hours a week in a little research lab that Raytheon had in Cambridge, Massachusetts, and one of the important persons in Raytheon at that time was C. G. Smith. Perhaps there's someone here who knew the man; I think he is still alive. Anyway, one day he asked me to look at the theory of essentially the cyclotron, but he called it a device to accelerate electrons. So I looked at the theory and saw that it would work for fairly low energies, but that you couldn't get very high energy electrons with it because of the relativistic effect. I guess, because of the many other things he had to do, that Smith did not go on with that thought, but anyway it was in the air.

The Neutron: The Impact of Its Discovery and Its Uses
Eugene P. Wigner

INTRODUCTION: *H. H. Barschall*

Introducing Eugene Wigner is a great honor but also a difficult task. The problem is that his contributions to physics, and the honors he has received, are so numerous that I don't know which ones to mention. The list of honors includes about every distinction and prize that a physicist can get. His list of honors as of 1973 includes nineteen honorary degrees, and I know he has received several since then. His contributions to physics have had an enormous impact on many fields of physics and chemistry, but today we are concerned only with nuclear physics. His name has been mentioned many times in the last couple of days, and we hope that after his talk he will be able to answer some of the questions raised then.

Perhaps I should mention that Professor Wigner was trained as a chemical engineer, and he just told me that a few months ago he visited the factory in Hungary where he was first employed as a chemical engineer. My first contact with Professor Wigner was when I took a course from him, when I was a student. He has talked with me and helped me ever since. One of the first occasions when I received Professor Wigner's help was after I mentioned to him my difficulty in getting some krypton for filling a counter, which I needed for my thesis work. Professor Wigner was quite surprised that I had any difficulty. He told me that in Hungary the light bulbs are filled with krypton. So he wrote to one of his friends in Hungary, and pretty soon a tank of krypton arrived, which I then used in my thesis work.

After I joined the staff of the University of Wisconsin, Professor Wigner was a frequent visitor to Madison, and much of the work in nuclear physics which has been carried out at the University of Wisconsin was inspired by Professor Wigner's suggestions and explanations, particularly in the area of the understanding of resonances and the importance of isotopic spin. Professor Wigner's work and inspiration has profoundly influenced both experimental and theoretical nuclear physicists everywhere. I think that only Hans Bethe has had a comparable influence on all of nuclear physics. We are fortunate to have Professor Wigner here this morning and to hear him talk on the impact of the discovery of the neutron on nuclear physics.

The Neutron: The Impact of Its Discovery and Its Uses

Eugene P. Wigner

Thank you very much for the very kind introduction. It really coincides a little bit with the first sentence I wanted to say, namely, that I am no expert on history. I forget very easily very much. In particular, I completely forgot this krypton incident.

Next, I want to compliment the organization of this meeting. I am not an expert on the history of anything, and certainly not of nuclear physics or the history of neutrons. But there is a certain advantage to ask not historians, who know history and dates and authorships wonderfully, but to ask physicists, because they may not know the dates, may not know all the papers, but they know a bit of the spirit, and they do remember the spirit even of times long passed. And the spirit of the time had a very great influence, as we learned from the previous speech.

Perhaps I should also mention that it gave me a great deal of pleasure to reread articles that I had read a long, long time ago. In many cases, my recollection was entirely in error. But on the whole, it was a pleasure to reread these articles, to relive what I had learned, what I had digested at one time, and what had given me pleasure then. Two circumstances did not give me pleasure. First, I found

Nuclear Physics in Retrospect, edited by Roger H. Stuewer, copyright 1979 © by the University of Minnesota.

many names and many articles almost completely forgotten, hardly ever mentioned, such as those of Guido Beck, K. T. Bainbridge, Moon and Tillman, Kopfermann, even W. Elsasser, and Dunning and Pegram. Second, I saw that things are also forgotten. I will compliment highly the article of Bethe and Bacher when I come to it, but even this forgot something. It forgot, for instance, that the neutrino was discovered, or was predicted, by Pauli. But let me get to that later.

I hope you will excuse my knowing best the articles I wrote myself; I will comment on both the good and bad sides of them. Another reason why it is a little strange for me to give this address is that I won't be able to say anything that is not known. Usually, if a physicist speaks, he tries to say things the audience does not yet know, things he has discovered three weeks ago. But this is not possible in the present case. I also want to be excused for speaking not only about the neutron but also about its relation to other particles and to other nuclei.

Well, as to the spirit, I would like to make one more remark. A few weeks ago I gave a talk on the great effect that the development of science has had on all of our lives, on the way we think and the way we feel. I am now tempted to discuss a little bit the effect of science on itself. The disorientation in physics which existed about thirty years ago is hard to imagine. We can remember it hardly more than we can remember the time when man's chief purpose was to hunt some animals, when he did not have chairs to sit in, did not have beds, did not have covers, did not have a jacket, did not have anything, not to mention that he did not have a radio to speak into. It is hard to remember that.

But it is also hard to remember how little we knew about nuclei thirty years ago. Originally, we knew that there are protons and electrons. This created a great difficulty for the understanding of nuclei in general. Originally, of course, people thought that those are the elementary particles, that everything is composed of protons and electrons. But they also knew that this can't be true, and this led to a serious disorientation. It can't be true for at least one obvious reason, namely, that if the energy of an electron in a helium nucleus is calculated by the uncertainty principle, it turns out, even assuming the too large radius which was believed at that time, to be 260 MeV. That this should be contained in the nucleus, the total binding energy of which is 28 MeV, was hard to believe. But there was another reason which came out very soon. And this was an observation of Heitler and Herzberg.[1] In 1929 they noted that the molecule N_2, if it is composed of protons and electrons, must have altogether 14 protons and 7 electrons. And it would therefore obey Fermi statistics, or exclusion principle statistics. Perhaps I should mention that I myself published an article[2] a little bit earlier than that, giving a rigorous proof that a composite particle obeys Bose statistics if it has an even number of particles, Fermi statistics if it has an odd number of particles. Heitler and Herzberg observed that the N_2 spectrum shows clearly that the N nucleus obeys Bose statistics. But this was only one of the many reasons causing disorientation.

Coming back to the history of the neutron, I should note that the idea of such a thing existing is very, very old, very much older than anything I mentioned before. Many people felt it should exist, but they were uncertain, and it was not possible for scientists of that time to make a real effort to use the neutron concept until the neutron's existence was demonstrated. As far as I know, the first suggestion of the existence of the neutron was made in 1920 by Rutherford,[3] and almost at the same time by a chemist. Chemists were in those days very much interested in isotopes, isotopic effects. The chemist who also proposed the existence of a neutron, William D. Harkins, was a professor of chemistry at the University of Chicago.[4] But nobody had the courage to take that entirely seriously. (As my friend, Dr. Bayman, told me, all that was only speculation; it had no significance.)

The next development in this direction was Pauli's suggestion of the existence of the neutrino. I mention this, because it illustrates the same point. Pauli stated already in 1928, both in his private conversations and in his letters, that he thought that a particle which we now call the neutrino exists. But he did not want to publish this idea, because he felt that it was speculation. It later turned out to be very real, in ways I will mention.

The next event on my list is that in 1932 the energy principle was again seriously questioned by Bohr.[5] He said the energy principle is not valid; it does not apply to nuclei. I mention these things because they show the general disorientation and the lack of courage to derive farther-reaching conclusions from any of these suppositions.

All this happened before what many people call the beginning of nuclear physics in 1932, when Chadwick wrote the paper interpreting the article of the previous year by Curie and Joliot, who bombarded beryllium with helium and found that somehow as a result of this reaction protons with an energy of 10 MeV were manifest in their cloud chamber close to the bombarding region. This was not understood by Curie and Joliot,[6] but it was correctly and very vigorously interpreted by Chadwick in a paper in *Nature*,[7] which I am sure was mentioned before. The paper appeared on February 27. And it immediately changed the attitude of physicists enormously. They suddenly thought: it is not speculation; we can believe it; we can develop ideas on the structure of the nucleus on the basis of the existence of a neutron. I will come back to the question whether these ideas were entirely correct. I am not convinced, and in that respect I disagree a little bit even with Chadwick's very excellent article which I mentioned before. Let me also mention that, in the same year, Urey, Brickwedde, and Murphy found the deuteron,[8] and the existence of the deuteron, as was mentioned fifteen times I am sure in these meetings, made experiments with nuclear physics very much easier.

Very soon thereafter, on June 7 — you see, just a few months later — Heisenberg published his first article on nuclear physics.[9] Heisenberg's articles are always inspiring and always interesting, but reading it, you realize that you did not remember it correctly, because what he says in detail often doesn't hold water. In particular, I seemed to remember that the proton-neutron charge symmetry, that the nuclear proton-proton interaction is as strong as the neutron-neutron interaction, was contained in that paper. It is not. He introduced the idea of the isotopic spin notation, but he said the proton-proton interaction is electrostatic, the neutron-neutron interaction unknown. In addition, he did not believe in Pauli's idea of a neutrino. He really treated the subject as if the neutron were a proton-electron compound. Of course, this was contradicted in a way by the spin of N^{14} and of many other nuclei, and by the Heitler-Herzberg observation of the statistics. He described the neutron-proton interaction as the interaction of two molecules, and in this way he satisfied the saturation requirements. In other words, he had essentially, not entirely, but essentially, exchange forces, which lead to the result that if there are many particles in the nuclei, the total binding energy does not increase with the square of the number of particles but only with the first power.

He then discussed in a beautiful way the structure of heavier nuclei. He pointed out the fact that light nuclei have on the whole equal numbers of protons and neutrons—how he got that when he didn't assume that the proton-proton and neutron-neutron interactions are the same is not clear. He also showed that the odd-odd nuclei have higher energies than the even-even nuclei, and his graph of this was later reproduced many times. Altogether, it is a pleasure to read the article even though one can't always agree with it. (I hope you don't mind my saying this honestly, but if we talk about history we should be honest and not deny tiny, tiny weaknesses in very, very, inspiring papers.)

Heisenberg wrote three articles,[10] all of which are very inspiring and very interesting, and in a way I have summarized them already. He was the first person who took seriously the concept that the nucleus consists of protons and neutrons, even though he disagreed, in this article at least, with Bohr, that the energy principle is not valid. He applied the energy principle and applied it very impressively.

The next article I want to mention is, I am afraid, my own article,[11] which shows that at one time I was really a chemical engineer. I thought that the best thing to do was to apply the neutron idea to the lightest nuclei that exist, and there existed already the deuteron and the α-particle. The article was in a way stimulated by a paper of Dr. Peierls,[12] which many of us have forgotten. Peierls had shown that in a one-dimensional space an attractive potential—no matter how weak the attraction—always creates a bound state. Well, the first point I made in my article was that in a three-dimensional situation the opposite is true. There must be a reasonably deep potential to have any bound state, and therefore the existence of the deuteron gives a definite connection between the width of the proton-neutron potential and the depth of it. It gives a very good approximation, a good connection between the two. The fact that He^4 has so much more binding energy shows us that in this case the total interaction is very much stronger. From the two binding energies one can arrive at a definite width and a definite depth of the interaction potential.

The numbers calculated in this article were off by about 30 to 40 percent from their present values, mainly because it was believed in those days that the neutron mass is equal to the proton mass. As we now know, the neutron mass is larger, and this makes the deuteron binding energy smaller than it was assumed in that article. Let me say one more thing to criticize that article. It considers the forces between a proton and a neutron as ordinary forces. As Professor Bethe I am sure mentioned, this is a gross error, because it would lead to binding energies which for heavier nuclei are proportional to the square of the number of particles. In my own defense, I might mention that my idea was—and I am not entirely sure I have abandoned it completely—that the heavier nuclei are something like a crystal, that at very short distances the interaction is repulsive. At present some people think this is true. I will come back to this, because all these ideas have changed very significantly in the course of time. Anyway, all this showed that after Chadwick's paper people suddenly started to take the existence of the neutron seriously; it was no longer speculation to talk about nuclear structure and about the nucleus as consisting of protons and neutrons. I remember that in my article I still said that perhaps from chlorine on there are electrons and neutrinos present in the nucleus, which was less absurd than it was to assume it for helium. I believed this because the idea of the neutron's disintegration was not accepted. In fact, for a long time it was not experimentally observed. It was assumed to be a stable particle. The β-decay, if serious thought was given to it, was considered to be the emission of electrons and neutrinos present in the disintegrating nucleus to start with—no disintegration of a very light nucleus was yet observed at that time.

This picture was changed fundamentally by Fermi's theory of β-decay.[13] I do not remember when I first read it, but it was an inspiration to me. I then knew that electrons and neutrinos need not be in the nucleus. And we certainly learned that it is possible to make calculations and describe disintegration better without assuming that electrons and neutrinos are present in the nucleus to begin with. Let me make one more remark. Most of the articles, and most of the progress made in those days, were due either to the laboratory work of Chadwick and his collaborators, that is, in Cambridge, or to that of Fermi in Rome. It is really surprising how much work was concentrated at these places.

The next important observation comes from Chadwick and Goldhaber,[14] and I hope that Dr. Goldhaber mentioned it. They had disintegrated deuterons by γ-rays of energy 2.52 MeV. And they found, I think to their surprise, that the masses of the neutron and proton are different, the mass of the neutron being larger. They had two masses for the deuteron, and they gave therefore two masses for the neutron. One of the two masses, 1.0090 (the mass of the hydrogen atom is 1.0081), agrees to the last decimal with the present value of this mass, which is really quite remarkable.

One more article which is almost forgotten and which I think was very important was by Crane, Delsasso, Fowler, and Lauritsen,[15] who questioned Bohr's idea that the energy principle is not valid, that the spread of the energy of the β-rays is due not to neutrino emission but to a violation of the energy principle. What they did — I mention only one of the two experiments, the one that is the more convincing — was to determine the energy of B^{12} by the reaction by which it is produced, deuteron plus B^{11}. They then knew the mass of B^{12} with reasonable accuracy. They had some trouble, because B^{10} also reacts with deuterons and also creates γ-rays, but they found that most of the reaction comes from B^{11}. They then measured the average energy of the B^{12} disintegration and found that it is 4 MeV. If the energy principle is valid only on the average, this average energy would have to be about 10½ MeV. The maximum energy of the β-rays was found to be 11.2 MeV or close to this, so that it was in good agreement with the energy principle if one assumed that the β-disintegration energy is more than 10 MeV, but in miserable agreement if it was assumed that it was 4 MeV. Therefore, they were very strongly supporting the idea that the energy principle is valid, that neutrinos, which were observed directly only very much later, are emitted (with an average energy of 6 MeV). And everybody believed it.

Let me mention that in Princeton the idea of the nonconservation of energy was not taken very seriously. The reason was that Einstein had such a strong conviction of the energy principle that he didn't want to question it. Much earlier when he reported — he was the one who reported—on the Bohr-Kramers-Slater theory,[16] that the energy principle is not valid, he said at once that we should make experiments to contradict this, and indeed the Bothe-Geiger and the Compton-Simon experiments contradicted this.[17] Einstein somehow had the idea—his greatest accomplishment in my opinion—that the symmetry principles, the invariance principles, are of basic importance and of complete validity. And the energy principle follows from the invariance principle. Therefore, he believed in that. Let me say also—this is not neutron physics, but I mention it just the same because it is very interesting—that he had the idea of the guiding field much before Schrödinger. The guiding field was the field that guides light quanta, guides electrons. In other words, it is the electromagnetic field, or something like the de Broglie field for electrons. But he realized that if this guiding field is taken seriously, and if a collision occurs between, say, a light quantum and an electron in the Compton effect and both follow only the guiding field, energy is not conserved, because one of the colliding units doesn't know which way the other follows the guiding field. To conserve energy and momentum, the guiding field must be in configuration space, and he did not conceive this possibility. The field guides not the particles but the configuration. Schrödinger's wonderful accomplishment was that he introduced the guiding field, not for particles but for the configuration. Einstein actually never published his idea of the guiding field because the way he imagined it, it was not in conformity with energy conservation. But those of us who knew him realized this and knew that he believed more firmly in energy and momentum conservation than in any suitably chosen counterexamples.

    Let me now come back to 1934 and mention the really amazing papers, also coming from Fermi's laboratory in Rome, of Amaldi, D'Agostino, Fermi, Pontecorvo, Rasetti, and Segrè.[18] It was truly surprising what they did—how much they did, and how much they discovered about the behavior of neutrons. First, they discovered that neutrons can be slowed down, and as a matter of fact the papers of Chadwick and Goldhaber, which I have mentioned, already used the fact that if neutrons are slowed down it is much easier to detect them. Not only did Fermi and his group discover that the neutron cross section of most nuclei is much greater at low energy, but they even demonstrated that the cross section at very low energies is inversely proportional to the velocity; they obtained the l/v law. They also investigated the interaction of neutrons with practically every element of the periodic system, and they had a long list of cross sections, of the thermal cross sections, of almost every element. If one looks at them now, when one has accurate values for the cross sections, one realizes that, yes, sometimes there is an error of 10 percent or even 15 percent, but on the whole their data were excellent. They also observed that the largest cross section per unit mass is that of B^{10} — it disintegrates into Li^7 and He^4 — and the largest nuclear cross section is that of cadmium, about 60,000 barns. There is at least one nucleus with a much higher cross section, 2½ million barns, and that is a xenon isotope which was found much later. But altogether this idea of the slow neutrons and neutron physics was largely due in those days to Fermi and his school. There are many other papers — I am sure you heard about the paper by Bethe and Peierls[19] — but I won't go into that because I hope you have heard about it.

I might say one more thing, that the idea of charge symmetry, in other words, that the proton-proton interaction (apart from its electrostatic component) is as great as the neutron-neutron interaction, was present already at this time, even if it wasn't published in so many papers. The fact that this makes some β-disintegrations very much faster than others for the same energy was also already known. But the general assumption was, even though this was not contained in Heisenberg's article — I assumed it as a matter of course — that the proton-proton interaction is electrostatic, and nothing else, and the neutron-neutron interaction is negligible. That made charge symmetry; it explained the fact that the most stable nuclei at low masses are the nuclei with equal numbers of protons and neutrons.

The next great discovery in my opinion was an experimental discovery which contravened this completely. The first person who observed a very strong proton-proton interaction was Milton G. White.[20] He observed proton-proton scattering and found that it deviates enormously from that due to Coulomb interaction, particularly at 45°. This experiment was published in 1936, but very soon afterward Tuve, Heydenberg, and Hafstad also published an article[21] on the same question, on proton-proton scattering and the proton-proton interaction, which clearly showed that it is very strong. The interpretation of these experiments was given in an article immediately following the article of Tuve, Heydenburg, and Hafstad in the *Physical Review* by Breit, Condon, and Present.[22] They showed that the proton-proton interaction is as strong as the interaction between a proton and a neutron in the singlet state. I did not mention—though perhaps I should have—that the interaction between a proton and a neutron is stronger in the triplet state than in the singlet state, which follows from the fact that the deuteron is in the triplet state. It's so much weaker that it is not sufficient to create a bound state; though the so-called virtual state is at a rather low energy, around 100,000 volts, and is responsible for what was thought at that time to be the very large scattering cross section of neutrons by protons—not 60,000 barns, but 10 barns, which was already very large. Breit, Condon, and Present evaluated the proton-proton interaction, and of course the proton-proton system is at 0 angular momentum in the singlet state, because the wave function is anti-symmetric. They found, and they said so in their article, that within experimental error it is as large as the proton-proton interaction in the singlet state. This then naturally created the idea that the neutron-neutron interaction is equally large, in other words, that the interaction of a proton pair, a neutron pair, and a proton-neutron pair is very nearly the same. This was the so-called isotopic spin independence, which led to many interesting conclusions, in particular to the idea of super-allowed transitions,[23] but also to many other results.

The next article I want to mention is that of Bethe and Bacher in the *Reviews of Modern Physics*.[24] As I now reread it, I found things with which I don't agree—for instance, they don't mention Pauli's contribution of the idea of the neutrino; there are some other things of a physical nature with which I also do not agree. Just the same, the article helped the development of theoretical nuclear physics so much; it is hard to remember its influence today because we forget that even then nobody could read all the articles in nuclear physics. Well, they at least read more articles in nuclear physics than many suitably chosen counterexamples, and this article was enormously helpful in the development and understanding of nuclear theories.

I am afraid that the next thing I have on my list is again something connected with my name, and that is the Breit-Wigner theory of neutron resonances.[25] In my opinion, this contribution is often overestimated. Fundamentally, Breit and I said that we all know how light quanta act; they create excited states, and that is what the neutrons create, and the form of the neutron resonance is the same as had been known for about forty years—the Lorentz shape. Still, the courage to say this clearly was very helpful and initiated many applications. It also explained, I think for the first time reasonably clearly, the so-called l/v law of neutron absorption (I am sorry to say I don't agree entirely with the explanation of it in the Bethe-Bacher article, but there are worse things than not to agree with everything one has read). Just the same, the existence of these resonances, their explanation, furthered nuclear physics a good deal, because it is often useful to say even the obvious things.

I now come to something that played a very great role. The next great discovery, in 1938, was the discovery by Hahn and Strassmann of nuclear fission.[26] I remember that this represents the correction of the most significant error that one can find in the articles of Amaldi, D'Agostino, Fermi, Pontecorvo, Rasetti, and Segrè. They thought that the uranium absorption of neutrons created heavier particles than uranium. To some extent it creates these, but most of the radioactivity does not come from these. Most of the radioactivity comes from the fission products, and this is what Hahn and Strassmann, with the collaboration of Meitner, established. Again it was important that some chemists engaged in this business, because they knew chemistry; they found the element barium in the products and then realized that this is not an element heavier than uranium. It would not fit into the periodic table that way.

The development during the next few years is not so easy to describe, because fission was investigated in many, many ways. The news was transmitted to the United States by Bohr, who visited for half a year at Princeton and who collaborated with Dr. Wheeler in developing a theory of fission, which was on the whole an extremely successful theory.[27] In particular, they realized that fission induced by slow neutrons comes from U^{235}. Another Princeton man, L. A. Turner, predicted that when neutrons are absorbed by U^{238} this would eventually lead to an element, plutonium, which would also be fissionable;[28] he therefore predicted much of the present knowledge on fissionability. That this can lead to a chain reaction was, I think, evident to every theoretical and probably experimental physicist.

I remember that the summer following Bohr's visit to Princeton, and the work of Bohr and Wheeler, there was a theoretical physics conference in Washington. It was organized by Gamow, and at that conference the possibility of a chain reaction was amply discussed. Essentially everybody believed it. But the year after 1938 was, not very surprisingly, 1939. And in 1939, as you know, the Second World War started and everybody was worried, everybody was excited, over what would happen. The first German victories made many people worried, including your speaker, that Hitler would conquer the world. Also, people were very worried that since nuclear fission had been discovered in Germany, the Germans would start working on it and would realize the possibility of nuclear weapons, just as much as most every theoretical physicist at the Gamow conference realized it. This, therefore, started a lot of work on neutron-induced fission and on the secondary emission of neutrons.

I don't think I need to go into the description of that, but perhaps I might mention that the first chain reaction was instituted by Fermi in 1942, just ten and a half years after the establishment of the existence of the neutron by Chadwick. The idea of a large-scale reaction had been discussed much earlier at Princeton and probably many other places, but I remember I pushed for it. I remember that Dr. Wilson, whom I pushed to work on the resonance absorption of U^{238}, once told me, "Well, Eugene, you are pleasantly disagreeable." Well, I was insisting that the possibility of, and the conditions for, a large-scale reaction be investigated. The first large-scale chain reaction was in the Oak Ridge graphite reactor in 1943; in 1944 a really large chain reactor went into operation.

Let me go through the next things reasonably fast. The decay of the neutron was first observed very much later, in 1948 by Snell and his collaborators,[29] and I could tell a funny story about that. I was at the time the research director of the Oak Ridge National Laboratory, which fortunately I abandoned after a year. But Snell and Miller wanted to investigate the decay of the neutron, because they had a very good facility for doing this: a reasonably large neutron-producing reactor. Well, when the military head of the project heard about it, he simply wanted to forbid it, because he said that if the neutrons are disintegrating, the chain reactions will come to a stop! This shows how different the knowledge of these things was at that time. I should have mentioned that Fermi always started his more public speeches this way: "The neutron is a tiny particle; the neutron is a tiny particle." The first shield design that somebody submitted to us was a strong cage. A fox could not have come out of it, a squirrel easily could have, because a squirrel is a small animal. But the neutron is even smaller than a squirrel. Yes it is, but that was not clear to people. Well, Snell and Miller observed neutron decay; the neutron lifetime was reasonably well measured, first in 1950 by Robson.[30]

This is a wonderful story of the neutron, but the neutron had, so to say, a semi-death. Up to that time, there were three elementary particles of finite mass: the proton, the electron, and the neutron. In 1964 Gell-Mann, Ne'eman, and other people degraded the neutron to become one of the many, many what they called elementary particles. Well, I don't know what the concept "elementary particle" means, and if we had a physics meeting, I would go into that. But surely there is a more or less continuous transition between the neutron with a mean lifetime of enormous length — 15½ minutes — and the other elementary particles with much shorter lifetimes, and there is a continuous transition to lifetimes at which nobody could call a particle elementary; they would be called resonances. So the question whether the neutron is an elementary particle is a matter of definition to a very considerable degree. In 1964 the neutron was degraded, and that is where I want to finish my discussion. Thank you very much.

REFERENCES

1. W. Heitler and G. Herzberg, "Gehorchen die Stickstoffkerne der Boseschen Statistik?" *Die Naturwissenschaften*, 17 (1929), 673.
2. E. Wigner, "Összetett Rendszerek Statisztikája a Quantummechanika szerint," *Magyar Tudományos Akadémia Matematikai és Természettudományi Értesitö*, 46 (1929), 576-582 (German summary, "Statistik Zusammengesetzter Systeme nach der neuren Quantenmechanik," on page 583).

3. E. Rutherford, "Nuclear Constitution of Atoms," *Proceedings of the Royal Society of London*, 97[A] (1920), 374-400.

4. See for example, W. D. Harkins, "The Nuclei of Atoms and the New Periodic System," *Physical Review*, 15 (1920), 73-94.

5. Niels Bohr, "Chemistry and the Quantum Theory of Atomic Constitution," *Journal of the Chemical Society* (1932), 349-384, especially 382-383.

6. Irène Curie and F. Joliot, "Émission de protons de grande vitesse par les substances hydrogénées sous l'influence des rayons γ tres pénétrants," *Comptes rendus*, 194 (1932), 273-275.

7. J. Chadwick, "Possible Existence of a Neutron," *Nature*, 129 (1932), 312; "The Existence of a Neutron," *Proceedings of the Royal Society of London*, 136[A] (1932), 692-708.

8. Harold C. Urey, F. G. Brickwedde, and G. M. Murphy, "A Hydrogen Isotope of Mass 2 and Its Concentration," *Physical Review*, 40 (1932), 1-18.

9. W. Heisenberg, "Über den Bau der Atomkerne. I," *Zeitschrift für Physik*, 77 (1932), 1-11.

10. W. Heisenberg, "Über den Bau der Atomkerne. II," *Zeitschrift für Physik*, 78 (1932), 156-164; "III," *Ibid*. 80 (1932), 587-596.

11. E. Wigner, "On the Mass Defect of Helium," *Physical Review*, 43 (1933), 252-257.

12. R. Peierls, "Über die Existenz stationärer Zustände," *Zeitschrift für Physik*, 58 (1929), 59-62.

13. E. Fermi, "Versuch einer Theorie der β-Strahlen. I," *Zeitschrift für Physik*, 88 (1934), 161-177.

14. J. Chadwick and M. Goldhaber, "The Nuclear Photoelectric Effect," *Proceedings of the Royal Society of London*, 151[A] (1935), 479-493.

15. H. R. Crane, L. A. Delsasso, W. A. Fowler, and C. C. Lauritsen, "The Emission of Negative Electrons from Boron Bombarded by Deuterons," *Physical Review*, 47 (1935), 887-888.

16. N. Bohr, H. A. Kramers, and J. C. Slater, "The Quantum Theory of Radiation," *Philosophical Magazine*, 47 (1924), 785-802. For the German version, see *Zeitschrift für Physik*, 24 (1924), 69-87.

17. W. Bothe and H. Geiger, "Experimentelles zur Theorie von Bohr, Kramers und Slater," *Die Naturwissenschaften*, 13 (1925), 440-441; A. H. Compton and A. W. Simon, "Directed Quanta of Scattered X-Rays," *Physical Review*, 26 (1925), 289-299.

18. E. Fermi, E. Amaldi, O. D'Agostino, F. Rasetti, and E. Segrè, "Artificial Radioactivity Produced by Neutron Bombardment," *Proceedings of the Royal Society of London*, 146[A] (1934), 483-500; E. Fermi, E. Amaldi, B. Pontecorvo, F. Rasetti, and E. Segrè, "Azione di sostanze idrogentate sulla radioattività provocata da neutroni," *La ricerca scientifica*, 5 (1934), 282-283.

19. H. A. Bethe and R. Peierls, "Quantum Theory of the Diplon," *Proceedings of the Royal Society of London*, 148[A] (1935), 146-156.

20. Milton G. White, "Scattering of High Energy Protons in Hydrogen," *Physical Review*, 49 (1936), 309-316.

21. M. A. Tuve, N. P. Heydenburg, and L. R. Hafstad, "The Scattering of Protons by Protons," *Physical Review*, 50 (1936), 806-825.

22. G. Breit, E. U. Condon, and R. D. Present, "Theory of Scattering of Protons by Protons," *Physical Review*, 50 (1936), 825-845.

23. L. W. Nordheim and F. L. Yost, "On the Matrix Element in Fermi's Theory of β-Decay," *Physical Review*, 51 (1937), 942-947. E. P. Wigner, "On Coupling Conditions in

Light Nuclei and the Lifetimes of β-Radioactivities," *Physical Review*, 56 (1939), 519-527.

24. H. A. Bethe and R. F. Bacher, "Nuclear Physics. A. Stationary States of Nuclei," *Reviews of Modern Physics*, 8 (1936), 82-229.

25. G. Breit and E. Wigner, "Capture of Slow Neutrons," *Physical Review*, 49 (1936), 519-531.

26. O. Hahn and F. Strassmann, "Über die Entstehung von Radiumisotopen aus Uran durch Bestrahlen mit schnellen und verlangsamten Neutronen," *Die Naturwissenschaften*, 26 (1938), 755-756.

27. Niels Bohr and John Archibald Wheeler, "The Mechanism of Nuclear Fission," *Physical Review*, 56 (1939), 426-450.

28. Louis A. Turner, "Atomic Energy from U^{238}," *Physical Review*, 69 (1946), 366. Turner was persuaded by Leo Szilard to delay publication until after the war.

29. A. H. Snell, "Search for Beta-Proton Coincidences Associated with Neutron Decay," *Physical Review*, 76 (1949), 195. Arthur H. Snell, Frances Pleasonton, and R. V. McCord, "Radioactive Decay of the Neutron," *Physical Review*, 78 (1950), 310-311. The authors note that E. F. Shrader, D. Saxon, and L. C. Miller collaborated actively in the experiment in the past.

30. J. M. Robson, "Radioactive Decay of the Neutron," *Physical Review*, 78 (1950), 311-312.

DISCUSSION

Bromberg: In your memory, or perhaps in the memory of some of the other people here, what were the other fields in physics or in mathematics or in other branches of science that were interacting most strongly with nuclear physics in the 1930s and early 1940s? Were any other fields giving ideas or shaping nuclear physics, or was it very largely an autonomous branch of physics?

Wigner: This is a very difficult question. I, myself, was very much influenced by something which I probably should not mention, namely, that we worked at Princeton on the theory of mechanical cohesion, and this influenced me so strongly that I thought that the nuclei had some structure similar to a solid body. And that this is the reason for the saturation. But this was incorrect. You know, it is not easy to say this, but physics was so interesting, and so many new things became evident, that the devotion to nuclear physics was moderate in the beginning. It was in competition with atomic physics and molecular physics, and many people felt that nuclear physics —

yes, it exists but let me first think out what is the structure of metallic strontium, or what is the structure of the water molecule. In other words, they had an attitude, as we now have, toward those parts of the world which are completely unexplored. I don't think I answered your question very well, but that's the best I can do.

Malley: Otto Hahn, in his autobiography, states that after he discovered fission he was not interested in applications, and he was not concerned with or even aware of the possibilities with warfare. I am wondering, in the light of what you said about this having been discussed — the chain reaction — whether you think perhaps he is concealing a humanitarian motive, or if it was possible that he did not think of such things?

Wigner: I think among the German scientists such as Hahn there was a considerable antagonism toward the Hitler regime, and as a result they did not particularly want to make possible an easy victory of the Nazi regime over the rest of the world. Just the same, they had a nuclear project. Perhaps I should also mention, although it has little to do with neutrons, that when we were working in Chicago on the nuclear chain reaction, one day we received a telegram from Dr. Houtermans, whose name may or may not be familiar to you, with this text: "Hurry up. We are on the track." Now, apparently this was not correct, because the Germans were far from achieving a chain reaction, but it does indicate that every reasonable physicist realized that the chain reaction is possible, and it's easy to calculate that the energy liberated in the chain reaction is 200 million electron volts, compared with about 4 electron volts in an ordinary chemical reaction. And there is a significant difference between 200 million and 4.

Bethe: I have a number of questions and comments. As you certainly know, in my talk I mentioned very similar things to those in your paper. You mentioned the singlet S-state of the deuteron. It was clear that the singlet S-state must be higher than the triplet S-state. But it would have been perfectly possible that it was 10 volts higher. And to refresh your memory, you and I were coming from some meeting at Columbia University, probably it was the Physical Society, and on the subway train we discussed the question of the large cross sections, and you said: "But that's very simple. The single state is just very little bound, and that's the explanation."

I was very much impressed by that and included the remark in the article with Bacher which you quoted. Well, this is the smallest of the points.

In your 1933 paper, you made some remarks about the Heisenberg paper which were far from complimentary, and in fact I believe you . . .

Wigner: The scattering of neutrons by protons?

Bethe: No, no. In your 1933 paper on the binding of the α-particle and the deuteron.

Wigner: Oh, that's 1932, but that's irrelevant.

Bethe: It's in the *Physical Review* in 1933, so you probably submitted it in 1932. You said that Heisenberg uses electrons in the nucleus, and I think this is really not a fair statement.

Wigner: It is not?

Bethe: He was, after all, the first to say, "Now we have neutrons, so we have neutrons and protons, and we can have a quantum mechanics of the nucleus." But then I agree—and I said in my talk, and you mentioned it again—he still had the idea in his mind that the neutron might be a composite particle and that electrons exchanged between protons and neutrons. But he had pushed the electron into the neutron, and I think it is really quite legitimate to push the difficulty one step deeper and work with the things you have, in this case neutrons and protons.

Wigner: Yes, you are absolutely right. He was the first person who realized that now that we have neutrons, we can have a reasonably simple theory of the nucleus. I didn't put this forward strongly enough.

Bethe: Today I think you were quite fair about it. The next point concerns the Breit-Wigner theory. I gave praises to the Breit-Wigner theory, and I think you were unfair to it. It was, after all, the theory by which everybody was able to interpret experiments. I don't believe it was trivial. The point that people raised about the Breit-Wigner theory, and with which I don't agree, is that is did not contain the idea of the compound nucleus. In my opinion it did contain quite a lot of that idea. It did contain the idea that at least two particles, possibly more particles, can interact strongly to give a level of the nucleus as a whole, which is quite different from a single-particle level; and I emphasized that in your paper you said very strongly that the one-body theories of neutron reactions had obviously failed, because they give a lot of scattering and not so much capture, contrary to observation.

Wigner: Perhaps I dare say that you are absolutely right, and I never believed—I believed far too little—in the independent-particle model. We put that remark in, because I felt that the situation can be explained more easily if only two neutrons are put in, but I did not feel that way. [N. B. I did realize, and so did Breit, that at the actual level densities the single- or even double-particle models are inadequate. In fact, we believed in the independent-particle model too little; we did not realize fully its usefulness even at low energies of excitation.]

Bethe: I assumed you didn't feel that way, and I think you had essentially the compound nucleus model. In my opinion your paper and Bohr's paper are complementary and give different aspects, but you both have the concept.

Wigner: I think you are right.

Bethe: Now, the next point. I talked, among other things, about the shell model. I have two questions about that. One is, there is a paper by Feenberg and yourself which I mentioned, in which you use the shell model with only the orbital motion and then find that more important than shells is the interaction of the nucleons within the shell. How much did you believe in the shell model when you . . .

Wigner: Not terribly. It is surprising that the shell model gave the proper spins for the normal states. But even with atoms I am not terribly shell-model minded, independent-particle minded. Now, I may be wrong, but why don't we ask Dr. Feenberg what he thought? Of course we had no idea of the real shells, which were . . .

Bethe: That's the second question.

Wigner: . . . no idea, at least I assume we did not.

Feenberg: Of course, at that time we knew about the papers on the shell model by Elsasser and Guggenheimer, and there were other names which I don't recall at the moment. These papers were very impressive, but they used up all the experimental evidence, and so there wasn't much more that could be done with them. But after the war, there was such a flood of experimental evidence, on magnetic moments, γ-transitions, isomeric states, beta decay, that it became obvious that there was a shell model.

Wigner: I did not remember this. Thank you.

Bethe: The second question refers to the real shell model of 1949, when the spin-orbit interaction was included. There is a rumor, and I don't know any evidence for it, that you were against that shell model at the time, and if so, why? and if not . . .

Wigner: I was not against it. I was very impressed by it. I should not mention the person who was against it, but he compared the shells at 8, 20, 50 with the subway stops in New York, and pointed out that you can always find some numbers which have a special significance. That was a very, very excellent physicist and a very famous man: Niels Bohr.

Bethe: I see. In connection with the spin-orbit coupling, you sponsored some calculations quite early about spin-orbit coupling as produced by a tensor force. Could you say a little about that? This is my last question.

Wigner: Well, the existence of tensor forces was a shock to me — that they are so strong that they create a quadrupole moment for the deuteron was a real shock to me. I thought that the spin-dependent forces were very small. I was completely mistaken on that. But when it was evident that there are tensor forces, I thought perhaps the tensor forces are responsible for the spin-orbit coupling, and somebody wrote a Ph.D. thesis at Princeton on this subject. Of course, these forces made the supermultiplets a much less valid concept than I thought it was, but I thought that it's enough if we have one spin-dependent force, and let's have only the tensor force. I was mistaken about this. The forces between nucleons are much more complicated, and much more difficult to describe, than the forces, say, between electrons.

Barschall: My recollection is that you knew the order of the, say, $P_{1/2}$ and $P_{3/2}$ levels in nuclei before it had been actually measured. Is that recollection correct?

Wigner: I knew that Li^7 had spin $3/2$, and this made it evident, but I had no theoretical reason. Of course, it followed from the tensor forces, but it was a great shock to me that tensor forces exist, because I felt that it was good if we have a supermultiplet theory, and it had some good applications. But Dr. Feenberg collaborated on that. Isn't that right?

Bethe: The two papers followed immediately upon each other.

Feenberg: I can comment on this. They appeared in the same issue of the *Physical Review*, but some time elapsed between the writing of them, and the charge independence came in between, as a basic concept which you used. [N.B. After the conference I looked at the paper I wrote with Wigner on states in the first p-shell and learned that we had referred to what must be the earliest statement of charge independence, by L. A. Young, *Physical Review*, 47 (1935), 972. Unfortunately, this reference was forgotten and generally overlooked in later publications.]

The Development of Our Ideas on the Nuclear Forces
Rudolf Peierls

INTRODUCTION: Robert Serber

This morning during Professor Wigner's talk we heard several references to the difficulties of understanding the fact that the binding energies of nuclei went linearly with the numbers of particles. Of course, we are familar with other systems that have this property, such as this podium, and one wonders why one didn't say that the nuclear forces between nucleons were like the forces between molecules. I suppose the reason was that in those days people had thought that the nucleons must be elementary particles. They were not yet aware that they were constructed of quarks.

I was glad that Ed McMillan spoke in such detail this morning about accelerators, because I know that in teaching physics and I suppose in describing the history of physics, there is a very great temptation to make it sound as if the progress of physics were due to ideas and theories, when in fact it is to a very large extent, it seems to me, controlled by the technical means available. The experiments one does are determined largely by the state of electronics and accelerator design at the given time.

After the developments in accelerator technology during the early 1930s which Ed described, there was progress in learning something about the forces acting between nucleons. But with a limited technology, it turned out that one could learn just a few features of the properties of the forces because one could only carry out experiments below 20 MeV. There were brilliant interpretations of the results by Peierls, Bethe, Wigner, Breit, and on the basis of very fragmentary evidence, they drew far-reaching and correct conclusions. After Heisenberg's 1932 paper, there was a period of fifteen years during which one soon found out something about the properties of nuclear forces, and then people just speculated about them. But as far as information goes, nothing new came in for a long time. It wasn't until the new development in technology started by Ed's theory of phase stability and the completion of the 184-inch cyclotron at Berkeley, which could get into the 100 million-volt range, that really detailed and new information could start to be developed.

Development of Our Ideas on Nuclear Forces

Many of us in this audience got to know Professor Peierls for the first time when he came to Los Alamos and played a very important role there. His earlier exploits have been mentioned many times already in these meetings. I would like to say that in addition to his contributions to theoretical physics he is a great teacher. After the war, the school of theoretical physics that he originated and led in Birmingham made very great contributions to physics in England and in the rest of the world. Also, we are pleased that at this time he is not only a professor at Oxford but also at Seattle. Professor Peierls will now tell us of the development of the ideas about nuclear forces.

The Development of Our Ideas on the Nuclear Forces
Rudolf Peierls

The nature of this talk differs a little from some of the preceding ones, which were given by people who had themselves contributed substantially to their topics. I did not really contribute to the subject of nuclear forces anything that would be worth being modest about. I was mainly a spectator. My task in presenting this subject was made easier by previous speakers, particularly Hans Bethe and Eugene Wigner, who covered some of the ground. Wigner also already explained what I had intended to stress about what we are doing here: We are not trained historians, but we were there, and we remember how it felt, and so we are trying to pass on the spirit rather than the historical details. Again, we remember only partially, as Wigner also already said. Memory is unreliable, and we have to go back to some of the original literature and there find many surprises.

On the subject of memory, I regret to discover that Eugene's memory is as good as it is, because I had hoped that my paper about the existence of stationary states, which he mentioned, was forgotten by now. It makes a correct statement about the problem in one dimension, but in its last paragraph goes on to say that, since the

---

Nuclear Physics in Retrospect, edited by Roger H. Stuewer, copyright © 1979 by the University of Minnesota.

Schrödinger equation in three dimensions looks the same as in one, the same result would apply. This, as he rightly pointed out, is far from the correct answer, because of a different boundary condition.

You will note the predominance of theoretical papers in my references. This does not indicate that I fail to appreciate the crucial importance of experimental developments, but in discussing the interplay, and often the confusion, of ideas, one has to refer to the theoretical interpretation.

It was already stressed several times in this symposium that no sensible discussion about nuclear forces was possible before the discovery of the neutron. It was also mentioned how difficult it was at the time to accept the idea that the neutron was as elementary as the proton. I have a few quotations which, I think, illustrate this point in an interesting way. In his first paper on the neutron in the *Proceedings of the Royal Society*, Chadwick says: "It is, of course, possible that the neutron may be an elementary particle. This view has little to recommend itself at present, except the possibility of explaining the statistics of such nuclei as N^{14}."[1] In other words, he saw the possibility but did not take it much further.

Probably it was clear to everybody that there must be neutrons in the nucleus; but it was not so easy to accept that there were only neutrons and protons, particularly that there were no electrons. There were at the time a number of papers discussing the constitution of the nucleus in the light of the discovery of the neutron. The earliest one I could find that mentions the neutron explicitly as a constituent was Fournier's,[2] with ideas later expanded by Francis Perrin.[3] But they still have electrons, α-particles, and deuterons (called "semi-helions") on an equal footing with protons and neutrons. Iwanenko made a similar remark early, but he, too, had a lot of electrons in the nucleus. However, in a later paper[4] (August 1932) he actually says that β-electrons must be created at the moment of emission, similarly to photons emitted by an atom. As far as I could discover, that is the first mention of the idea in print.

Heisenberg[5] took a very complicated view, as was already mentioned. He did have electrons in the nucleus, mainly (or entirely) inside the neutrons, but a telling phrase in his third paper is: "the possibility of separating clearly the quantum mechanical factors from those new ones which are characteristic of nuclear physics."

Development of Our Ideas on Nuclear Forces   185

By the latter he meant the electron losing its individuality, its statistics, and perhaps the conservation of energy. But he knew that for describing the dynamics of a nucleus one could forget this. Another interesting remark in the same paper is: "The disintegration experiments permit us to regard, for example, the neutron as an immutable elementary particle, like the proton and the electron. This assumption would be expressed in the Hamiltonian in the first instance by the absence of exchange terms." He refers to "the heavy nuclei, which include β-emitters and, on the assumption of immutable neutrons would also have to contain electrons." These remarks make strange reading in retrospect, but they show how hard it was to get adjusted to the new point of view.

On the other hand, it was immediately accepted by everybody that a new, and very strong, force was required to hold the nucleus together. There was never any doubt that there was some new force at work. What one wonders about in retrospect is why, in the case of the atom, before there was a properly working quantum mechanics, nobody ever speculated that there could be a force other than the Coulomb force (and the related magnetic forces). It was always taken for granted that there could only be electromagnetic forces in the atom. With the nucleus it was different.

In his first paper Heisenberg had a first shot at the force problem, based on the analogy with the hydrogen molecule and molecule ion, which led him to the idea of exchange forces. There is no indication in his three early papers that he saw the connection between exchange and saturation, i.e., the idea that exchange forces could prevent a nucleus from collapsing to a binding energy proportional to the square of the number of particles. But the notion of exchange forces was picked up by others because of saturation.

This is particularly clear in the paper by Majorana,[6] with which we are all familiar. He discussed saturation very clearly and introduced exchange (in a form somewhat different from Heisenberg's) as a means of achieving it. In looking at Majorana's paper it may be worth quoting one sentence about his notation: "I shall label the neutron and proton as different, to avoid the inconvenience of the ρ coordinate" ("ρ" was Heisenberg's symbol for the isotopic spin variable). In other words, here was the brilliant innovation of avoiding the isotopic spin by treating neutrons and protons as distinct particles. This was, of course, entirely sensible at that time, because, although Heisenberg had invented the isotopic spin notation, it did not seem then to serve any visible purpose.

Let us look further at the problem of saturation. It had been known for a long time that the binding energies of nuclei rose only proportionally with the mass number, and there are various ways in which nuclear forces could lead to this behavior. In particular, four possibilities come to mind:

1. Exchange,
2. Repulsive forces at short distances ("Repulsive core"),
3. Velocity-dependent forces,
4. Many-body effects.

For many years all these possibilities were ignored, other than exchange. Why was that? Looking back, we can see that people were looking for a simple law because of the state of physics at the time. We thought, in the late twenties and early thirties, that physics was almost finished. I remember a conversation in Copenhagen, not as part of a meeting, but perhaps over lunch, when many people took the view that soon fundamental physics was going to be finished. There was only the relativistic behavior of electrons and the electromagnetic field which still needed to be understood. One did not know where the fine structure constant $1/137$ came from, but it seemed as if this was related to the unresolved problems of electron self-energy, etc. There was the proton-electron mass ratio to be explained, but that appeared to be related somehow to the fine structure constant. There was, of course, the nucleus. That did not seem such a large problem, because one did not have much information that called for explanation. In any case at that time nuclei still contained electrons, and this was evidently a highly relativistic problem. In the development of quantum mechanics fundamental problems had been resolved so quickly that it was natural to think that the next step might be the last. Some started to discuss what they would do after physics was complete — perhaps they might go into biology. Max Delbrück did change to biology about that time, but I do not know what was his motivation or whether he was present at the conversation I reported.

After 1932 things were different; we knew of the neutron and we had to find a new law of force. It was reasonable to hope that this would be something as simple and as fundamental as Coulomb's law. After all, Coulomb's law, and indeed the whole content of Maxwell's equations, can be demonstrated by a very limited number of experiments, and with the use of very few parameters. Therefore, we looked for something simple.

This was, I think, the background to the attitude of most theoreticians, when they rejected the idea of a repulsive core to the forces, which would make nuclear forces act like interatomic forces in molecules or solids. Majorana noted this quite explicitly: such forces would be aesthetically unattractive, he said. There was, also, a related point — I do not remember who expressed it — that a repulsive core would have to have a shorter range than the already short-range attraction, and to have any effect on the wave function it would therefore have to have enormous strength. Its potential might have to be as great as 1 BeV. It seemed hard enough to accept the idea that in a system in which binding energies per particle, and excitations, were usually a few MeV, one was dealing with attractive potentials of 50 to 100 MeV, and it seemed quite fantastic to postulate repulsive energies yet another order of magnitude higher.

Velocity dependence was not discussed at the time, although it could also help to get saturation, since by Fermi statistics compression increases the nucleon velocities, so that an interaction that gets less attractive at higher relative velocity would make higher densities less favorable.

The possibility of invoking many-body forces was also generally ignored, with some exceptions to which I want to return. So it was taken for granted, for many years, that the observed saturation required exchange forces.

In setting out to find what forces would account for the known properties of nuclei, we naturally started to look for a law of force that in some sense was simple. We made, deliberately or implicitly, a number of assumptions, to be revised in the light of experience. The simplifying assumptions from which almost everybody started were:

a. Two-body forces,
b. Central,
c. Static,

d. Short range (zero range?),
e. "Ordinary" (i.e., not exchange forces),
f. Spin-independent,
g. One sign only (i.e., no repulsive core),
h. Neutron-proton force only,
i. Charge symmetric (i.e., n-n force equals p-p force, if any),
j. Charge independent. (This simplification, which contradicts h above, came later.)

The interesting fact is that, as the consequences of these assumptions were studied, and particularly as new experiments provided additional information, almost all these simplifying assumptions had to be abandoned one after the other, so that the nuclear force turned out to be almost as complicated as could be imagined. In fact, of all these simplifying assumptions the only ones to survive were charge symmetry and charge independence. It is amusing to think that charge independence was deduced, or rather guessed, on the basis of extremely weak evidence, as we shall see.

The belief in ordinary forces was dropped quite soon, because as I discussed, exchange forces had to be invoked to account for saturation. The idea of zero-range forces, proposed as a possibility in Wigner's paper of June 1933,[7] had to be abandoned when L. H. Thomas[8] showed, in a very clever paper, that for given deuteron binding energy, the binding energy of three nucleons, H^3 or He^3, rises indefinitely as the range is decreased, so that zero range would lead to an infinite binding energy for mass number A = 3. The relation between the deuteron binding energy and that for A = 3 and 4, had already been discussed in Wigner's paper of February 1933,[9] in which he pointed out that the very much stronger binding, particularly of the a-particle, was to be expected for short-range forces. The calculations by which he illustrated the point were carried down only to a certain finite range, and it is not clear whether he realized at the time that the range could not be made too short. Probably this was not yet understood, because in his June 1933 paper Wigner regarded zero range as a possibility.

The spin independence of the force had to be abandoned because of the slow neutron scattering results, as discussed, we heard, by Wigner and Bethe in the subway.

At this time there was therefore a provisional picture of the forces, which had already many applications. I was involved with Hans Bethe in one of them.[10] Perhaps I might add to what was already said a remark about how this work started. We were visiting Cambridge where Chadwick had just done (I do not know whether at this point with or without Goldhaber) the photodisintegration experiment.[11] He did not tell us the result, but when there was general talk about neutrons and protons and their interaction, Chadwick said: "I bet you could not calculate the cross section for photodisintegration of the deuteron." We felt this did not look too hard and thought we might try it. This resulted in a paper in which some answers were right, but others, such as the capture cross section for slow neutrons, were quite wrong, because we did not yet know about the spin dependence of the force. The approach of this paper foreshadowed the effective-range approximation, which later proved a very convenient way of handling low energy scattering data, though we did not fully develop the generality and simplicity of this technique.

However, at that time most people still thought of forces between unlike particles only. Perhaps the motivation was that we knew from the existence of the deuteron that there must be a force between neutron and proton, and there was no equally direct evidence for a neutron-neutron or proton-proton force (other than the Coulomb repulsion in the latter case). It was also believed that the dominance of the n-p attraction explained the tendency of nuclei to have equal numbers of neutrons and protons, modified for the heavier nuclei by the electrostatic repulsion of the protons.

There had already been many suggestions that this was not so. Guggenheimer[12] concluded that there must be a force between like particles of the same order of magnitude as between unlike particles. His arguments were based on a rather peculiar model and one might doubt, therefore, how seriously they should be taken. L. A. Young[13] reached the same conclusion by a much clearer argument. Much more quantitative support for this idea came from the calculations of Feenberg[14] on the binding energies of the three- and four-nucleon nuclei. We know today how hard it is to get good values for these nuclei, but his variational results seemed reliable enough to show the need for an appreciable interaction between like nucleons.

Late in 1936 came direct experiments on proton-proton scattering. The first qualitative results of White were already mentioned, as well as the later, quantitative measurements by Tuve, Heydenburg, and Hafstad.[15] These showed that the interaction was strong, and the analysis of these results by Breit, Condon and Present,[16] and the discussion of the implications by Cassen and Condon[17] and by Breit and Feenberg[18] led to the suggestion that the p-p force in the ^1S state (which for like particles is the only S wave allowed by the Pauli principle, and therefore the only state observable at low energy) is the same as that of the n-p system in that state.

This was the first mention of the charge-independence hypothesis. It was then a guess, prompted by very limited evidence. This looked, perhaps, a little more weighty at the time because we did not yet have a clear command of the effective-range approximation.

We now know that at the energies then available one is well within the range of validity of that approximation, so that for each state only two parameters are needed, the scattering length and the effective range. The scattering lengths for the ^1S state in neutron-proton and proton-proton scattering were similar (actually they are rather different, but both are very large, and it is the inverse that should be looked at) but the effective range for the n-p ^1S state was hardly known at all. So all the relevant information was that one number had similar values in the two cases. In the absence of the effective-range theory we had to make numerical calculations for each postulated shape of potential, and compare the resulting curves for cross sections as functions of the energy. There was even some discussion about which shape of potential gave a better fit to the data, although we now know that any two-parameter family of potentials can give the same fit within effective-range theory. There is a remark in the 1939 paper by Share, Hoisington, and Breit[19] that the Yukawa shape fits the proton-proton data better than other forms, and in a companion paper the same authors say that, with this shape, the p-p interaction potentials agree more closely than otherwise.

In fact, the confidence in this conjecture of charge-independence was so strong that people started finding explanations for the small difference in the two numbers whose near-equality had given rise to the hypothesis. What is so amazing is that it all turned out to be correct, and that, as we now know from very extensive evidence, the nuclear forces are indeed charge-independent, except for small corrections, which can be understood.

That was when Wigner[20] wrote his important paper studying the symmetries of the problem. He considered a number of situations of varying degree of symmetry, culminating in the most symmetric case in which the forces were charge-independent. As motivation for this he quoted the experimental results of Tuve and his collaborators and their theoretical analysis. He discussed applications, which showed good agreement, but he did not emphasize that this supported the most symmetric scheme, and hence the postulate of charge independence. Similar schemes were developed independently by Hund.[21]

These methods were soon adopted by others; the first spectroscopic calculation using Wigner's symmetry scheme was probably that of Inglis and Young in March 1937.[22]

There were also many attempts to explain the binding energies of the heavier nuclei. Most people, including v. Weizsäcker,[23] used for this the Thomas-Fermi approximation, which ignores all quantum effects, except for Fermi statistics. Heisenberg[24] showed that a better approximation could be obtained from the Hartree-Fock approach, but even then the binding energy of the heavier nuclei did not come out right if the forces were adjusted to fit the two- and three-body data.

The clearest test came in calculations of the binding energy and density of "nuclear matter," i.e., of a hypothetical nucleus so large that surface effects would be neglible. For this one had to neglect the Coulomb force between the protons, otherwise such a system could not be stable. The properties of such nuclear matter can be estimated by extrapolating the data from real nuclei.

The most ambitious calculation of this kind in the prewar days was that of Euler,[25] working in Heisenberg's group. (The same problem was studied in 1949 by Huby,[26] using Yukawa-type forces, with similar conclusions). This is worth describing in some detail, because it reveals a very interesting situation.

Euler calculated the energy per nucleon of nuclear matter as a function of specific volume in the Hartree-Fock approximation. In this limit the problem of self-consistency becomes trivial, since the one-nucleon wave functions are simple plane waves. His results look qualitatively like the curve in Figure 1. At very high density the energy is strongly positive, as a result of the saturating effect of the exchange forces and the kinetic energy due to the Fermi statistics. For medium density the energy becomes negative, so there is binding. The energy minimum occurs in about the right place, but it is not deep enough, giving far too small a binding energy. The extrapolated value for nuclear matter is, by comparison, near the cross.

This disagreement could, of course, be due to the approximation being inadequate. The calculation assumes independent particles, thus it neglects correlations, and the strong short-range forces might indeed cause quite substantial correlations. To test this, Euler calculated the energy to second order in the nucleon-nucleon interaction. In second order the correlations are approximately taken into account, so the size of the second order term is a measure of the importance of correlations. The result is about as indicated by the broken curve in the figure. It gives a much increased binding energy but is still substantially short of the correct value.

Euler concludes from this that the correction is not small and that therefore the use of independent particles plus perturbation theory is inadequate. Today we would draw the opposite conclusion. We have to remember that the calculation gives separately the kinetic and the potential energy. The binding energy is a fairly small difference between these two large quantities, of which the potential energy is the more sensitive to the details of correlations. We should therefore look at the error not as a fraction of the net binding energy but as a fraction of the potential energy. In that light it is seen to be gratifyingly small, and we must conjecture that higher-order corrections would be even smaller, so that his calculation to second order should be quite close to the exact result. So the conclusion should really have been that the forces assumed in the calculation were wrong.

I remember my own reaction to that situation at the time, which was probably fairly typical of the many people who were concerned about the disagreement. I argued that, although the forces could be wrong, this could not be responsible for the result found by Euler, because we knew, for example from calculations such as Feenberg's, that the assumed forces were capable of giving the right binding energy for the α-particle. A substantial part of the binding energy

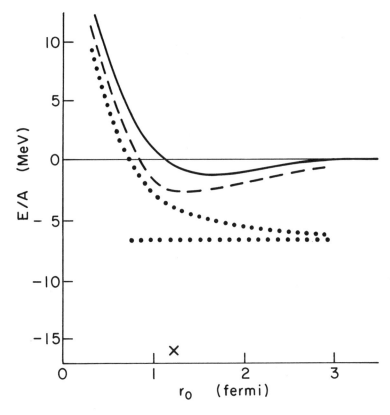

Fig. 1. Energy per nucleon for nuclear matter, with the "old" forces.
——— Hartree-Fock calculation, ----- including second-order correction.
· · · · · · (horizontal line) α-particle binding energy per nucleon.
· · · · · · (curve) conjectured shape of exact result with "old" forces.
The data plotted are those of Huby, but they are very similar to those of Euler discussed in the text.

of nuclear matter can be obtained by combining the nucleons into α-particles, and the additional energy gained by the attraction between the α-particles is relatively small. The α-binding energy per nucleon is shown by the horizontal line in the figure, and if the calculation gives a minimum above that, the evaluation must be incorrect.

The answer to this became clear only many years later. We can be sure that the postulated forces give the right α-particle energy only if we leave enough room for the a-particles to be separate, in other words we can be sure of the portion of the horizontal line to the far right. As we reduce the volume, real a-particles attract each other, but there is no guarantee that this feature is given by our assumed forces.

Obviously, the approximation does not work for high specific volumes, and this is not surprising. If the density is so low that the nucleons collapse into a-particles, correlations are evidently not negligible. On the other hand, the correlations are suppressed, and the effective interaction weakened, by the Pauli principle. This point was made clearly many years later, in a note by Weisskopf in 1951.[27] The point is that correlations require changes in the wave functions of the nucleons, and such changes can be described as adding to each wave function terms proportional to other wave functions. But according to the Pauli principle, one cannot use for one particle contributions from a wave function occupied already by another. The only remaining flexibility is to add small multiples of the unoccupied states above the Fermi level. But this costs much energy if the density is high, and therefore the changes in the wave function are small. In this sense the perturbation theory gets better for higher density. One would therefore conjecture today that the correct energy curve for the forces used by Euler would run like the dotted curve in the figure. His approximation was good, but the forces then fashionable were bad.

But this was not the interpretation accepted at the time. It was then believed that there should be very strong correlations, so that an approach neglecting correlations, or treating them as small perturbations, was misleading. This was one of the reasons why there was such strong skepticism about the shell model for the nucleus. The shell model, at least its customary application as in the atom, neglects correlations and proceeds from a Hartree-Fock approach like Euler's. If Euler was wrong, it seemed that the shell model had to be wrong. This is one reason why, as Bethe pointed out, there was a long period in which there was no work on the shell model.

Another reason was the success of Niels Bohr's compound nucleus model, coupled with his description in terms of the liquid drop model. This was misunderstood to give the impression that even in the ground state or low excited states of a nucleus very strong interactions occurred, causing strong correlations, which could not be ignored. On this view, the shell model could be justified for the nucleus no more than for a molecule or a drop of liquid. I believe, though this is only second-hand information, that when Bohr visited Japan shortly before the Second World War, he met Yamanouchi, who had done important work in group theory and who told Bohr that he planned a serious study of the nuclear shell model. Bohr convinced him that there was no hope for the nuclear shell model and dissuaded him from wasting his effort on it.

A new complication, disproving yet another of the conjectured simplifications, was the discovery by Kellogg, Rabi, Ramsey, and Zacharias[28] of the quadrupole moment of the deuteron, which demonstrated the existence of noncentral forces, the so-called tensor forces, between neutron and proton. So the forces were not central forces. It was not immediately clear that tensor forces would play a large part in determining nuclear states and their energies. Bethe[29] was the first to point out that they were likely to be very important in the deuteron ground state, and he suggested the possibility that the tensor force might be the only interaction in that state. A fuller study of the effects of the tensor force was made later by Rarita and Schwinger.[30]

This new surprise made the situation more complicated. One question was, How far were the very general conclusions about low-energy behavior, which had been derived on the assumption of central forces, invalidated by the existence of the tensor force? Hepner and I[31] showed that most of these results remained valid with some quite minor corrections. Today we know that the effective-range approximation can be derived even in the presence of a tensor force, though its statement becomes a little more involved.

Volkoff[32] realized that tensor forces were important for saturation. On a superficial view it appears that tensor forces would help saturation, because in a large nucleus each nucleon is surrounded by others of different spins and situated in different directions from it, and in that case the tensor force tends to average out. But it is also part of the saturation requirement that it should not be energetically favorable to pack nucleons at high density into needle-shaped or disk-shaped nuclei (according to the sign of the tensor force), with the nucleon spins aligned along the symmetry axis.

Volkoff showed that this makes the saturation requirements more stringent.

Up to the late 1940s the hypothesis of exchange forces was a pure conjecture. It was realized quite early, and pointed out, for example by Wick[33] in 1933, that exchange forces would show up in the angular distribution of neutron-proton scattering. If the de Broglie wavelength is comparable to the range of the forces, it becomes hard to deflect particles, and small deflections are more likely than large ones. But exchange forces interchange the nature of the particles, so the incoming neutron, which tends to continue in the forward direction, has become a proton and the outgoing neutron tends to go backward in the center-of-mass system. Hence ordinary forces favor forward scattering, exchange forces backward scattering.

The difference does not show up at low energies when the scattering is dominated by S waves, which give isotropic scattering. When the Berkeley experiments of Hadley et al. in 1949[34] gave the angular distributions of p-n scattering at 40 and 90 MeV, they showed a forward and a backward peak of the same magnitude. This was confirmation of the presence of exchange forces, but at the same time showed that the contribution of exchange forces to the total was not enough to satisfy the saturation requirements. In other words, although the exchange forces really existed, they were not adequate to serve the purpose for which their existence appeared certain. There had to be something else to account for saturation.

What this was became clear soon afterward, when the experiments reached energies as high as 340 MeV. The first curves found by Chamberlain and Wiegand[35] looked different from what was expected, and Jastrow[36] was the first to point out that they could be accounted for by assuming a repulsive core to the forces. Hans Bethe mentioned that Morse, Fisk, and Schiff[37] had considered a potential with a repulsive core as early as 1936, but this was in the course of a study of the most general potentials for which the Schrödinger equation of the neutron-proton system could be solved and gave the usual low energy behaviors, so they showed that a repulsive core was not ruled out by the low energy data; they gave no reason for believing in its existence.

Later, the phase shift analysis of the scattering showed that the S wave phase shift becomes negative at high energy, and that is a

Development of Our Ideas on Nuclear Forces   197

direct proof of the existence of strong repulsive forces, at least as long as the force is static.

Still later, and this is getting far beyond the period we are discussing at this symposium, the success of the shell model, which assumed a spin-orbit coupling, prompted the conjecture that there were also spin-orbit forces in the two-nucleon interaction, and this was confirmed also by the analysis of high energy scattering. Since this interaction contains the orbital angular momentum, the assumption of static forces has now failed, too.

By now nearly all the tentative simplifications which I had listed have had to go. The idea of a predominantly two-body force has not been questioned much. This is hardly surprising in view of the difficulty of obtaining clear evidence for many-body forces. The only way to show their existence empirically would be to determine, from the two-body data, the most general class of two-body forces which would fit these, and then to show that none of them are compatible with the data about larger nuclei. This would require a much greater computational effort than has been possible so far, and we cannot claim a final answer. We may hope that, for the sake of the sanity of nuclear physicists, we can confine ourselves to forces that are predominantly a sum of two-body terms, though it is likely that there exist some corrections involving several nucleons at a time.

There have been attempts to invoke a certain type of many-body force. These involve the idea that the nucleons interact with some kind of field, which can saturate, and each nucleon then interacts with the general field, rather than with its individual neighbors. Such a force could easily lead to saturation, and it would also be an attractive and simple way of accounting for the validity of the shell model. This possibility has been pursued particularly by Teller. In 1938 Critchfield and Teller[38] tried an electron-positron theory of the forces, assuming the field to saturate, and in 1955 Johnson and Teller[39] proposed a similar scheme based on the pion field. In spite of the attractive simplicity of this view, the impression is that information from experiments on the interaction of pions with nuclei and on pion production provide strong evidence against it, since they tend to show up the high momenta and strong correlations expected during close encounters between individual nucleons in the nucleus. The analysis of such experiments is not simple, but there is a firm impression that the strong two-body interactions are essential for understanding the results.

I have taken the story of the phenomenological exploration of the nuclear force well beyond the time in 1935 when Yukawa published his famous paper[40] on the meson theory of nuclear forces. There was some delay in the general recognition of the significance of this idea. The delay was due in part to the fact that not many physicists in the West had easy access to the Japanese journal in which the paper was published, and in part to the fact that it was regarded as very speculative to postulate yet another particle.

The climate of opinion changed rapidly when Anderson discovered the μ-meson, which was then mistaken for the particle playing the part postulated by Yukawa. There developed then a great industry of papers on meson theory and their application to the force problem.

Here, it seemed, we had at last found the simplicity we were looking for. The interaction was mediated by the meson field in precisely the way in which electromagnetic interactions were mediated by the electromagnetic field and its photons. It took a long time to find some very elementary difficulties. The coupling constant between the nucleons and the meson field had to be so strong that a treatment using the first order of perturbation theory is not applicable, different from the situation in electrodynamics, where the small value of the fine structure constant makes perturbation theory very rapidly convergent.

The earliest theories assumed a scalar meson, but Serber[41] discovered that this would give a repulsive force in ^1S states. Papers by Kemmer,[42] and by Fröhlich, Heitler, and Kemmer[43] in 1938, generalized the approach to include mesons with other spins and parities. Limiting themselves to the nonrelativistic approximation, they concluded that only vector mesons had a chance of giving the right kind of forces. In this analysis the pseudoscalar meson was missed, because it gives no effect in the nonrelativistic approximation, and all the resulting force then comes from relativistic effects.

A further, interesting difficulty arose because it seemed at first sight that meson theory could not easily account for charge independence. A charged meson would, in first approximation, give a force only between neutron and proton, since only a neutron could absorb the positive virtual meson which the proton could emit, and vice versa. To account for the n-n or p-p force, the existence of a neutral meson had to be postulated. But this would give an interaction between unlike as well as between like particles, and therefore the unlike particles would interact by two mechanisms, the like ones only by one. This made it seem very unlikely that their interactions would be the same.

This trouble was resolved by Kemmer, who had tried already in 1937[44] to use the concept of isotopic spin in the obsolete β-interaction theory of the nuclear forces, showing that this could give charge independence, although the magnitude of the resulting force was many orders of magnitude too weak. In his 1938 paper he applied the same idea to the meson field and showed that the interaction could be formulated in such a way that it would conserve isotopic spin, and hence the forces derived from it would evidently be charge independent. In simple terms, the neutral meson would give a force which, like the Coulomb force in the electromagnetic case, was of opposite sign for like and unlike particles. This disposed of the argument that the force between unlike particles should necessarily be stronger than the other.

Most versions of meson theory resulted also in a tensor force, but this was very singular at short distance, and would not allow any finite solution to the two-nucleon problem. A workable scheme was proposed by Møller and Rosenfeld[45] in 1940 and elaborated in a paper published in 1943. This assumed two types of mesons combined in such a way that the most singular part of the tensor force canceled, but a finite amount remained. This served as a pattern for many studies until the high energy scattering data showed the existence of a repulsive core and of other features not compatible with the scheme.

My account of the story of nuclear forces is far from complete, but I think I have covered, and indeed exceeded, the period that was to receive attention at this symposium.

Today it is still true that most of our quantitative information comes from the analysis of experimental data, mostly on the two-body system but supplemented and tested by the knowledge of heavier nuclei. However, as I tried to show, we face a picture of great complexity, and therefore it is very hard to extract from the data all details of the law of force. It is, in fact, a fairly common experience, on looking at the present state of the phenomenological analysis, to become sufficiently disappointed to conclude that we shall never reach a complete answer until the field theory of the forces is in better shape. Yet when the present state of the theory of the pion field and other fields is examined, with its strong coupling, its need for renormalization, and other problems, it still seems safer to rely on the phenomenological approach.

REFERENCES

1. J. Chadwick, "The Existence of a Neutron," *Proceedings of the Royal Society of London*, 136[A] (1932), 692-708.
2. Georges Fournier, "Sur la composition des noyaux atomiques," *Comptes rendus*, 194 (1932), 1482-1483.
3. Francis Perrin, "La constitution des noyaux atomiques et leur spin," *Comptes rendus*, 195 (1932), 236-237.
4. D. Iwanenko, "Sur la constitution des noyaux atomiques," *Comptes rendus*, 195 (1932), 439-441.
5. W. Heisenberg, "Über den Bau der Atomkerne. I," *Zeitschrift für Physik*, 77 (1932), 1-11; "II," *Ibid.*, 78 (1932), 156-164; "III," *Ibid.*, 80 (1932), 587-596.
6. Ettore Majorana, "Über die Kerntheorie," *Zeitschrift für Physik*, 82 (1933), 137-145.
7. E. Wigner, "Über die Streuung von Neutronen an Protonen," *Zeitschrift für Physik*, 83 (1933), 253-258.
8. L. H. Thomas, "The Interaction between a Neutron and a Proton and the Structure of H^3," *Physical Review*, 47 (1935), 903-909.
9. E. Wigner, "On the Mass Defect of Helium," *Physical Review*, 43 (1933), 252-257.
10. H. A. Bethe and R. Peierls, "Quantum Theory of the Diplon," *Proceedings of the Royal Society of London*, 148[A] (1935), 146-156.
11. J. Chadwick and M. Goldhaber, "The Nuclear Photoelectric Effect," *Proceedings of the Royal Society of London*, 151[A] (1935), 479-493.
12. K. Guggenheimer, "Remarques sur la constitution des noyaux. II," *Le journal de physique et le radium*, 5 (1934), 475-485.
13. Lloyd A. Young, "Note on the Interaction of Nuclear Particles," *Physical Review*, 48 (1935), 913-915.
14. Eugene Feenberg, "Neutron-Proton Interaction. Part I. The Binding Energies of the Hydrogen and Helium Isotopes," *Physical Review*, 47 (1935), 850-856; "Part II. The Scattering of Neutrons by Protons," *Ibid.*, 47 (1935), 857-859; Eugene Feenberg and Julian K. Knipp, "Internuclear Forces," *Physical Review*, 48 (1935), 906-912.
15. M. A. Tuve, N. P. Heydenburg, and L. R. Hafstad, "The Scattering of Protons by Protons," *Physical Review*, 50 (1936), 806-825.

16. G. Breit, E. U. Condon, and R. D. Present, "Theory of Scattering of Protons by Protons," *Physical Review*, 50 (1936), 825-845.
17. B. Cassen and E. U. Condon, "On Nuclear Forces," *Physical Review*, 50 (1936), 846-849.
18. G. Breit and E. Feenberg, "The Possibility of the Same Form of Specific Interaction for All Nuclear Particles," *Physical Review*, 50 (1936), 850-856.
19. S. Share, L. E. Hoisington, and G. Breit, "The Exponential and Meson Potentials" [Abstract], *Physical Review*, 55 (1939), 1130.
20. E. Wigner, "On the Consequences of the Symmetry of the Nuclear Hamiltonian on the Spectroscopy of Nuclei," *Physical Review*, 51 (1937), 106-119.
21. F. Hund, "Symmetrieeigenschaften der Kräfte in Atomkernen und Folgen für deren Zustände, insbesondere der Kerne bis zu sechzehn Teilchen," *Zeitschrift für Physik*, 105 (1937), 202-228.
22. D. R. Inglis and L. A. Young, "Stable Isobars," *Physical Review*, 51 (1937), 525-526.
23. C. F. v. Weizsäcker, "Zur Theorie der Kernmassen," *Zeitschrift für Physik*, 96 (1935), 431-458.
24. W. Heisenberg, "Die Struktur der leichten Atomkerne," *Zeitschrift für Physik*, 96 (1935), 473-484.
25. H. Euler, "Über die Art der Wechselwirkung in den schweren Atomkernen," *Zeitschrift für Physik*, 105 (1937), 553-575.
26. R. Huby, "Investigations on the Binding Energy of Heavy Nuclei," *Proceedings of the Physical Society*, 62[A] (1949), 62-71.
27. Victor Weisskopf, "Nuclear Models," *Science*, 113 (1951), 101-102.
28. J. M. B. Kellogg, I. I. Rabi, N. F. Ramsey, Jr., and J. R. Zacharias, "An Electrical Quadrupole Moment of the Deuteron," *Physical Review*, 55 (1939), 318-319; "An Electrical Quadrupole Moment of the Deuteron. The Radiofrequency Spectra of HD and D_2 Molecules in a Magnetic Field," *Physical Review*, 57 (1940), 677-695.
29. H. A. Bethe, "The Meson Theory of Nuclear Forces," *Physical Review*, 55 (1939), 1261-1263.
30. William Rarita and Julian Schwinger, "On the Neutron-Proton Interaction," *Physical Review*, 59 (1941), 436-452.
31. W. Hepner and R. Peierls, "Non-Central Forces in the Nuclear Two-Body Problem," *Proceedings of the Royal Society of London*, 181[A] (1942), 43-57.
32. G. M. Volkoff, "Tensor Forces and Heavy Nuclei," *Physical Review*, 62 (1942), 126-133; "A Note on Exchange Tensor Forces in Heavy Nuclei," *Ibid.*, 62 (1942), 134-136.
33. G. C. Wick, "Über die Wechselwirkung zwischen Neutronen und Protonen," *Zeitschrift für Physik*, 84 (1933), 799-800.
34. J. Hadley, E. Kelly, C. Leith, E. Segrè, C. Wiegand, and H. York, "Experiments on N-P Scattering with 90- and 40- Mev Neutrons," *Physical Review*, 75 (1949), 351-363.
35. Owen Chamberlain and Clyde Wiegand, "Proton-Proton Scattering at 340 Mev," *Physical Review*, 79 (1950), 81-85.
36. Robert Jastrow, "On Charge Independence and High Energy Scattering," *Physical Review*, 79 (1950), 389; "On the Nucleon-Nucleon Interaction," *Physical Review*, 81 (1951), 165-170.
37. Philip M. Morse, J. B. Fisk, and L. I. Schiff, "Collision of Neutron and Proton," *Physical Review*, 50 (1936), 748-754.

38. C. Critchfield and E. Teller, "On the Saturation of Nuclear Forces," *Physical Review*, 53 (1938), 812-818.
39. M. H. Johnson and E. Teller, "Classical Field Theory of Nuclear Forces," *Physical Review*, 98 (1955), 783-787.
40. H. Yukawa, "On the Interaction of Elementary Particles. I," *Proceedings of the Physico-Mathematical Society of Japan*, 17 (1935), 48-57.
41. R. Serber, "On the Dynaton Theory of Nuclear Forces" [Abstract], *Physical Review*, 53 (1938), 211.
42. N. Kemmer, "Quantum Theory of Einstein-Bose Particles and Nuclear Interaction," *Proceedings of the Royal Society of London*, 166[A] (1938), 127-153.
43. H. Fröhlich, W. Heitler, and N. Kemmer, "On the Nuclear Forces and the Magnetic Moments of the Neutron and the Proton," *Proceedings of the Royal Society of London*, 166[A] (1938), 154-177.
44. N. Kemmer, "Field Theory of Nuclear Interaction," *Physical Review*, 52 (1937), 906-910.
45. C. Møller and L. Rosenfeld, "On the Field Theory of Nuclear Forces," *Det kgl. danske videnskabernes selskab. Mathematisk-fysiske meddelelser*, 17, no. 8 (1940), 1-72; "Electromagnetic Properties of Nuclear Systems on Meson Theory," *Ibid.*, 20, no. 12 (1943), 1-66.

DISCUSSION

Wheeler: Looking back on the history of the meson theory of nuclear forces, if one had to think today of what the history would have looked like if we had never had the meson theory of nuclear forces, what would we have lost, as you see it from the perspective of today?

Peierls: I think different people may give different answers to that question. My answer would be that we would first of all be less confident about charge independence. That's one thing, although we might have accumulated enough evidence from isotopic spin conservation. Perhaps the most important point relates to the fitting of phase shifts to high energy nucleon-nucleon scattering experiments. Because of many ambiguities which would otherwise arise, it has been found possible to do this fitting sensibly only by using the device, first proposed I think by Breit, of taking it for granted that the long-distance behavior of the forces, and therefore the scattering in the high angular momentum states, is given by the one-pion approximation. I think that has been a vital step in the analysis. Without that, the analysis of high energy scattering would be very much more complex and difficult.

Bethe: I have a couple of questions and a couple of comments. First, a question: when did we become aware of charge symmetry? You mentioned that in Feenberg's papers it was necessary to assume like-particle forces. I remember that in the summer of 1936, in lectures at the summer school in Michigan, it was generally assumed—everybody assumed—that there were forces between like particles, and I believe in our *Reviews of Modern Physics* article we assumed like-particle forces and assumed charge symmetry. Am I right?

Peierls: With my incomplete recollection of what is said in your paper, I think that is so. I think the question of charge symmetry, and the fact that you couldn't explain the equality of neutron and proton numbers with neutron-neutron forces which are substantially different from proton-proton forces—I think that's there.

Bethe: That I am sure is there, and like-particle forces are there. I just don't remember who first talked about them. Because they surely didn't take Guggenheimer . . .

Serber: May I interrupt here? The mirror nuclei were first discussed in 1936 by Fowler, Delsasso, and Lauritsen. I think they gave a fairly complete explanation. Willy [Fowler] will no doubt remember this—that was 1936.

Bethe: But that was already too late, because our *Reviews of Modern Physics* article was written in 1935, since it had to be published in early 1936. So it must have been clear to us in 1935.

Peierls: I think that it was clear to very many people, but I cannot say who pointed it out first.

Wigner: I thought it was Heisenberg who claimed charge symmetry. You know, one doesn't read an article fully, and I read it as assuming charge symmetry. I did not realize he did not assume charge symmetry until a couple of weeks ago when I reread it.

Bethe: His first article certainly does not have charge symmetry; it has neutron-neutron forces . . .

Feenberg: Perhaps I could add something. In the summer of 1936, I was at the summer session in Ann Arbor, and I received a short letter from Breit in which he proposed charge independence; so we had some correspondence on that, and he wrote up a little paper which appeared in the *Physical Review*. But Bernard Cassen and Condon were also at the summer session, and one day Cassen accosted me and said he had something he wanted me to read. So I read the little paper that he and Condon had written, in which they proposed charge independence and expressed it in terms of the isospin variable, which made it a very elegant presentation. This was published in the *Physical Review* in the same issue in which the paper of Breit's and mine appeared, and the experimental work on proton-proton scattering and the theoretical analysis. I think the two proposals are essentially equivalent, but the Cassen and Condon method of expressing it was much more refined and sophisticated, and led to interesting consequences. They had it all there, but they did not invent the new quantum numbers which were implicit in charge independence. Wigner came along with that shortly, perhaps a few months later.

Bethe: But they did have isospin, didn't they?

Feenberg: Yes, they used the isospin formalism. As they used it, it was a convenience and not a source of physical ideas or results.

Bethe: This is exactly what I remember. Charge independence certainly was a very new and unexpected thing. On the other hand, I have the feeling that charge symmetry was assumed by all of us implicitly, without our ever saying that this is an interesting assumption. It is certainly in our article, but it must have arisen somewhere. It was in the air.

Wigner: Yes. As I said, I thought that Heisenberg had it . . .

Bethe: That was certainly not true.

Wigner: I misread Heisenberg in a most favorable way.

Bethe: Do you [Fowler] have something to add on this subject?

Fowler: Only to say to you what I said to you before. I concede, without a doubt, that we knew, we implicitly believed in charge symmetry. But as Bob [Serber] says, as far as I have been able to ascertain by looking in the experimental record, the regular increase with charge Z of the β-decay energies in the mirror nuclei was first noted in the thesis of a person whom in all modesty I cannot mention except to say that his first initial was W. In the paper I wrote with Delsasso and Lauritsen in the *Physical Review* of 1936, we, as far as I can tell, first made the explicit statement, because it could be calculated that the general increase was due to the Coulomb force taking into account the mass difference of the neutron and proton, that this was an experimental proof of the charge symmetry. And, as far as I know, there was no other experiment before that, but I agree with you that it was implicit in our thinking that charge symmetry was there.

Bethe: My second subject is the Euler paper. I don't know whether anybody has done a modern type of nuclear matter calculation with Euler's forces, and only if such calculations were done would I agree with the statement that Euler's force did not give attraction beyond the α-particle. Do you know whether there is such a . . .?

Peierls: No, I don't. I believe the ingredients for it exist in a paper you wrote which was the first application of the Bethe-Goldstone equation to forces that were well behaved, so that one might just as well have applied ordinary perturbation theory. Now, what forces precisely were assumed there and what the answers were, I do not remember. But inserting appropriate numbers in that calculation would just about do it. But I don't know of any calculation . . .

Bethe: I think it is not clear whether the Euler paper failed to give enough binding energy because of an insufficiently good method or because the forces were not strong enough.

Peierls: I don't quite agree there, because I think we now realize that the convergence of such a method should be looked at in terms of the correction, not to the binding energy, which is a small difference, but to the total potential energy, and in that sense, the next correction in Euler's case is remarkably small.

Bethe: Yes. Well, I think one really should do this. Then, a small comment on the many-body forces. Gerry Brown studied this to some extent on the basis of the theory of T. D. Lee with Σ-forces. It becomes very messy; mostly, the many-body forces become far too big. There was always the danger that they are attractive, in which case you get into worse trouble, and the only thing that Gerry Brown could conclude is that the parameter used in T. D. Lee's papers cannot be correct. But I think that any such field theory as T. D. Lee's Σ-theory leads to the danger of too big many-body forces, and many-body forces very likely will give more attraction and less saturation than will just two-body forces.

Peierls: Could you not visualize many-body forces that are predominantly repulsive?

Bethe: I could visualize them, but it is hard to get them.

Wigner: My first question is closely related to Professor Bethe's question, although it is not about many-body forces. Is it at all conceivable that the coherence of the nuclei is best described not in terms of forces but in terms of a picture similar to that which obtains in metals, for instance, so that there is a cloud of mesons; and that nuclei move in this cloud, and that two-particle or three-particle forces are not the most adequate description? I don't know what the situation is, but perhaps you know something about this.

Peierls: No I don't. I have only an impression, which is that this is the kind of picture exemplified, for example, by the Johnson and Teller approach, so it doesn't have to be precise.

Wigner: By whom?

Peierls: Johnson and Teller.

Wigner: I did not know that.

Peierls: I think that what you described is essentially their picture. My impression, very crudely speaking, is that the evidence from various high energy reactions involving either fast protons or pions, the evidence on the momentum distribution of the nucleons (and nucleons have a very high momentum if they happen to be close to another one), would be hard to explain on a picture where the individual force between two nucleons approaching closely was very much less than it is on the conventional picture. Now, that's only an impression, and I don't think any attempt has been made to see whether these two things can be reconciled.

Bethe: There is a further calculation on this picture by Walecka, with others. He tried this with the view particularly to very high density nuclear matter, such as occurs in neutron stars. The result is essentially that you cannot fit parameters which will do justice to all of the limiting cases. In particular, one is likely to get insufficient repulsion at high density, and insufficient attraction at low density, if one takes this continuum assumption. One can get a fairly good description, but never fit all density regimes well enough.

Wigner: How do you have different densities in stars?

Bethe: In neutron stars versus ordinary nuclear matter. If you fit the parameters so that ordinary nuclear matter is sufficiently bound and has the right density, neutron stars will collapse.

Wigner: Thank you very much, and that is a paper of Walecka and some other people?

Bethe: Correct.

Wigner: Thank you very much. May I ask my second question? This is not so closely related to what you have told us about, but it is a question in which I am very interested, namely, what do you think about the concept "elementary particle"?

Peierls: I think it is a very good question, but I don't think I know an answer to it. I'm not sure there is an answer, or it may well be that "elementary particle" is a limiting case. I would think that the description of a neutron or a proton or electron as "elementary particle," in spite of the limited lifetime of the neutron, is probably a very sensible idea, but it's only a limiting case, and other particles deviate more and more from that limit. That's, however, not a clear answer to your question.

Wigner: But it represents what I believe also.

Serber: I would like to give a much more definite answer. It may not be more correct, but it's at least more definite. That is, I think the experience of very high momentum transfer reactions on Bob's [Wilson's] machine and SLAC give an indication that, as far as we can see, point field theory is correct. Momentum transfers occur, for instance, in electron scattering, with $1/q^2$ as far as you can go. If point field theory is correct (presumably there's a Lagrangian field theory), the particles (more strictly, the fields) that interact at a point are the elementary ones. I mean, they are certainly not the protons and neutrons, but let's say the partons, to use a general term.

Wigner: Let me quickly say one more thing. If the neutron is an elementary particle, the state of a neutron is perpendicular to all states of proton-electron-neutrino. Now, whether this is correct is of course an experimental question. It is terribly difficult to demonstrate directly, but it seems to be a meaningful question. Do you think the neutron is an elementary particle?

Serber: Oh, no, certainly not.

Wigner: No? Then I misunderstood you.

Serber: No, I said that it's clear that the nucleons are made up of something more elementary, from the really high energy, high momentum transfer experiments.

Wigner: Thank you very much. I misunderstood.

Segrè: I would take a very pragmatic attitude to this question, which has been asked hundreds of times, namely, whenever one is asked what is an "elementary particle," I would reply with another question: What means do you permit me to use? I think in a fairly true sense that if you are permitted to use only chemical means, the elementary particles are the 92 elements. If you are only operating at low energies, the neutron and proton, and if you wish the electron, are the elementary particles. And so on and so forth. So the question is really not well defined unless you specify the means you use. Otherwise you go into very vague speculations.

Wilson: I would like to add what Abdus Salam said, which I believe is consistent with what has been said here: "All particles are elementary, but some particles are more elementary than others!" To change the subject, it seems to me that perhaps implicit has been some criticism of how the field has developed. I would like to ask for some critical remarks, perhaps from you Rudi [Peierls], about your view of the way things might have developed had there been much less theoretical activity — a few phenomenologists such as Breit and perhaps Hans [Bethe] writing a review article from time to time. Then had the experiments just proceeded as they all had, all of what we know of the nucleus would have been made clear with much less fuss and little difference in the end. Could you comment on that?

Peierls: I will try. I'm afraid I'm probably biased by the fact that I happen to be a theoretician. I think it is probably true that in this field I was reviewing there were rather more theoretical papers than was absolutely necessary. However, this field does not hold by any means the record in that respect, and I can give very specific examples. Now, of course, one can criticize many of the prejudices we had, and the wrong starts, but I think that the most important thing that one can ask of a theory is that it should not impede the progress of the experiments, and second, that it should occasionally stimulate experimental questions that lead to informative and profitable experiments. Now, I think nuclear theory and the theory of nuclear forces certainly in most cases have satisfied the first criterion. I think it may be that things like the exaggeration of the relevance of the liquid drop model of Niels Bohr has retarded certain types of approaches, maybe experimental approaches also, but in general I think the viscosity created by that has been small. And I think that one can quote quite a few examples where in fact the experimental research was stimulated by these theoretical ideas, inadequate as they undoubtedly were at the time.

Swenson: I'd like to ask a more general question that I have encountered over the years in trying to do a little teaching to spectators on the side. It has to do with the question of nomenclature and philosophy: the question of "strong" and "weak" nuclear "forces" versus nuclear "interactions." Could you say something about how these two words, "forces" and "interactions," were used in your early years, and how the problem of defining the usage of the words "forces" and "interactions" came to their present state of usage?

Peierls: I am afraid that I will not be able to cover all relevant things in this quick answer, but I would say, approximately, that "force" and "interaction" are used interchangeably, with two exceptions: First, one is reluctant to use the word "force" when one means "interaction potential," particularly for those who were brought up on classical physics, where there is an inverse square law "force" and an inverse distance law "interaction potential." Therefore, people are sometimes reluctant to use the word "force" where they mean an energy or potential. Second, "interaction" has two uses, one is in the terms "interaction potential" and "interaction energy," the other is when one talks about the "interaction" of a neutron with a nucleus, meaning a collision or reaction. If one wants to avoid that misunderstanding, one must go back to the term "force." But other people may have different things to say.

I think the distinction between "weak" and "strong" interactions was invented by the elementary particle physicists: "weak" interactions are those responsible for the β-decay and other interactions discovered later of about that order of magnitude—one now means forces violating a certain number of conservation laws; "strong" interactions are the interactions between pions, nucleons, baryons, and other particles of that kind, with electromagnetic interactions being somewhere in between.

Fowler: I wish he were here to say it himself, but one of my distinguished colleagues, Murray Gell-Mann, has frequently taken a very skeptical view of the whole picture of protons and neutrons as nuclear consituents, as nucleons, and in fact he rather teases those of us who still work in classical nuclear physics about our "primitive" beliefs that the nuclei we play with consist of neutrons and protons. For example, he said, and I don't think I'm misquoting him, "Well, the deuteron itself only looks like a proton and a neutron maybe half the time." And then I understand there recently have been attempts to build a deuteron on a six-quark model. Is that right? So I guess my question would be: In the long run, how fundamental is this development that we've been talking about, the nuclear forces between nucleons, going to be?

Peierls: I'm glad that you raised that point, because I think we all agree that if you want to describe things with sufficient accuracy, the presence of quark constituents, if any, the presence of virtual mesons, and whatever else will have to be brought in. Therefore, in that sense, the nuclear forces as we have discussed them this afternoon are only an approximation. Perhaps I was at fault in not making that reservation at the very beginning of my talk. I think at the same time we hope, and it is only a hope, that there is a range of energies and a range of accuracies in which the use of the idea of nuclear interaction potentials is a sensible approximation and is profitable and permissible.

Feshbach: Can I add to that a little bit? I think you made this point in the course of your talk. If one asks what is the nuclear interaction when the nucleons are far apart, one has very little difficulty. Difficulties increase at smaller distances, smaller separations, between the nucleons. I think the quark hypothesis that Professor Fowler mentioned will become very important at very short distances between the nucleons, but when nucleons are further apart I think our present methods are quite adequate.

Bethe: I want to reemphasize what Segrè said, namely, it depends entirely on the means used to investigate the nucleus, or matter in general. Depending on the energy, one picture or another is the most useful. Surely in the case of the deuteron if we want to know its size and its wave function at large distances, the neutron-proton picture is the only one that makes any sense; if we want to know how the deuteron behaves when interacting with an electron of 100 Gev, it's a different matter.

Peierls: But it also depends on the accuracy one means. If one wants to calculate even the binding energy of the deuteron by dead reckoning to one part in 10^6 or something, all sorts of things not implicit in the neutron-proton picture will have to be allowed for.

Some Men and Moments in the History of Nuclear Physics: The Interplay of Colleagues and Motivations

John Archibald Wheeler

INTRODUCTION: R. R. Wilson

We come to the final session of this delightful and instructive and nostalgic voyage through time and space, as Willy Fowler has called this symposium. It is entirely appropriate that it has been held at the University of Minnesota, for Minnesota is one of the strong centers of physics and of nuclear physics that has been the focus of this meeting. It has been a personal pleasure for me to have the opportunity to visit this university about which I have heard so much from my many friends who have worked at Minnesota. I think one thing that has gone unremarked has been the *Physical Review*, which seemed to me to reside here because this was where Professor Tate was. When you submitted a manuscript, I recall that you would get a postcard back from Madeline Mitchell, and I think that probably, in some historic way, Tate did have something to do with the excellence of the *Physical Review* and thereby imposed some discipline on American physics, which probably did make some difference to at least American physics. I also believe that it has been an inspiration to have this meeting between the physicists who carried out much of the early work, and the emminent historians of science who are assembled here.

It's my privilege and pleasure and honor to introduce Professor John Archibald Wheeler—Johnny. There are many prosaic things I could say about Johnny. I first knew him when I went to Princeton as an instructor in 1940. He was an ancient by then, from my point of view, for he had been an assistant professor for all of two years. At Princeton, it was enough to have to put up with the overwhelming and relentless politeness and gentility of Professor Wigner, but to add to that the gentle manners of this perfect southern gentleman, John Wheeler, was almost too much to bear. It was made possible only because of Pauli's presence there at the same time and because of his somewhat abrasive personality. However, behind that gentlemanly facade of John Wheeler there resided—a perfect gentleman. And behind that facade there was—another perfect gentleman. He was a gentleman through and through.

However, somewhere among those polite facades there was a tiger loose; a reckless buccaneer, a bold adventurer, a fearless and intrepid explorer, a man who had the courage to look at any crazy problem. And somehow, after he had thought about that crazy problem, things were less crazy than when he started. This adventuresome thinking was done in the context of a great amount of extremely solid work. However, he did think about advanced potentials, gravitational collapse, geons, neutron stars, black holes, gravitational waves—all sorts of way-out things like that—things that I believe nowadays are the subject of serious books and serious thought; things that are almost commonplace. And of course at the same time he has played a prominent role and made important and solid contributions to the nuclear physics which we are talking about today. His subject is "Some Men and Moments in the History of Nuclear Physics"—he knew the men, he was present at the moments. No one is better qualified to discuss this subject, and I am waiting eagerly to hear what he has to say. Professor Wheeler.

Some Men and Moments in the History of Nuclear Physics: The Interplay of Colleagues and Motivations

John Archibald Wheeler

Thank you, Bob, and my appreciation also to so many wonderful colleagues for staying indoors this beautiful day when they could just as well be out-of-doors or on their way home. At the April 26–28, 1934, meeting of the American Physical Society in Washington, one of the many ten-minute papers[1] was given by a National Research Council Fellow who had got his Ph.D. with Karl Herzfeld in Baltimore the year before, his quantum mechanics thesis[2] dealing with the scattering and absorption of light by the helium atom and the connection between these two processes. At the time he gave his report he was finishing a year of postdoctoral work in nuclear physics with Professor Gregory Breit in New York. A few months later he would be taking a fifty-five dollar freighter trip across the Atlantic for a second fellowship year with Niels Bohr in Copenhagen. The title of the Washington report was "Interaction between Alpha Particles." Let me read the abstract in its entirety.

From measurements by Rutherford and Chadwick on the anomalous scattering of alpha-particles by helium, H. M. Taylor deduced a law of force between alpha-particles strongly attractive at small distances. More recent experiments do not agree well with Taylor's original interpretation. The present

Nuclear Physics in Retrospect, edited by Roger H. Stuewer, copyright ©1979 by the University of Minnesota.

work has shown that it is not sufficient to take account only of the wave function of zero angular momentum. A method has been developed by which the distribution in angle of the scattered particles determines the Faxen and Holtsmark phase shifts for angular momenta L = 0 and 2. The resultant curve of phase shift plotted against velocity is very different from Taylor's and indicates that no modification of the coulomb field at distances less than 5×10^{-13} cm can account for the anomalous scattering. Departures from inverse square forces must then extend to unexpectedly large distances or the concept of a potential barrier must be considered inadequate for treating the scattering of complex particles. This conclusion cannot be considered final until more complete scattering experiments have been made.

The "plan" of nuclear physics as one could envisage it in those days, though not obvious from this abstract, was in the oral report a central theme: Experiments give scattering, scattering gives phase shifts, and phase shifts give law of force. The law of force between individual nucleons would come first. Then the interactions between more complex nuclei would be worked out from measurements of scattering, or from many-body calculations based on the law of coupling between the elementary particles, or both.

For this program of "interaction from scattering" to succeed it was essential to find a workable method to deduce phase shifts from scattering. The report described such a method and applied it to a case—a-a-scattering—where the data were as good as any at that time available.

What a change of goals and outlook that participant in nuclear physics was to experience in the next five years! On the Washington program of April 27-29, 1939, a second ten-minute paper by him appeared, written jointly with Niels Bohr and entitled "Mechanism of Nuclear Fission."[3] This oral report and the final published paper[4] developed a simple picture of the mechanism of nuclear fission. It also predicted rates of spontaneous transformation and cross sections for fission and for competing reactions. Yet not one reference did it make to any detail of the law of force between nucleon and nucleon, nor to any hope, desire, or need to find that law.

The nucleus is not celestial dynamics, it is condensed matter: that was the change in view. What experiences impelled the young worker to that shift? How did his view of physics in its larger sweep alter in the same 1934–39 period? And how did he get into physics in the first place? As that question touches on his university years, some words about 1927–33 may make an appropriate prelude to an account of 1934–39.

Johns Hopkins

What was the state of physics in America in the 1930s as seen through the eyes of a student at Johns Hopkins University? Hopkins, from the day it went into action, October 3, 1876, had been a pioneer in this country as a research-oriented university. The spirit of the great founding president, Daniel Coit Gilman, lived again in the handshake and voice of Joseph Sweetman Ames, president in my time, early authority in physics and dominant creative member of the National Advisory Committee on Aeronautics. This freshman, meeting him by chance on the campus, got a friendly inquiry about how I found engineering—and this encouragement, "No matter what it is, everybody who is anybody at Hopkins goes to work on something that deeply interests him."

Counted more important than any class or lecture were the laboratory, the library, and the seminar. Reports by the students themselves often made the backbone of a course. Freed of the traditional regimentation of seats, we sat around a long table like board members, interrupting, participating, discussing. Such seminar courses contributed to an atmosphere of freedom, a sense of working together in the search for understanding, a feeling of colleagueship between students and professors. The very word "professor," which at many other places had for its number one meaning "teach," at Hopkins meant "profess": make a profession of allegiance to the great issues of one's subject.

Henry Augustus Rowland, the greatest experimentalist in the United States in his day, had died in 1901, but his ruling engine, his gratings, and his Hopkins successors, Robert W. Wood, A. H. Pfund, and Gerhard Dieke, kept the tradition of spectroscopy very much alive. Wood, famous for his research on resonance radiation, for his work on every aspect of spectroscopy, and for his pioneer book *Physical Optics*,[5] gloried in not having known enough advanced mathematics to get a Ph.D. degree. His loss was our gain.

Not in mathematics but in clear words he gave us the story of diffraction, the theory of the concave grating, and the machinery of resonance radiation. He provided diagrams, order-of-magnitude estimates, and, above all, the living reality of the experiment itself, put through its paces and made to do tricks, explained tricks, before our eyes. Physics was physics, not pencil and paper exercises taken out of a book. Wood's puckish and sometimes sardonic sense of humor is too well known[6] to bear recounting here. It ranged from his best seller, *How to Tell the Birds from the Flowers and Other Woodcuts*,[7] to the unmasking of many a medium and hoax, including Blondlot and the N-rays;[8] and from the greening of Old Faithful Geyser to persuading a student to drink—or himself drinking—a little liquid air.

It is difficult to recall a single lecture by Wood that did not contain some surprise designed to make fun of pomposity in physics. The circulation theorem? A line of vorticity can never come to an end within a fluid? Nothing better illustrates the theorem than a smoke ring? Then, with words about people who pontificate to this effect, he proceeded to put together a smoke-ring generator before our eyes and get out of it half a ring. The white semicircle traveled a meter or more before dissipating. How he did it was up to us to figure out. We finally realized that he had inserted a partition within the generator to separate air above from smoke below.

On Wood's behalf Gerhard Dieke—with whom I was then setting up a discharge tube—asked me one day if I would contribute to the next edition of *Physical Optics* some diagrams of the motion of an electron in a hydrogen atom under the influence of electric and magnetic fields. Fascinating as I rate the project today, at that time I begged off, attitude typical, I am ashamed to say, of one—and possibly more than one—graduate student with too much hidden pride to be willing to do anything unless it might lead to something new.

In no one did the joy of experimenting bubble up more happily than in A. H. Pfund, cigar smoker, teller of stories about men of physics past and present, busy always with putting together equipment and making measurements. "Read Rayleigh" was each time his advice to me, whatever the subject of discussion, and I have learned to repeat his words. To have Pfund show one in an advanced laboratory the channeled spectrum or how to silver a mirror or how to see the fine structure of the lines of the mercury arc was ever afterward to think of optics as one of the most attractive of all the attractive fields of physics.

These and other riches of physics at Hopkins were yet to open out to me. I had come from the Rayen High School in Youngstown, Ohio. There I had had the same teachers in English, Latin, and especially mathematics (Miss Doerschuk and Miss Lida F. Baldwin) that my sometime Princeton physics colleague, M. L. Goldberger, was later to have. I had entered Johns Hopkins in September 1927 as a freshman in the School of Engineering. I was thinking of a practical profession. I was inspired by a schoolboy love for mechanisms:[9] from adding machines, combination safes, and automatic pistols, their gears and parts whittled from wood, to a wood-turning lathe built from an old washing machine motor, a Morse-code telegraph to my friend Rushworth Steckel half a block away, and a crystal radio to hear that pioneer broadcasting station KDKA in Pittsburgh, 100 kilometers away. I spent the summer between freshman and sophomore years rewinding electric motors used to pump water out of the deep workings of the Pittsburgh Veta Grande Mining Co. silver mine at Zacatecas, Mexico.[10]

Physics had already begun to fascinate me. The appeal of the subject came in part from good books. Among them were J. Arthur Thomson's *Outline of Science*,[11] Charles P. Steinmetz's books on dynamos, motors, mathematics[12] and on relativity,[13] and especially H. A. Lorentz's wonderful *Problems of Modern Physics*,[14] where gravitation theory and the quantum were central topics of attention, as they are today. In part the appeal came from a visit, before my university days, to the National Bureau of Standards laboratory of Paul R. Heyl, then making that determination of the Newtonian constant of gravitation[15] destined to stand supreme for years.[16] He showed me the moving masses, quartz fiber, mirror and scale, and explained how resonance increased the sensitivity of his measurement. Taking me from his laboratory to his office, he operated for me his analyzer of the tides, built from sliding centers of rotation, pulleys turning on those centers, and a cord winding around them, that ever afterward gave special charm both to the theory of the tides and to Fourier analysis.[17] Partly a sense of the centrality of physics derived from my Saturday-night experiences manning the central Science and Technology Department of the Enoch Pratt Library, Baltimore's public library. There I had to dig out information for patrons on everything from the size of Jupiter to the principles of design of a heat exchanger and from the frictional wear of steel to the proper adhesive for bakelite. I had been instructed to secure for every member of the public the best available thinking relevant to his needs, whatever the topic and whatever the effort it took.

At Johns Hopkins, too, the libraries of engineering, mathematics, and physics were one and the same. As I was looking up there the properties of an alloy steel to calculate the strength of an I-beam for a course in the strength of materials, or working out with trigonometry how far out from closure my partner and I had been in our afternoon of surveying a circuit around the campus, I would find myself tempted by the latest issue of the *Zeitschrift für Physik* —then the great journal of physics—and the mysteries of electron orbits, angular-momentum coupling diagrams, and half- and quarter-quantum numbers. It did not help in resisting such temptations to walk and talk with Robert T. K. Murray, my classmate and closest friend, who had a magic pipeline to everything new and a contagious enthusiasm for physics.

Our section man in chemistry, H. M. Smallwood, gave fortunate freshman a broader survey of the rush of new ideas in physics than any conventional course in physics itself then afforded. Our section man in physics, R. Bowling Barnes, I found one day in the waiting room of an optometrist's office. He explained to me that with atoms one was now coming to the point where one could do the mechanics for them, quantum though it was, in much the same comprehensive way in which the solar system had been treated in earlier times. To me that quarter of an hour was an inspiring glimpse into the possibilities that might lie ahead: to bring everything one lays one's eyes upon into the world of the understandable.

The so-called New Plan at Johns Hopkins allowed those who so desired to move gradually from undergraduate work to graduate work with no sharp transition, a "nonstop flight" in six years from entering freshman to Ph.D., with no intermediate degree. I took advantage of this opportunity in the latter part of my sophomore

Some Men and Moments in Nuclear Physics   223

year and switched from engineering to physics, continuing, however, an interest as well in English and history as in French and German.

Mathematics made a happy tie between the old concentration and the new. I never knew a greater illuminator of the methods of mathematical physics than Francis D. Murnaghan, gift of Dublin to Baltimore, charming, intense, distinguished, onetime president of the American Mathematical Society, author of *Vector Analysis and the Theory of Relativity*,[18] of *Hydrodynamics* with Dryden and Bateman,[19] of *Theoretical Mechanics* with Joseph S. Ames,[20] and of eight other books. He was interested in applications as well as principles. One intending Baltimore brewer, preparing for the 1933 repeal of prohibition, came to Murnaghan and on Murnaghan's advice to me. What was the right design for the heat exchanger he was going to need? Working out the answer proved good practice for the days in 1942 when du Pont and Chicago colleagues and I would be occupied with the cooling of the first high-level nuclear reactors. A treatment of electromagnetism that Murnaghan had mentioned to me he worked into one of his courses at my request. Capitalizing on Cartan's principle that "the boundary of a boundary is zero," it was later to find itself expanded and expounded in one chapter of a book with which I was associated.[21] Abraham Cohen introduced us to Goursat;[22] John R. Williamson, to some of the deeper sophistications of matrices, determinants, and invariants;[23] and Aurel Wintner, to functional analysis, the problem of three bodies, ergodic theory, and his *Spectral Theory of Infinite Matrices*.[24]

I began my closer identification with graduate work in physics at a moment when the faculty of the department had come up with a wonderful new idea. Each student was to spend a month or more working with a professor of experimental physics in the department, then a month with another, and so on, until one had a handful of arrows in one's quiver. Following work with John C. Hubbard on acoustics, and with Gerhard Dieke on getting a discharge tube to work, I spent some time with Joyce A. Bearden in his X-ray laboratory. He was then doing his work on the determination of the charge of the electron, or the standard X-ray wavelengths, from the dispersion of electrons,[25] an interesting project with which I soon found myself also associated on the theory end.[26] This was not the total of my happy association with spectroscopy. For several summers I worked as a student assistant to William F. Meggers at the National Bureau of Standards,[27] at $40 a month. Meggers was helpful to a greenhorn both in the photographic darkroom and at one or another of those instruments for measuring line positions where I—with C. C. Kiess, W. J. Humphreys, and B. F. Scribner—spent so many hours. He told me stories of the great men in the field—Kayser and Runge—introduced me to what was then the most influential book in spectroscopy, Friedrich Hund's *Linienspektren*,[28] and invited me to collaborate with him on a paper on the band spectra of three diatomic molecules.[29]

The photographic darkroom was not the darkest room I met in these welcome associations with older experimental colleagues. Sit in a room with no light at all, I was instructed by the newcomer. Wait until the pupil becomes fully dilated. Then with a simple eyepiece look at a zinc sulphide screen. Count the flashes. In this way determine how many α-particles are arriving per minute from a radon source right behind the screen. See how this number varies from hour to hour and from day to day—and how it depends upon the distance from source to screen. This and more was my introduction to nuclear physics from preposterously young Norman Feather, freshly arrived from Cambridge's Cavendish Laboratory September 1929. The Johns Hopkins physics department had seen the handwriting on the wall. They had charged Robert Wood to recruit for the faculty a promising nuclear physicist on his next trip to a meeting of the Royal Society. I was a direct beneficiary, not only in practical experience, but also in two courses I took with Feather. One, "Radioactivity," was based on the great book of Rutherford, Chadwick, and Ellis.[30] The other took for text that 1903 classic *Conduction of Electricity through Gases*,[31] described by Lord Rayleigh as a summary of "the work of Thomson's great days at the Cavendish laboratory," work which had won "J. J." the 1906 Nobel prize and which was a beautiful combination of simple experiment and simple theory. Would that one of our colleagues of today might follow this model and make an even more delectable, and more timely, book out of the path-breaking 1934–38 experiments of Fermi and his group on slow-neutron physics.[32]

Quantum mechanics? That one learned in those days mainly on one's own—I from Hermann Weyl's deep and panoramic *Gruppentheorie and Quantenmechanik*,[33] a book for me filled with deep happy mysteries, studied day after day in the summer sunshine in a chair-with-lap-board set out in a Vermont cow-pasture. Where better than there could one also work out all the problems in Lewis and Randall's *Thermodynamics*[34] and read Slosson's *Creative Chemistry*?[35] Or come to appreciate from Compton and Allison,[36] and from Ruark and Urey,[37] the great sweep of "the elementary processes of atomic physics," enjoy John Livingston Lowes's *The Road to Xanadu*,[38] and be thrilled by the great story of Hutton,[39] Lyell,[40] and Darwin?[41] But the strongest impact came from participation in academic-year seminars, among them the seminar on X-rays that Bearden presided over, and on new developments in quantum mechanics guided jointly by Karl Herzfeld and Maria Mayer. Of the ten or so topics and papers I reported on during my graduate years, the one most pregnant with the future was the 1930 discovery by Bothe and Becker of a highly penetrating radiation emanating from beryllium bombarded with α-particles.[42] This finding, puzzling to me and to us all, was left puzzling and even more challenging by the observation of Joliot and Curie that this "beryllium radiation"—which they interpreted as γ-radiation—knocks high-speed protons out of hydrogen-containing materials.[43] When the next step came, and Chadwick discovered the neutron,[44] the excitement was great among students and faculty alike in our laboratory. Are there any neutrons in the earth's atmosphere? Can neutrons be liquefied?[45] Is a neutron radioactive? These are a few of the questions that Robert T. K. Murray and I and fellow-students found ourselves discussing.

Maria Goeppert-Mayer often, and from time to time her husband Joseph Mayer, crossed the lawn from chemistry to physics to participate in a seminar or consult Herzfeld. The higher administration had denied her any proper position. She did not teach. She nevertheless taught as effectively as anyone—by her questions and comments in seminars and colloquia. Her concern for whatever the issue was made her an example to us all. It is impossible to forget her quiet firmness, the rising emphasis at the end of her last sentence, and her inability to leave a problem until it was clarified.

Karl Herzfeld was for me the great attraction of the department, by reason of his breadth of knowledge and his interest in the great problems. Already in my second undergraduate year I was sitting in in his graduate course. In whatever year, and in whatever field, he began with a wonderful opening survey of physics. He used it to frame and set into perspective the central features of his course of lectures, whether they were to deal with Brownian motion, or kinetic theory, or electricity and magnetism, or hydro- and aerodynamics, or the dielectric properties of matter. Physics for him was not a secular but a religious calling; it aimed, in his view, to make clear the structure and beauty of God's creation. The great men of physics were for him great men, and he spoke of them as great men. To queries he replied with humility and honesty. "How stupid I am," he would often exclaim; and often ask, "Is not my question completely without meaning?" Great piles of unread manuscripts and unanswered letters littered his worktable, the sign not of too little conscience but of too much—because every letter of commentary he wrote, I later learned, was for him a major judgmental undertaking.

Herzfeld was author of treatises on kinetic theory,[46] on the properties of condensed matter,[47] and on dispersion theory;[48] and was later to occupy himself with electronic orbitals in polyatomic molecules and—with T. A. Litovitz—to write a book on the dispersion and absorption of ultrasonic waves.[49] Neither in discussion nor, later, in studying his bibliography did I ever find an area of physics in which he did not have interesting comments and long-standing questions. I am grateful that he suggested to me a thesis topic connected with dispersion theory[50] because that topic binds together so much so meaningfully.

Physics—who cared about physics, or knew what it was? A depression was on. A neighbor down the street killed himself because his business had failed and he saw no way of supporting his wife and children. Men, fine looking men, would ring the doorbell and ask for work, any work at all, no matter how menial, to help their families pull through. How many of those who came to the Washington meeting of the American Physical Society looked like lost souls, lost except for their love of science—"I'm a Nobody. I come from Nowhere, where the Nobodies live on their Nothing a Day."[51]

Of the handful of those who received their Ph.D.'s at about the same time as I, Lawrence Hafstad would have liked to have a teaching position, but none was available. He took a minor laboratory job at the Carnegie Institution's Department of Terrestrial Magnetism. He went on to make a distinguished name for himself in nuclear physics and later become director of research for the General Motors Corporation, an organization with a research-plus-development budget of over a billion dollars a year. Allen Astin (Ph.D., New York University, 1928), a National Research Council Fellow at Johns Hopkins, also got no teaching job and took a minor laboratory position at the National Bureau of Standards, rising to become its director and build it into an organization of more than five thousand people. R. Bowling Barnes, who had received his Ph.D. in physics at Johns Hopkins and had been a National Research Council Fellow at Berlin and Breslau and then my freshman physics instructor at Johns Hopkins, had no long-term teaching post. He joined the American Cyanamid Company, built research there, and went on to organize his own successful Barnes Engineering Company. John W. Mauchly (Ph.D., Johns Hopkins, 1932) did get a teaching position—at Ursinus College, near the railroad tracks in the outskirts of Philadelphia. There, with J. P. Eckert, Jr., he pioneered a large-scale electronic integrator and computer and invented Eniac, Edvac, Univac, and Binac electronic computers, foundation of today's multimillion-dollar position of the Sperry Rand Corporation in the computer industry. James D. Hardy (Ph.D., Johns Hopkins, 1930), deeply in love with the infrared, liked to show one his beautifully simple wire-wound grating for sorting out the infrared spectrum; but he found that absolutely nobody, either in industry, university, or government, was about to fund any work in infrared. After months of despair he suddenly realized that the Johns Hopkins Hospital exists, that people are sick in bad times as in good, and that by infrared it was possible to study the body conditions of patients. He ended up as professor of physiology at the Yale School of Medicine and director of a great Yale laboratory for the physical analysis of physiological problems.

What examples these careers were of the "Great Pump," the alternation of bad times and good, of ice age and tropical conditions, that drives animals and man to fill every known ecological niche and make their way into new ones.[52] Of all levers that workers in the field of physics have for prying themselves into new fields of usefulness, none—I came to realize—has ever surpassed physics-based instruments. They reach from medicine to communication, from biology to technology, and from the great to the small. It is no wonder that the number of kinds of such instruments rose from 1200 in 1963 to 2400 in 1976.[53]

How a truly new kind of instrument develops was known to all Hopkins students who followed the progress of their colleagues' theses; who piled into a car to go to the Atlantic City meeting and hear Arthur Compton and Robert Millikan report—and debate—their latest measurements; and who could picture thirty-two-year-old Albert Michelson sitting in the front row in Lord Kelvin's 1884 Johns Hopkins lectures,[54] inspired there by Kelvin's questions to take up anew, and by better means, the search for an ether drift: push the possibilities to the limit!

For a view of one American university at the beginning of the nuclear age it is good to have a glimpse of theses and jobs as well as traditions and professors, books and students, and seminars and experiments; and perhaps also good to end with a fleeting look at two of the many visitors who helped give the feeling that physics, wherever done, is one.

It was the custom to assign to the graduate students, in turn, the responsibility to meet the visitor at the train and look after him during his visit. It fell to me to be the student host for Peter Debye, the same Debye who had put Erwin Schrödinger on the track of the wave equation[55] and who had done so much on polar molecules, electrolytes, specific heats, and scattering theory. Never before had I met a deep thinker who so resembled a well-dressed efficient business executive. Neither had I encountered until then a person of such wide acquaintance with people and laboratories, university and industrial, American and European, nor one with so witty a way of putting things. He had just visited a university blessed with a new physics building. After being conducted around the gleaming laboratories he had asked his hosts, "And now show me where you do your work."

The other visitor, Paul Ehrenfest, was the guest of his old friend, Karl Herzfeld. He had a marvelously infectious enthusiasm and a gift for promoting discussions of principle. One day Herzfeld tried to get Ehrenfest to lead the seminar discussion. Despite much urging, Ehrenfest declined, insisting that Herzfeld himself should take the floor. Herzfeld finally did. He explained why one might hope for a "separation of variables" even in a wave equation that is nonseparable by any standard criterion. Look at a given proper function, he proposed, and look where its nodes lie, and pick new coordinates to conform to those nodal surfaces. Then, he suggested, the wave equation might be separable. Ehrenfest smilingly broke in, "My dear Herzfeld, you are completely crazy"; and the liveliest discussion developed between the two friends.

I was never to see Ehrenfest again, but I was to have close relations over the years with the two leaders of theoretical nuclear physics in America who had gone to Leyden to study with Ehrenfest—Robert Oppenheimer and Gregory Breit. Ehrenfest wrote one paper jointly with each: with Breit on the remarkable features of space quantization;[56] and with Oppenheimer on whether a given nucleus obeys Fermi or Bose statistics.[57] Of any memory of Ehrenfest I did not have the good fortune to hear Oppenheimer speak; but Breit has told of a discussion in which Ehrenfest was doing all the talking until he interrupted himself to say sternly, "Breit, behaupten Sie etwas"—Breit, take a position, stick up for something!

I had applied for a National Research Council Fellowship for the year 1933–34; but if my application was granted—as by extremely good fortune[58] it finally was—with which man—Oppenheimer or Breit—should I work? I had corresponded with each, had talked with each at an American Physical Society meeting, and could have worked with either. Both showed me many kindnesses over the years. In personality they were utterly different. Oppenheimer saw things in black and white and was a quick decider. Breit worked in shades of gray and could be described in those words that Charles Darwin used in speaking of his own most important qualities: "the love of science—unbounded patience in long reflecting over any subject—industry in observing and collecting facts—and a fair share of invention. . . ."[59] Being temperamentally uncomfortable with quick decisions, and attracted to issues that require long reflection, I chose to work with Breit.

New York

Anyone in the United States who before World War II contributed more importantly to more fields of physics than Gregory Breit would be difficult to name. Who earlier than he recognized the potential of artificially accelerated particles to advance nuclear physics?[60] On this and other issues Breit, born July 14, 1899 (Ph.D., Johns Hopkins, 1921), worked 1924–29 at the Department of Terrestrial Magnetism of the Carnegie Institution of Washington. There he enjoyed the collaboration of Merle Tuve—later to become the distinguished director of that laboratory—and Otto Dahl, who after daring adventures in mountains and deserts had concluded that the real frontiers lie in the world of the small. By 1929 Breit had concluded that artificial sources could and must be used to disintegrate nuclei,[61] and the three men together had proposed a laboratory method to produce high potentials.[62] At this point Breit moved to New York (1929–34), from where he was later to go to the University of Wisconsin at Madison (1934–47) and then to Yale (1947–58) and the State University of New York at Buffalo. Tuve and his group modified their proposed approach to high voltages, adopted the idea of Van de Graaff,[63] and by 1934 were achieving nuclear disintegrations with 600 keV ions.[64]

Earlier, contributing both theory and observation, Breit had taken a central part in establishing the existence and properties of that layer of ionization in the upper atmosphere which is known today as the Heaviside layer.[65] He was even more concerned with questions of principle. His work on the effect of retardation on the interaction between two electrons is widely known.[66] No one could be more conscientious than he in treating the many subtle ramifications of fine structure, hyperfine structure, magnetic moments, and nuclear isotope shift, as seen in the many papers by Breit and his collaborators on the subject. With I. I. Rabi he had developed the principles for a standard method for the measurement of nuclear spin.[67] His great *Reviews of Modern Physics* treatise on dispersion and absorption is still a standard reference.[68] Capitalizing on earlier experimental work of Robert W. Wood, he pursued the idea of determining nuclear magnetic moments by observing the scattering of resonance radiation as influenced by magnetic fields easily available in the laboratory. On that and closely related topics there was close contact during the year between Breit and Irving S. Lowen, Jenny E. Rosenthal, Lawrence A. Wills, and especially Norman Heydenburg,[69] then also a National Research Council Fellow, close colleague and sharer with me of a rented room. Still to come was the great 1936 work of Breit and Wigner on the resonance theory of the capture of slow neutrons[70] and on the saturation requirements for nuclear forces,[71] Breit's subsequent and long continuing involvement with the interpretation of nucleon-nucleon scattering experiments, and his collaboration[72] with Gerald E. Brown, his then Yale student and now leader in nuclear physics.

At University Heights in New York's Bronx Borough was the campus of the uptown division of New York University. On that campus stood a building which as a house was large and distinguished but which was cramped when transformed into a physics laboratory. A good fraction of the ground floor was taken up with the continuing enterprise of Breit—with his new New York collaborators—to produce high voltages and induce nuclear reactions. Space on the next, office, floor was so limited that I shared a small room with Breit. Others who were in the building or visited it included Richard Cox, William Crew, Harvey Hall, Otto Halpern, Montgomery Johnson, Lloyd Motz, Edward Salant, and Hugh Wolfe. Years later through William Crew I met delightful Henry Crew, founding member of the American Physical Society, then in his nineties, visiting Nassau Hall for his last Princeton reunion.

Often four or five of us would have lunch together. I cannot forget that lunch in which Cox told us of measurements of the angular distribution of nuclear β-rays that violated what today would be called the principle of the conservation of parity. Two of the younger theoretical physicists took turns explaining to a defenseless experimentalist why those results could not be right. It is not surprising that in the paper as ultimately published[73] there appeared this caveat,

It should be remarked of several of these suggested explanations of the observations that their acceptance would offer greater difficulties in accounting for the discrepancies among the different results than would the acceptance of the hypothesis that we have here a true polarization due to the double scattering of asymmetrical electrons. This latter hypothesis seems the most tenable at the present time.

This statement convinced nobody of anything. Not until 1957 was the community to learn that parity is not conserved in β-decay.[74]

An especial privilege were the two-to-three-hour Saturday afternoon walks of Heydenburg, Lowen, myself, and sometimes one or two others, with Breit, talking of problems of the day and discoveries of the past. It was an even greater treat one fall afternoon to take the train with Breit to Princeton for a carefully unannounced seminar by Einstein, whose report on his current work came shortly after he had established himself in this country. It marked for me the beginning of an ever closer contact with Einstein's life and work.[75] The same visit gave me the chance to meet E. U. Condon, H. P. Robertson, and Eugene Wigner, with all of whom I was to have such happy relations over the years.

Those were the days when new discoveries came so fast that almost anything seemed possible. Or was it all a deception? One experimental nuclear physics colleague from Washington whom I have always admired confessed his skepticism to me about all those new artificially produced radioactivities that were being found. We were walking wintry Amsterdam Avenue while returning from lunch to the afternoon session of the American Physical Society meeting at Columbia. "It is just too improbable," he said, "that all those periods should be so conveniently suited to measurements—seconds, minutes, hours, days. Some kind of psychological self-deception must be involved." Not much later he admitted me via lock and key to the attic of his Washington laboratory, and showed me the equipment he was using to test whether the so-called Allison effect[76] was real. In the end it turned out that not everything *is* possible!

I. I. Rabi and Gregory Breit together conducted a joint Columbia-New York University seminar. Afterward we often adjourned to Rabi's apartment for a late evening discussion. Molecular beam work was in full swing at Columbia. Rabi's wonderfully productive and down-to-earth approach was evident in the topics under way and in his favorite advice to anyone tempted by the "great questions": "Shoot low to hit the mark."

Breit, too, believed that one does not have to do everything to do something. Yes, the resonances in nuclear transformations were puzzling; and puzzling too why for any given energy of the bombarding particle the outgoing particles, classified by energy, should fall into such a small number of groups. Nevertheless, don't worry; do the doable; calculate the penetration factors for the entry of the primary and the escape of the secondary. This proposal was attractive. I soon fell in love both with Coulomb wave functions and with the Wentzel-Kramers-Brillouin method of getting quick insights and good estimates—one of Leyden's gifts to Breit, and one of Breit's gifts to me. From our modest work together on this topic[77] I went on later to develop a so-called amplitude-phase method for determining wave functions,[78] and still more recently to exploit the WKB method in the problem of finding a potential from scattering data, the problem of "inverse scattering."[79] For our immediate applications and for the getting of numerical results it fell to me to train a friendly middle-aged out-of-work New Yorker in the use of the usual mechanical desk calculator of those days. The work was financed by the Works Progress Administration, set up by Congress to revive employment. Much as we learned, and much as we discussed resonances in nuclear transformations, we arrived at no all-encompassing theory of them. However, one by-product of our work on Coulomb wave functions was a small result on the fine structure in the spectrum of the Li^+ ion.[80]

I was much more on my own in another problem that Breit suggested to me, a study of the scattering of a-particles in helium. This topic, the subject of my ten-minute report at the April 1934 meeting of the American Physical Society,[81] lent itself to analysis by graphical methods which, in their mechanism of action, almost compared with that analyzer of the tides which Paul R. Heyl had shown me long before. Unhappily the scattering data then available were not definitive; but during my Chapel Hill period much more precise scattering cross sections were measured and reported. They were relevant to one part of the problem, to go from scattering to phase shifts; but for the more difficult part, to connect phase shifts with interactions, I had much thinking to do in the four years to come, with results I was to find most instructive.

I am especially grateful to Breit for his introducing me to pair theory, and for inviting me to collaborate with him to evaluate the cross section for the collision of two quanta to produce a pair of electrons out of empty space.[82] Far away though the process at that time seemed from realizability, I was to be at Copenhagen only a short-time—on a second fellowship year—before I realized how central it was to the theory of electron showers and even to the great test that quantum electrodynamics itself was about to pass.

I said goodbye to the Gregory Breit who had taught me so much about nuclear physics, atomic physics, and pair physics and who had supported me in my move to go and work with Bohr. Insufficiently appreciated in the 1930s, he is today the most unappreciated physicist in America. If he did not happen to have the empathy to resonate with the community, that was a small tragedy for him and a great tragedy for the community. Let who will take what problem he has to Breit and see what wise advice he will get.

To Copenhagen—to London—to Copenhagen

What happened in World War II to one of my three cabin mates on the freighter crossing, that young German who had thrown up his job as an American farm worker to return and "join Hitler"? In Antwerp, the port of debarkation, did ever anyone in the troubled decade to come seek out for serenity Christophe Plantin's printing establishment, with its union of home and work, of beauty and making beauty? In Köln the hostile glare of that booted brownshirt with his shoulder swastika, pacing up and down as I, foreigner and lone patron, briefly off the train, took my breakfast in the station restaurant: how many others was it destined to chill? And by what act would it at last be extinguished? In downtown Copenhagen, investigation of an attic room for rent took me past the scarlet sentries of Amalienborg Palace; but I did not foresee that ten years later Denmark would be an occupied land, that from that palace gateway day after day the king would ride forth alone, on horseback, among his people, to keep up faith and hope, nor that he would wear the yellow arm band—imposed on Jews by Hitler—to declare the principle that all men are brothers.

The Danish board-and-room house at Strandvaenget 8 and Bohr's Institutet for teoretisk Fysik at Blegdamsvej 15 had known me for only a few weeks when I took some of my slender savings to board the train-boat-train from Copenhagen to London with a few other Institute visitors and members. One was Georg von Hevesy, distinguished radiochemist of the Institute. With Dirk Coster in 1922, guided by the predictions of Bohr's atom model, he had discovered Copenhagen's own element, hafnium. Originally of the landowning nobility, he was full of stories of his native Hungary and of science, and happy to joke about the effects of drinking heavy water.[83] Others I met on the boat included twenty-six-year-old Edward Teller and his charming and vivacious bride of a few months, Mici, with both of whom I was later to have many rewarding associations. Teller was going from a Göttingen career and a few months refuge in Copenhagen to a lectureship at the University of London, destined to give way within a year to a professorship at George Washington University.

Under the joint auspices of the International Union of Pure and Applied Physics and the (British) Physical Society, an International Conference on Physics[84] took place in London and Cambridge September 1934. It was dedicated to the two fields of nuclear physics and the solid state. It took place at an exciting time, only two years after the *annus mirabilis*[85] in which the deuteron was detected and positively identified,[86] the nucleus was first disintegrated by artificial means,[87] the neutron was discovered,[88] the first nuclear transmutations were made with the 1 MeV Berkeley cyclotron,[89] and the positive electron was discovered.[90]

No one starting a new period of life and work in nuclear physics could have had a better survey of the problems and prospects in his field:[91] great men, great moments, great ideas. Lord Rutherford opened the meeting. A few nights later, at a reception in the rooms of the Royal Society, I, one of the few not in evening dress, found myself in a circle pressed around this towering and vital man. He lived only three more years. Sir J. J. Thomson, seventy-seven, frail and white haired, to remain as master of Trinity College, Cambridge, until his death six years later, was host at a reception there. Max Born, deprived of his position in Göttingen, and newly arrived in the U.K., opened the London sessions on nuclear physics by writing in huge letters on the blackboard, "NUCLEAR PHYSICS," then with eraser and chalk—to laughter—altering the title to read "UNCLEAR PHYSICS." Thirty-three-year-old Enrico Fermi reported results of the Rome group on radioactivities produced in a variety of elements by neutron irradiation. Less than a month later (11 A.M.,

October 27, 1934) came their great discovery that hydrogenous substances moderate neutrons and that this moderation of the neutrons "increases the activation intensity by a factor which, depending on the geometry used, ranges from a few tens to a few hundreds."[92]

Neither in nuclear physics nor in solid state do I remember a single paper—whether on the program or brought up in the discussion—that was not significant. Especially interesting to me were the results of Chadwick and Goldhaber[93] on the binding of the deuteron, of Bethe and Peierls[94] (updated by Bethe and Bacher[95]) on the photodisintegration of that two-nucleon system, the check on the balance of energy in β-decay—and therefore on the Pauli[96]-Fermi[97] concept of "β-decay as neutrino emission"[98] — by Ellis and Mott[99] and by Ellis,[100] the anomalous scattering of γ-rays in lead observed by Gray and Tarrant,[101] the cloud chamber studies of individual cosmic ray particles by Anderson and Neddermeyer,[102] the observations by Rossi of a penetrating component and of showers in the cosmic rays,[103] and the Bethe[104]–Bethe-Heitler[105] treatment of the degradation of electrons to photons and photons to electrons —the last four items (Gray-Tarrant, cosmic rays, Bethe-Heitler) to be central items of attention in the coming months at Copenhagen.

Copenhagen

The first sight of Bohr's Institute surprises anyone aware of how important it has been to the development of physics over the years. How could so much come out of a building so small, smaller than many a house? The impression of surprise was only increased to this young arrival in September 1934 by seeing how few there were inside. Only once or twice a year at times of conferences did the attendance climb as high as twenty-five. Naturally there was the familiar, active experimental group in nuclear physics working in a laboratory addendum to the building. It included T. Bjerge, K. J. Brostrøm, G. von Hevesy, J. C. Jacobsen, and E. H. Rasmussen. Old collaborators stopped in at various times during the year, including Werner Heisenberg, Oskar Klein, Harrie Massey, Ivar Waller, and C. F. von Weizsäcker. There were also special visitors from time to time in consequence, direct or indirect, of Hitler's January 30,

1933, accession to power and the growing shadow over Germany. They came for a few days or a few weeks or a few months on their way from positions in Germany to positions elsewhere, often still to be found. Among them were Hans Bethe, Max Delbrück, James Franck—a prince of a man—Hilda Levi, George Placzek, and Edward Teller. Learn Danish in one month? That was one of the achievements of Otto Robert Frisch, visitor for the year who had quickly made himself another highly productive member of the experimental group. Theoretical colleagues there for most of the year were few but hospitable: Bohr himself, Fritz Kalckar, Christian Møller, Milton S. Plesset, S. Rozental, E. J. Williams, and myself, supplemented for extended periods by Bohr's wonderfully scholarly and long-term collaborator, Léon Rosenfeld. The library was quiet, the cleaning woman each day cleaned the neat gray trim and aired it all out no matter how cold this operation made the user. It was the standard place for work and for occasional two- or three-person discussions.

Williams was mentor and guide in my first months at the Institute. He was of less than average height, compact, vigorous. Coming from Aberystwyth, he had used only his native Welsh until near the end of his days in secondary school. At Cambridge he had done experimental work until Rutherford forced him into theory by his rule that no one should be in the laboratory beyond the end of the afternoon. Williams explained to me about the ways of life of the Institute and of the Niels Bohr whom I had first seen and heard speak at the "Century of Progress" Chicago World's Fair of 1933. He told me of the agony Bohr had experienced on his arrival at Cambridge as a young postdoctoral worker, shy, and at that time ill at ease in English. Not knowing what to say when he had his first meeting with J. J. Thomson, director of the Cavendish Laboratory, and having read thoroughly and appreciated a book of Thomson (his *Electricity and Matter*[106]), in his embarrassment he could only blurt out a catalog of the principal errors in the book—foretaste of Bohr's disastrous May 16, 1944, encounter with Winston Churchill.[107] It was no wonder, Williams explained, that close relations did not develop between the two men. The twenty-seven-year-old Bohr found it more stimulating to go to Manchester and work with the magnetic forty-year-old Rutherford. At Copenhagen, Williams told me, it had been Bohr's habit each year to start to give a course of lectures. Each time, however, an exciting issue had developed in the course of the first session or two, and Bohr had put off further lectures until the question should have been entirely cleared up. A few months before I arrived Williams had visited the Soviet Union in company with Bohr, and regaled me with an account of his own exciting motorcycle trip in the outskirts of Leningrad. Eleven years later, at age forty-two, his heart failed, and the man of so much warmth and drive was gone.[108]

In the summer before my arrival in Copenhagen Niels Bohr had lost his oldest son Christian overboard from their sailboat in a storm. He and the family were too deeply affected for him to go to the London conference, or to show up regularly at the Institute the first part of the fall. One day, however, as I arrived on bicycle at the usual hour at the Institute, I noticed a workman tearing down the vines that had grown too thickly over the gray stucco front, and on closer view saw that it was Bohr himself, following his usual modest but direct approach to a problem. I had talked with him briefly on a couple of earlier occasions; but now full Institute activity was to start. How did he pick the topic? How did he make headway with it? And how did he communicate his conclusions?

"He sees farther ahead than any man alive": that was why—with the backing of Gregory Breit—I had asked to spend with Bohr the second of my two fellowship years. Bohr distilled the central issue out of dialogue with those who were themselves distillers of issues, former collaborators and special visitors. The single-hearted attention that Bohr gave to such a colleague showed nowhere better than in the way the two men walked up and down outside the Institute. They might share a less private discussion at the lunch to which so many of us brought the open-faced sandwiches—smørrebrød—purchased down the street. However, soon the talk would focus again more sharply on the issue that had been, or was in the course of being, "smoked out." Bohr would take the visitor away to his office, often carrying along his "right-hand man" as of the moment— Léon Rosenfeld or E. J. Williams in my time. Bohr went round and round the table as he talked or joked, expostulated or reflected, his whole soul taken up in the action. He stopped to make an especially strong point—or to listen briefly. His words were forceful. His voice was soft. His glance was piercing as he looked up from time to time and stared into one's eyes. His mood changed from time to time as dictated by the discussion itself: for making a point, "How could one possibly believe . . ." or "There is not the slightest evidence that . . .". If in doubt, his head tipped to one side as he spoke to one position; to the other, as he spoke to the opposite position.

Explanation was never dry pedagogy, but a one-man tennis match in which Bohr hit the ball from one court, then ran to the other fast enough to hit it back—the more times, the more enjoyable the game: "Such-and-such an effect leads one to expect thus-and-so. . . . Indeed one sees thus-and-such, but then so-and-so observed such-and-such. . . . Then we were in immense difficulty. Just at this point so-and-so pointed out that the proper formulation of the principle is thus-and-such. . . . This discovery brought the whole subject into order. But then so-and-so realized that this extended principle stands in absolute contradiction to the stability of such-and-such. . . . We were all lost until we found that the new formulation was really absolute nonsense. . . . What fools we have been! We have only to recognize such-and-such and we see that absolutely everything has to be exactly as it is." From time to time, to make a point or lighten the atmosphere, there would come a joke of which a favorite was Bohr's definition of a "great truth": A truth whose opposite is also a great truth.

Of all ways to tell the visitor of some new finding at the Institute, and convince him of it, and tell him how to convince others, it would be difficult to imagine a single one at the same time more modest, and more effective, than this "explanation-by-tennis"; but for the real colleague it was only a warm up for the real tennis match. In it he himself knocked the ball back and forth with Bohr. The spirit might have been the game for the game's sake; but the excitement came from expectation of the unexpected. The best witness to the level of the dialogue was the level of the participants, from Kramers to Heisenberg and from Bethe to Pauli. How else could a man hold up his end in such an encounter except to make his own point of greatest concern?

Complementarity had a place in almost every discussion. "*Contraria sunt complementa*" Bohr later made his heraldic motto.

Years after his death Heisenberg told me how, right after the time of his own paper on the uncertainty principle, he was out sailing with Bohr and was explaining to their sailing partner, Niels Bjerrum, the contents of the article. After hearing him out Bjerrum turned to Bohr and said, "But Niels, this is what you have been telling me ever since you were a boy." Bohr's lifetime advocacy of complementarity owed much to the philosophy of his great teacher, Harald Høffding,[109] as Einstein's continuing rejection of complementarity, and 1917–29 rejection of the big bang, were influenced by his youthful admiration for the thought of Benedict Spinoza,[110] advocate of determinacy and of a universe that goes on from everlasting to everlasting.

Sometimes a week or two would go by without a meeting. When it came, three or four of us would gather with Bohr in his office or another room to discuss some then worrisome point. To bring about a seminar it took a visitor, perhaps an experimental physicist and former collaborator from Poland, or a new and important paper reported by someone at the Institute, perhaps Jacobsen or Rosenfeld. The attendance ranged from one to two dozen. The language was usually German, occasionally English. The joy was to have something that "wouldn't fit." The central idea of the Institute[111] was clear, "No progress without a paradox." Most seminars were successful in the sense that Bohr broke in halfway through or sooner to solve the puzzle or explain the central point at issue. He would get to his feet and, reflecting as he kept talking and pacing up and down before the blackboard, encourage himself by saying every now and then, "Now it comes, now it comes." Suddenly it really would come, and he would give the explanation to us all as another tennis match. It was more reminiscent of football, however, where he had been a national hero, to see the way he plunged into the middle of things, found the central point, seized on it, and delivered it with great force to all assembled. Only rarely was it that the worst happened: nothing came up that surprised anyone, and Bohr had to utter those dreaded words, "It was an interesting seminar."

Usually the new issue became a focal point of discussion in the next days. Those days could almost have been numbered odd and even. One day was a day of building. If so-and-so is true, such-and-such follows. That will give us the chance to understand thus-and-so. That means it will be absolutely central to measure this-and-this cross section. Then we will be able to predict such-and-such with great assurance. No criticism. That was reserved for the next day. If at its end anything survived, that battle-tested core became the starting point of yet another day of building—and so on,[112] up to a conclusion that could be played out as a complete tennis match.

Progress with an idea I never saw Niels Bohr make except in dialogue or dictation or sudden revelation out of the depths of the subconscious. Always the end desired was a harmonious account of a wide range of experience. For this purpose he kept a continuous slow fire under about fifteen topics. They ranged from the angular momentum of light to dispersion relations for reaction cross sections in the continuum and from stopping power to superconductivity. Preliminary drafts of papers on each he stored in a little cabinet at Carlsberg in his home office, just off the Pompeian court, where he did his dictation on any issue of importance.

Is Quantum Electrodynamics Valid?
Way Station to the Discovery of the Meson

Of all the issues that had come up at London, and of all the issues that had been recognized as central at Copenhagen, Bohr took the most central as obviously this: Is quantum electrodynamics valid for high-energy electrons?

At the December 29, 1933, Chicago meeting of the American Physical Society I had heard Robert Oppenheimer give a lecture[113] on the status and problems of quantum electrodynamics before a great and enthralled audience. In it and a subsequent paper with Wendell Furry[114] he had expressed the conviction that current predictions[115] and predicting power would fail when the energy of the electron exceeded a value of the order of $137 mc^2$. Hans Bethe, speaking at the London conference in the discussion following the paper of Rossi, had said,[116]

The experiments of Anderson and Neddermeyer on the passage of cosmic-ray electrons through lead are extremely valuable for theoretical physics. They show that a large fraction of the energy loss by electrons in the energy range round 10^8 volts is due to emission of γ-radiation rather than to collisions, but still the radiative energy loss seems far smaller than that predicted by theory.

Thus the quantum theory apparently goes wrong for energies of about 10^8 volts, and it would be of special value for any future quantum electrodynamics to know exactly at which energy the present theory begins to fail, in other words to have much more experimental data on the energy loss of fast electrons (energy 10^7 to 5×10^8 volts) passing through matter.

Carl Anderson later summarized the situation as he saw it at this time[117] (and as it was seen in the fall of 1934 in Copenhagen):

Professor Seth H. Neddermeyer and I were continuing the study of cosmic-ray particles using the same magnet cloud chamber in which the positron was discovered. In these experiments it was found that most of the cosmic-ray particles at sea level were highly penetrating in the sense that they could traverse large thicknesses of heavy materials like lead and lose energy only by the directly produced ionization which amounted to something like 20 million ev per cm of lead. . . .

There were difficulties, however, with any interpretation in terms of known particles, as was pointed out as early as 1934 in a paper presented to the International Conference on Physics held in London of that year. . . .

. . . [To] interpret these particles as protons would mean *assuming the existence of protons of negative charge* since these sea-level particles occurred equally divided between negative and positive charges. . . .

There were difficulties also in interpreting these sea-level penetrating cosmic-ray particles as positive and negative electrons. . . .

. . . [The] paradoxical character of our data could be reduced if one assumed that the Bethe-Heitler theory, although correct for electrons of energies below a few hundred million ev, in some way became invalid for electrons of higher energy, thus permitting high-energy electrons to have a much greater penetrating power. . . .

In the face of these pervasive doubts Bohr—with Williams—sought and found a convincing way to establish the correctness of quantum electrodynamics at high energies. This great development, spread rapidly by word of mouth in the winter of 1934-35, and fully developed later in a comprehensive paper by Williams,[118] created a climate of opinion which, together with the further experimental results of Anderson and Neddermeyer,[119] established the existence of the meson.

The central idea was simple and powerful, "the method of the equivalent radiation field." Bohr had developed it twenty years earlier to analyze the stopping of a-particles.[120] With good results he had treated the electric field of the fast particle as equivalent in its effects to a radiation field with the same Fourier components. The probability that the atom will be excited or ionized by the passing particle had been identified with the probability for ionization or excitation by the equivalent radiation field. That field he had resolved by Fourier analysis into its monochromatic constituents. A decade later Fermi had applied this method of the equivalent radiation field to excitation of atoms by electrons, and to further details of the ionization produced by a-particles;[121] and Bohr thereupon commented[122] on the problems still left in the application of the method. These questions, insofar as they concerned excitation and ionization of atoms in distant collisions, and could be treated within the framework of the new wave mechanics, were dealt with in a 1933 paper of E. J. Williams,[123] who then came to the Institute to work with Bohr on the use of the method of the equivalent radiation field for a wider range of problems.[124] C. F. von Weizsäcker, visiting Copenhagen earlier and later, contributed further to extending the realm of applications, especially to bremsstrahlung,[125] or "splash radiation"; and Bohr's original method has become known in the age of wave mechanics as the method of Williams and Weizsäcker.[126]

Bohr and Williams extended to relativistic encounters this method of the equivalent radiation field. The shielded Coulomb field of the nucleus, looked at in the original inertial frame of reference of the oncoming electron, contains equivalent photons of all energies, up to the kinetic energy of the electron itself. When one of these photons is Compton-scattered by that electron, it is not only altered in direction but also transformed from "virtual" to "real." Modest though its energy might be in the moving frame, it typically has enormous energy in the laboratory frame, an energy equal to a substantial fraction of the kinetic energy of the electron. This then was the new view of the bremsstrahlung, the splash radiation, of Bethe and Heitler. It was sufficient for rederiving their result on these new grounds, Bohr and Williams showed, to limit attention to the low-energy end of the spectrum of the equivalent radiation, where photon energies are of the order of the electron rest energy and less. First, the cross section for scattering of higher energy photons falls off with energy. If theory gave that cross section wrongly, the true result could only be more scattering in the electron's frame, hence more bremsstrahlung in the laboratory frame, hence even more trouble for the ultrarelativistic electron to penetrate the lead shielding of Anderson and Neddermeyer. Second, for photons of the relevant energies the cross section for the Compton-knockon of an atomic electron, as given by the Klein-Nishina formula,[127] had been well confirmed by observation.[128] In brief, no room for doubt remained. The penetrating particles, with their comparable content of positives and negatives, could not be electrons. When Anderson and Neddermeyer gave the meson to the world,[129] penetrating power—as judged by momentum loss and ionization—was the key factor in establishing its identity.

Is Quantum Electrodynamics Correct?
Proof of the Foundations of Shower Theory

Bohr and Williams's relativistic extension of the method of the equivalent radiation field gave an equally firm foundation for understanding the other major part of the cosmic radiation, the "soft" or "shower" component. Their treatment had established that a fast electron necessarily radiates at least as strongly as Bethe and Heitler had calculated. However, did the high-energy photon thus produced have so high a cross section as Bethe and Heitler had predicted for forming a new pair of positive and negative electrons?

Williams had been entranced by the paper Breit and I had written on the production of a pair of electrons out of the vacuum by the collision of two photons.[130] Now he found a way to put our result to use. He considered the encounter of a high-energy photon with the screened Coulomb field of the nucleus. He analyzed the interaction in a reference frame moving in the direction of the photon. That photon was still present, though with reduced energy, in the moving frame; but flying toward it were now all the virtual photons of the equivalent radiation field. The probability for producing a pair out of the vacuum could be calculated easily out of our formula, and the Bethe-Heitler formula for pair production by a high-energy photon was confirmed.

A closer look is required to see the most important part of the reasoning. Only those photon-photon collisions contribute importantly to the pair production in which the product of energies of the two photons, a Lorentz-invariant quantity, is of the order of magnitude of a very modest quantity, the square of the rest energy of the electron. Extreme relativistic energies never come into question, no matter how high the energy of the incoming photon may be. If one believes quantum electrodynamics for any energy, one has to believe it for these energies. However, Bohr wanted to make doubly certain of this foundation point in the reasoning—and did. For this purpose he developed an extremely simple order-of-magnitude way of estimating from first principles the cross section for production of pairs by collision of two photons with energies of the order of the rest energy of the electron. Brill and I later provided a written description of his method in the course of applying it to other processes.[131]

Its foundations established, the theory of showers,

$$\begin{pmatrix}\text{highly energetic}\\ \text{electron passing}\\ \text{through matter}\end{pmatrix} \to \begin{pmatrix}\text{photons}\end{pmatrix} \to \begin{pmatrix}\text{pairs}\end{pmatrix} \to \begin{pmatrix}\text{more}\\ \text{photons}\end{pmatrix} \to \begin{pmatrix}\text{more}\\ \text{pairs}\end{pmatrix} \dots ,$$

"took off" in a series of papers by Oppenheimer,[132] Bhabha and Heitler,[133] Carlson and Oppenheimer,[134] and others. Another great part of the cosmic radiation was reduced to the rule of reason.

Is Quantum Electrodynamics Correct?
Heavy Nuclei and the "Anomalous Scattering" of γ-rays

Challenges to quantum electrodynamics were far from ended. When 2.6 MeV γ-rays from a radioactive source of Thorium C″ fall on lead, Gray and Tarrant had found,[135] γ-radiation comes off in the backward half-sphere, harder and more intense than anyone had any reason to expect. This result was no bolt from the blue. They had got into this problem because other investigations[136] had "shown that γ-radiation of 2 to 3 million volts quantum energy is much more strongly absorbed than would be expected, in elements of high atomic number." Gray and Tarrant attributed the additional absorption provisionally to interaction with the nucleus. They found that the secondary γ-radiation is

quite unlike the radiation scattered by the electronic system of the outer atom, in that the wave-length of this radiation is neither the same as that of the primary radiation, nor varies with angle in accordance with the Compton equations. Moreover, the quantum energy of the secondary radiation appears to be independent of that of the primary radiation, provided this exceeds a certain threshold value, which by two independent methods, has been found to lie between 1.5 and 2.0 million volts. These results suggest that the absorption process consists in the excitation of the nucleus, which subsequently emits "characteristic radiations."

The absorption curves of the characteristic radiations of lead, tin, iron, and oxygen have been found to be very similar. A unique determination of wavelength is not possible by absorption methods, but the simplest interpretation of the curve is that all four elements emit the same two radiations (having quantum energies of 0.5 and 1.0 million volts) but in different proportions, leading to the view that the radiations are characteristic of some unit of nuclear structure present in all nuclei.[137]

Others confirmed[138] the anomalous "scattering," or, better name as it proved in the end, the anomalously large secondary radiation. C. C. Lauritsen and Robert Oppenheimer commented on it.[139] So did Max Delbrück,[140] remarking—in the words of the abstractor[141] —that a "discrepancy arises in the anomalous scattering of hard γ-radiation," where "the predicted values are about 1000 times smaller than those observed."

Could not the incoming γ-ray, quite apart from creating a real electron pair in the electric field of force of the nucleus, create there temporarily a virtual electron pair? Could not that pair, upon annihilating, send off a γ-ray of the original energy into a new direction? This proposal was taken very seriously by Delbrück. It could only be evaluated by calculating the cross section for what in the language of quantum electrodynamics is a fourth-order process. That analysis required use of the technique of "subtracting off infinities" or "renormalization" in a far more sophisticated form than anyone had yet attempted. Delbrück, in the course of the year destined to become an exile from Germany, but at the time associated with Heisenberg in Leipzig, labored month after month at these for those days incredibly difficult calculations. It was widely believed that the anomalously large secondary radiation could be understood as such a "Delbrück effect," catalyzed by the field of force of the atomic nucleus.

In a seminar one of the senior experimentalists of the Institute, the pipe-puffing and always cheerful J. C. Jacobsen, reported the existing experiments, the idea of Delbrück, and his own measurements[142] confirming the anomaly. Lise Meitner visited and, telling about the results that she and her collaborators had obtained, reinforced the concern of all of us at the Institute.

Did it really require an unexpected new effect to explain the "anomalous scattering"? Was it not some unexpected combination of simple and well-known effects that conspired to produce the secondary radiation? No question seemed more natural than this to someone like myself who had had the privilege of working out and publishing with Gregory Breit in the preceding year at New York some of the elementary processes of electron pair physics. Roughly speaking, a 2.6 MeV γ-ray incident on a sheet of lead has a 20 percent chance to produce there a photoelectron, a 60 percent chance to knock on a Compton electron, and a 20 percent chance to create an electron pair. There is nothing anomalous about any of these processes. Moreover, not one of them produces anything like a high-energy scattered γ-ray. However, the electrons produced by the γ-ray captured my attention. A little investigation made it clear that they must experience strong scattering in the lead. In addition, they were energetic enough to have a strong probability of emitting splash radiation in the γ-ray region on their passage through the lead. Moreover, the positive electrons had a significant chance to annihilate in flight and produce energetic annihilation quanta. In brief, though the primary photon had minienergy on the scale of cosmic rays, did it not nevertheless produce a minishower? And did not that minishower, being of low energy and strongly scattered, retain a minimemory of the direction of the primary? What else could be more natural as an explanation of the strong back-directed secondary radiation?

Numerical calculations showed that this combination of three processes—absorption of the primary γ-ray, scattering of the resulting positive or negative electron, and radiation by that electron—accounted as well as could be expected for the observed secondary radiation. Thus there was no reason to believe in any anomaly or any failure of quantum electrodynamics in the 2 to 3 MeV region.

The word spread. The experiments stopped. The fourth-order calculations stopped. Delbrück arrived at Copenhagen on his way to a new and unknown life. Lunch with him at the Institute during his several-week visit was always stimulating because of his wide interests and liveliness in discussion. From those times when Bohr joined the lunchtime sessions and talk turned to the challenges of biology, it was easy to see the direction in which Delbrück's interests were turning, foreshadowing a career filled with great discoveries.[143]

At about this time the laboratory of Robert Watson-Watt burned down, destroying all his research and research records—and there was a change in management. He would never have been able to move ahead so fast in the new field, with all it meant for the future of the world, had it not been for those developments, he tells us.[144]

Recently, by pure accident, I had the honor and pleasure of sitting as one-day visitor in Delbrück's graduate course in theoretical biology at the California Institute of Technology on the occasion of his last lecture. It dealt with Gödel's great theorem about undecidable propositions.[145] A little later this announcement arrived, reminder of warm personal associations that grew out of that great issue of Copenhagen days, "Is quantum electrodynamics valid?": "The Caltech Board of Trustees has created a new position to honor professors of unusual distinction after full retirement from their regular professorships. . . . Two faculty members, both Nobel laureates, have been named to the new title of Board of Trustees Professor Emeritus. They are Carl D. Anderson . . . and Max Delbrück. . . ."[146]

Testing the minishower hypothesis had forced me to investigate the quantum electrodynamic cross sections for all relevant interactions of high-energy electrons and photons, systematizing the known ones and calculating new ones. This part of the work had brought me into collaboration with Milton Plesset. He, the only other National Research Council Fellow in Copenhagen, had worked with Oppenheimer on electron-pair theory,[147] as had I with Breit. I was frequently to enjoy his and his wife's hospitality. He and I together calculated the cross section for the inelastic scattering of quanta with the production of pairs,[148] and on the way found further applications for the method of the equivalent radiation field.

Nuclear Scattering by Dispersion Theory

We made an abortive beginning on a development of potentially much greater importance: the use of dispersion theory to calculate unknown cross sections from known ones. I had already seen the power of dispersion theory in my Ph.D. thesis on the scattering and absorption of light by the helium atom. Why go to the most difficult area of quantum electrodynamics to calculate the Delbrück scattering? Why not make a straightforward application of dispersion theory? Why not evaluate the amplitude for forward scattering by integrating the known or easily estimated absorption cross section—divided by the appropriate resonance denominator—over all frequencies? If the cross section has to include truly nuclear photoabsorption as well as photoabsorption by the bound electrons and pair production in the Coulomb field of the nucleus, so much the better for the believability of the results. The scattering thus calculated was of course too small to be relevant to the results of Gray and Tarrant and other workers in the field. However, it is interesting in its own right. It has been investigated experimentally by today's greatly improved techniques. We pursued this line of reasoning not only for scattering by an atom but also for scattering by an electron. However, we never published the resulting paper. Bohr was skeptical of the conclusions. What right had we to assume that dispersion theory was correct for relativistic results? Neither Kronig[149] nor Kramers[150] had ever given any reason to believe that the domain of validity extended so far. Nor had we.

Not until 1949 did I return to this issue, in association with John S. Toll. It soon became clear that the validity of dispersion theory could be extended to the relativistic domain by a proper application of the principle of causality. Engulfed though we soon were in a demanding project, we managed to state our results in one area of pair theory.[151] Subsequently John Toll, by then day-and-night associate director of Project Matterhorn, by working incredible hours put out a comprehensive thesis, "The Dispersion Relation for Light and Its Application to Problems Involving Electron Pairs."[152] It inaugurated a flood of papers, which have carried the subject far beyond its beginnings.

Plesset was at Liège during part of the period of our collaboration. I stretched my visit to him to include two days in Paris. There I renewed the rewarding acquaintance with Louis Leprince Ringuet and Pierre Auger that I had made in London, met for the first time two other distinguished experimental physicists, Maurice de Broglie and Irène Joliot-Curie, and ended with a visit to the tomb of Abélard and Héloise in the Père Lachaise cemetery.

The Copenhagen that I came back to had a leadership limited neither to physics nor to Carlsberg beer. However, the beer helped. The son of the owner of the brewery was J. C. Jacobsen, a man interested in history, archaeology and science. Having been for a time a student of Pasteur, he appreciated what a proper understanding of chemistry and fermentation could do for Carlsberg. He made the brewery a thriving and ever growing enterprise. In 1876 he gave the revenues to the control of a Carlsberg Foundation, closely associated with the Royal Danish Academy of Sciences and Letters. These revenues helped give Denmark a position in science altogether out of proportion to its three million population. Among the leaders there in my time were Niels Erik Nørdlung, brother of Margarethe Bohr, mathematician, director of the Danish Geodetic Institute, and mapper of Greenland and Iceland; Niels Bjerrum, great physical chemist and close friend of Niels Bohr; the large and warmhearted family—as distinguished in Baltimore as in Copenhagen—of Bernhard Bang (1848–1932), discoverer of the bacillus responsible for contagious abortion in cattle; and Elis Strømgren, director of Copenhagen's Observatory, and his son Bengt. Today Bengt Strømgren occupies the House of Honor, as Niels Bohr had in my time, that beautiful residence built by J. C. Jacobsen.

Stars and Nuclei

Bengt Strømgren and I had many discussions with Subrahmanyan Chandrasekhar, then a fellow of Trinity College, Cambridge, making a visit of some weeks to Copenhagen. Chandrasekhar had recently shown[153] that white dwarfs cannot have a mass in excess of about 1.2 times the mass of the sun. He had found it impossible to convince Arthur Eddington of the inescapability of this result, of the arguments of special relativity that underlie it. We both took a deep interest in Chandrasekhar's reasoning and accepted it.

We were less aware of the entirely new domain of compact objects that was being opened up by the discovery[154] of the neutron. Already a year before Baade and Zwicky had advanced "the view that supernovae represent the transitions from ordinary stars into *neutron stars*, which in their final stages consist of extremely closely packed neutrons."[155] In these objects densities are $\sim 10^{13}$ to 10^{16} g/cm^3, as compared with $\sim 10^6$ to 10^8 g/cm^3 of white dwarfs.

In the analysis of them general relativity has to be taken into account, as it did not in Chandrasekhar's white dwarfs. Oppenheimer and Volkoff were to be the first to give that analysis,[156] in 1939. For this purpose they approximated the nuclear matter of the star as an ideal neutron gas.

Chandrasekhar's treatment gave white dwarfs, with no hint of neutron stars; whereas the work of Oppenheimer and Volkoff gave neutron stars, with no hint of white dwarfs. The two fields of physics seemed to have no meeting place. Would a proper analysis yield both families of compact objects, and the transition between them? When I later found myself concerned with the issue of gravitational collapse, I could not let this question go unanswered any longer. To deal with it, Harrison and I developed an approximate equation of state[157]—since updated[158] and much improved upon[159] —for "cold catalyzed matter, at the absolute end-point of thermonuclear evolution," giving pressure as a function of density over the whole range from 10 g/cm^3 to 10^{16} g/cm^3—from ordinary matter to nuclear densities, $\sim 2 \times 10^{14}$ g/cm^3, and beyond. Using that equation of state Wakano and I integrated the general-relativity hydrostatic equation of stellar equilibrium.[160] The three of us found displayed before our eyes the whole spectrum of cold objects as we considered first one central density, then another. The spectrum began with planets. Then came small white dwarfs, then larger white dwarfs up to a limit of mass that we called the "first crushing point," because beyond this point pressure crushed protons and electrons together to make neutrons. Higher central density brought a continuation of configurations afflicted with this form of instability. The instability ended for a calculated central density of $10^{13.4}$ g/cm^3 and a mass of 0.15 solar mass—the beginning of a family of neutron stars. That second stable family was calculated to end at a "second crushing point" with a central density of $10^{15.8}$ g/cm^3 and 0.8 solar mass—a critical mass substantially increased in more recent times by more realistic allowance for that repulsion between nucleon and nucleon nowadays expected for these supranuclear densities. A neutron star beyond the second crushing point collapses to a black hole.

As a consequence of this excursion to a frontier of nuclear physics, I could not help reviewing the physics of neutron stars. Baade and Zwicky had reasoned that a neutron star should be left at the center of the debris from a supernova explosion. I knew that the Crab Nebula is the best studied example of supernova debris. Therefore I pored over Jan Oort's analysis—with Walraven—of this object.[161] It could not keep radiating from flickering wisps of luminosity that move from week to week, as it does, he concluded, unless something is supplying power to it. In my 1966 report[162] I analyzed six conceivable sources for this energy, had to reject four as incompatible with the evidence, and about the remaining two concluded (p. 423) that: "Energy of rotation of the star itself or energy of bulk motion of the ion clouds ejected at the time of its formation seems to be a stockpile of energy more rewarding for further investigation." Rotation it turned out to be, two years later, thanks not to this analysis, but to the wonderful discovery of pulsars by Anthony Hewish and Jocelyn Bell and their collaborators.[163] Those walks and talks in Copenhagen with Chandrasekhar and Strømgren had launched me on a sea of inquiry filled with endless beautiful surprises.

Nuclear Interactions

How can one make the most of the empirical evidence on the scattering of α-particles in helium? How can one connect the interaction of two systems of four nucleons each with the elementary interaction between nucleon and nucleon? This problem that I had brought to Copenhagen met no resonance there, and had to wait until Chapel Hill for significant progress. Nevertheless, I kept up a substantial effort on it in Copenhagen. A few weeks before my end-of-May departure a development occurred which was to have great consequences for the way I looked at nuclear physics.

Ever since Fermi and his group had made their October 1934 discovery[164] how to moderate neutrons, they had been exploiting this technique, and were to continue to do so right up to the end of 1938, when Fermi went to Stockholm, received his Nobel award, and began a new career in the United States.

On two occasions, in July 1936 and May 1938, I had exceptionally good opportunities to study the papers of the Rome group. My wife was in the lying-in ward of the Duke University Hospital, little occupied, and willing enough to translate those Italian words that I could not make out. In successive issues of *Ricerca scientifica* I learned about slow neutron resonances, and how it was possible to determine the energy of such a resonance from simple measurements of the distribution of activation of a metal foil moved from place to place within the moderator. I found out that the cross section for capture of slow neutrons in boron is inversely proportional to neutron velocity, and how this feature could be checked and exploited. I became aware how enormous the slow neutron cross section of cadmium and silver is compared with the geometrical cross section of the nucleus. These details of the developing work and more I was able to follow from month to month, but I will never forget the day I first heard about the results and what they meant.

The news hit me at a Copenhagen seminar, set up on short notice to hear what Christian Møller had found out during his Eastertime 1935 visit to Rome and Fermi's group. The enormous cross sections Møller reported for the interception of slow neutrons stood at complete variance to the concept of the nucleus then generally accepted. On that view the nucleons have the same kind of free run in the nucleus that electrons have in an atom, or planets in the solar system. Møller had only got about a half hour into his seminar account and had only barely outlined the Rome findings when Bohr rushed forward to take the floor from him. Letting the words come as his thoughts developed, Bohr described how the large cross sections lead one to think of exactly the opposite idealization: a mean-free path for the individual nucleons short in comparison with nuclear dimensions. He compared such a collection of particles with a liquid drop. He stressed the idea that the system formed by the impact of the neutron, the "compound nucleus," would have no memory of how it was formed. It was already clear before Bohr finished and the seminar was over that a revolutionary change in outlook was in the making. Others heard his thoughts by the grapevine before he gave his first formal lecture on the subject, before the Copenhagen Academy on January 27, 1936, with a subsequent written account in *Nature*.[165]

Bohr's liquid drop model provided a means to estimate some of the properties of his compound nucleus. At almost the same time appeared Breit and Wigner's theory of cross sections at and near resonance.[166] It gave tools to calculate in some details other properties of the compound nucleus. It is difficult to name any two papers which together, in shorter compass, brought more predicting power to nuclear physics.

Chapel Hill

Three days after my return from Copenhagen Joseph and Maria Mayer, assisted by Karl Herzfeld, gave a reception in Baltimore for Janette La Tounette Hegner. She and I were married five days after my landing. We went to live in Chapel Hill, our residence for the three years 1935-38.

Arthur Ruark was chairman of the physics department of the University of North Carolina. With Harold Urey, during the five years when Urey was at Johns Hopkins, he had written (1930) that bible of my student days, *Atoms, Molecules and Quanta*.[167] Ruark's probing interest in the famous paradoxes and models of physics made him a stimulating colleague. There were additional interesting associates both at Chapel Hill and at Duke University, including Walter Gordy, Walter Nielsen, and Lothar Nordheim. Among the graduate students with whom I interacted closely, besides Dudley Williams and Herman Parker, was one destined to become dear to the community of nuclear physicists for her own work in the field as well as for her contribution to theirs: Katherine Way.

Old themes long pursued lead to new themes. So it was with the two concerns I brought with me and explored at Chapel Hill. One was the physics of relativistic electrons and photons, the other was nuclear scattering and nuclear structure.

What is the Role of the Electron in Nuclear Physics?

Is the electron, the light particle, also the basic particle? Yes: so the successes of Dirac's relativistic theory of the electron[168] were beginning to make one think in the early 1930s. Moreover, the Dirac electron, by filling all negative energy states, had preempted even the vacuum for its own, and endowed that vacuum with a rich and interesting physics. The ostensibly infinite density of electric charge and infinite polarizability of this vacuum Dirac[169] and Heisenberg[170] in beautiful papers had taught us how to regularize.

Was not the dead dream of Lorentz[171] and Poincaré[172] — to give a purely electromagnetic account of the structure of matter[173] — almost at the point of reviving in a new form, an electron-theory account of matter?

"Why was it that you turned away from considering the electron as the truly simple particle," I once asked Heisenberg,[174] "while Dirac continued in that view?" "The discovery of the neutron[175] convinced me," he replied, "that the electron is not more fundamental than other particles."

To come so quickly to so firm a conclusion as Heisenberg's was not within my power. His hypothesis[176] — and Iwanenko's[177] — that nuclei are made of neutrons and protons I had used and I would continue to use wholeheartedly, as did other colleagues working in nuclear physics. However, in addition to that picture, not in contradiction to it, might not these neutrons and protons themselves be made of electrons? Would that not be a natural consequence of Dirac's concept of the electron as the basic principle? In default of convincing observational or theoretical evidence against this view, or some beautiful alternative, how could one keep from reflecting on what an electron theory of matter might mean?

Electrons? The basic entities in nuclear physics are not electrons at all but nucleons — according to a proposal made by Yukawa[178] in 1935 and soon generally accepted, far from being thoroughly definitive though that proposal is today, forty-two years and hundreds of research papers later. Not a dissenting voice was to be heard when lecture after illuminating lecture on nuclear physics began with the Standard Strong-Force Credo: (1) the nucleus contains no electrons; (2) the force that binds nucleons has nothing to do with electromagnetism and constitutes a new "strong force"; and (3) this force is transmitted by mesons.

A single miracle could be laid to the credit of this Strong-Force Credo: the correlation between the mass of the meson and the range of nuclear force. Was it impressive enough to justify the stilling of all doubt? A great part of physics was at stake. I would not have had a good conscience if I had foreclosed all thought of alternative interpretations of the evidence. Frankly I could not accept the Strong-Force Credo. Why not say neither yea nor nay until light should come? Coming from a family that had fled England in the great religious persecutions[179] of 1620–40, I had grown up in the tradition of silent dissent: silence in public, earnest discussion in private of where one should and where one should not agree with proclaimed doctrine. Therefore, from time to time I took council with Breit, Ruark, and one or two other colleagues about the weaknesses of the standard arguments. The nucleus contains no electrons? Nothing did so much as β-decay to make one think that the nucleus contains electrons, and nothing did so much as Fermi's theory of β-decay[180] to make one believe that the nucleus does not contain electrons. An advance copy of Fermi's paper had been sent to Gregory Breit. I still remember the interest and admiration with which he described Fermi's great idea and its implications. Everyone working in nuclear physics saw the force of Fermi's considerations. Until then doubt had existed whether the electron exists in advance in the nucleus, or whether it is formed at the moment of the β-transformation; doubt, even though no one could cite more promptly than Marie Curie[181] the remark of Aston[182] that the smoke does not exist in the pistol until the trigger is pulled. Moreover, no one could so misunderstand quantum mechanics as to think that the Lyman-α photon existed in the atom before it was emitted. However, a closer look made it apparent that the Fermi theory of β-decay is after all purely phenomenological. In and by itself it really says nothing about the presence or absence of electrons in the nucleus. Dirac electron theory itself was the best guide to the possibilites. It had been taken by many to provide the strongest single argument one had against the presence of electrons in the nucleus. Confinement in a space so small, of the order of the classical electron radius, implied momenta so great, ~ 137 mc, and binding forces so strong that those forces would inevitably produce positive and negative electron pairs in indefinite numbers.[183] As if this were not objection enough to the idea of electrons in the nucleus, one had two other powerful considerations. First, the spin and statistics of N^{14} had been determined[184] and, although they were compatible with the idea that this nucleus consists of seven neutrons and seven protons, they contradicted the idea that it consists of n negative electrons and $(n + 7)$ positive electrons.[185] Second, nuclear magnetic moments are very much smaller than one would offhand think reasonable for a system composed exclusively of electrons.

A closer look suggested that neither of these objections is as compelling as it first appears. One has only to turn through the pages of Mott and Massey[186] to discover that a sufficiently rapidly moving bound particle possesses only a fraction of its free-state magnetic moment. Moreover, researches both in spectroscopy in the early 1930s[187] and in particle physics in more recent times[188] provide striking instances where the apparent spin and statistics associated with the system are directly opposite to the actual ones and to what one would have deduced from straightforward application of the familiar elementary rules. On these grounds it seemed justifiable to delay judgment on the question of electrons in the nucleus.

What about the strength of the force that binds nucleon to nucleon? If it were 10^{39} times stronger, or even 10^6 times stronger, than the electric force, one could regard as compelling the argument for a new, "strong" force. However, the actual factor is only $\sim 10^2$. Moreover, if the nucleus actually contained electrons, they would be of necessity not only highly relativistic but also violently and frequently accelerated. Under these conditions the accelerative component of the force acting from particle to particle could be expected enormously to exceed the static component. Were these not exactly the right conditions to build up electromagnetic forces to the order of magnitude of the so-called strong forces?

How could one be conscientious about questioning the Strong-Force Credo if one did not give a careful look at this only evident alternative? And how then could one escape a careful look at the interaction between highly accelerated relativistic electrons? This was the motive for taking a fresh look at the theory of electromagnetic action at a distance of Schwarzschild,[189] Tetrode,[190] Frenkel,[191] and Fokker.[192]

An independent difficulty made it natural to think along the same lines: the so-called infinite self-energy of a point particle. How could one forebear from asking if this was not a difficulty of one's own making? In dealing with macroscopic bodies, one was accustomed to write the field energy in the form $(1/8\pi) \int E^2 d(\text{volume})$. However, evaluated for two-point particles at rest, where the electric field is

$$E = e_1 r_1 / r_1^3 + e_2 r_2 / r_2^3 ,$$

this integral contains not only an interaction term,

$$(2/8\,\pi) \int (\mathbf{E}_1 \cdot \mathbf{E}_2) d(\text{volume}) = e_1 e_2 / r_{12} ,$$

in full agreement with experience, but also two "self-energy" integrals which, being infinite, make no sense. Why not be happy with the part that agrees with experience and discard the part that doesn't? I kept a slow fire going under questions of this kind all through my three years at Chapel Hill; and sometime after mid-January 1939, several months after I had gone to Princeton I discussed these considerations with Léon Rosenfeld, then visiting Princeton. He objected that electromagnetic radiation seemed to have no place in this picture. Some time later, reflecting quietly at home one Sunday afternoon on the back of an envelope, I suddenly recognized that if there were enough absorber particles around to absorb completely the radiation from an accelerated source, it would make no difference how numerous were these particles, nor what their properties. However, I failed by a factor two to get the right result for the familiar force of radiative reaction. The next day I told Richard Feynman, then a graduate student, about my line of thought and about my results. Thanks to our usual lively discussion the factor two was cleared up along with many other ramifications of "interaction with the absorber as the mechanism of radiation."[193] We were not able to write up and publish this work[194] until after World War II; and we never did publish the larger ranging work we had been doing at the time on scattering theory.

Could one find some way to translate this classical theory of direct particle interaction into quantum terms? This question, Feynman tells us in his Nobel Prize Lecture,[195] drove him to his beautiful "sum over histories" way of doing quantum mechanics[196] with its later payoff in quantum electrodynamics.[197]

Why did Feynman and I give up direct interparticle interaction? It was not a failure at describing interaction. It was a failure at describing particles. Nowhere did this show more clearly than in the "zig-zag world line" description of pair annihilation which Feynman[198] attributes to me but which, unbeknownst to me, had been put forward by Stueckelberg.[199] From that picture one recognizes that the so-called particle in any proper quantum theory is in actuality a kind of cloud of virtual particles in which creation and annihilation are continually taking place. As a consequence the object that one calls a "particle" is immensely more complicated than is recognized by any single simple history of it as a world line.

Beyond Particles, beyond Fields

It is sometimes said that "no one gets religion like a reformed drunkard." In keeping with that observation, I who had given so much effort to the theory of direct particle interaction ("sweeping out electromagnetic fields from between the point charges"), and who later devoted the summer and fall of 1949 to exploring gravitational action at a distance ("sweeping out space and time from between the elementary particles") became as "field theoretical" as anyone can be. If everything cannot be reduced to particles and the interaction between them, is it not reasonable to think of everything built out of field, including the particles themselves? And not many fields, but only one field? And what field is more naturally taken as basic than geometry? Who that observes the marvelous depth and scope of Einstein's purely geometrical theory of gravity could think of any other field as primordial? Twenty years more (1953-73) brought many new insights but most of all the realization[200] that any such picture is too finalistic to be final. Surely no theory of physics that deals only with physics will ever yield the key to physics! Rather than particles, rather than fields, may not the building blocks of nature be something ever so much more tenuous: elementary quantum acts of observer-participant?[201] And what "machinery" do they fall together to make? However, that is another story!

Go anywhere. See anyone. Ask any question. Make headway by making all possible mistakes, but make them as fast as possible and recognize them:[202] this was the spirit with which I and many another colleague participated in "the search." One nevertheless knows that one will fall without hope of recovery if one tries to reach out too far without keeping one's feet on solid ground. In my own case, that solid ground was nuclear physics. That topic brings me back from relativistic electrons to the other subject of primary concern to me in my three years at Chapel Hill, the structure and interactions of nuclei with special reference to the scattering of one nucleus by another.

Resonating Group Structure

The four principal results of this period were the concept of resonating group structure, the concept of the scattering matrix and—with Edward Teller—nuclear rotations and the a-particle model of the nucleus. Not one of these developments did not, so far as I was concerned, take its origin in the problem that Gregory Breit had suggested to me in New York, to make what one could out of the experiments of Rutherford and Chadwick[203] on the scattering of a-particles in helium. That problem had led to the program that I outlined in my ten-minute paper at the April 1934 Washington Meeting. I had developed a systematic method to carry through the first part of the program, to determine phase shifts from scattering.[204] What was I to do about the second part: To get effective interaction from phase shifts? Most difficult of all, what was one even to mean by the term "effective interaction?" The very term conjures up the vision of a single degree of freedom and some kind of potential associated with that degree of freedom. But where was that degree of freedom to be discerned in a system of four particles impacting on and exchanging constituents with another system of four particles? Week after fruitless week went by while I attempted to get some grip on this question. Then an inspiration came, surely from Breit's beautiful treatment of the three-body problem[205] and from the application of the Ritz variational method by Hylleraas[206] and others to calculating the ground state of the helium atom. Why not apply the variational method to the eight-particle system? Why not write down a "preliminary trial wave function" for the system as a product of an internal wave function—assumed to be known—for the one four-particle system, multiplied by a similar internal wave function for the other a-particle and multiplied by an as yet undetermined function of the separation of the centers of gravity of the two a-particles? It was only necessary to antisymmetrize this wave function with respect to interchanges of neutrons and protons between the two systems in order to have a wave function which took due account of all interchanges which could go on in elastic scattering.

Once this trial wave function was in hand, the further steps were obvious:[207] use it to evaluate the expectation value of the energy of the eight-particle system. Extremize this energy with respect to variations in the sole undetermined function. Arrive in this way at a wave equation for this function. Separate out angle variables. Interpret the terms in the resulting radial wave equation. One is a "direct a-a interaction." Another is recognizable as an "exchange interaction," describable in mathematical language as "an integral operator with a symmetric kernel" and in physical language as a "velocity dependent coupling," such as had already been discussed by Dirac[208] in connection with the binding of electrons in atoms. Finally, still another velocity-dependent term or kernel appears in the effective reduced mass of the two-a-particle system.

This a-a wave equation made it evident what it meant to go from a-a interaction to phase shifts for a-a scattering. It also turned out that there is no unique way to go back from phase shifts to interaction. There are too many "pieces" (direct, exchange, and inertia) to be untangled.

It soon became clear that the variational approach was in no way limited either to a-particle clusters or to scattering. It could be applied to bound states; and the clusters could be as well deuterons or tritons as a-particles. Thanks to the interest of Katherine Way[209] and Herman Parker[210] it was possible to verify that this way of treating binding gave reasonable results for the lightest nuclei. On a December 1936 – March 1937 leave of absence kindly granted me by the University of North Carolina to work at the Institute for Advanced Study in Princeton, I asked Edward Condon for a good name for this approach to nuclear binding and scattering. He proposed the term "resonating group structure." I happily adopted it. In later times some investigators have given the name "clustering theory" to this way of analyzing binding and reactions. It was the direct ancestor of the "method of generator coordinates" for obtaining approximate wave functions for excited states of nuclei[211] subsequently to be further developed and applied.[212]

The Scattering-and-Reaction Matrix

What was one to do next with "resonating group structures," a theory that allowed one both to define and in principle to calculate the interaction between one deuteron and another, between one a-particle and another, between one cluster and another? Later days were to see a whole literature[213] and international conferences[214] devoted to such questions. However, at this early stage of nuclear physics the most immediately attractive application was also the simplest. One had for the first time a framework of ideas for calculating such a quantity as a phase shift in the collision of complex nuclei. That scheme did more. It allowed one to define for inelastic encounters and reactions a generalized scattering or reaction matrix.[215] I could see that: "The c's form a unitary matrix, for the equality which must exist between the numbers of incoming and outgoing groups in the [given] state . . . for arbitrary choices of the a's . . . is just the necessary and sufficient condition for unitary character." However, examples I looked into made it clear to me that this matrix, besides being unitary, is symmetric. Not seeing how to establish this symmetry on general grounds myself, I wrote for advice to Eugene Wigner, then at Wisconsin, and he promptly replied that time-reversal is the essential consideration and explained the consequences. Therefore I could add in the paper, with acknowledgment to him, "This relation shows that $\|c_{mn}\|$ is a symmetrical matrix, and . . . demonstrates that each element of the matrix is determined up to a factor ±1." The theory of the scattering matrix was "on the road"—and was thereafter given a great forward push by Heisenberg.[216]

Interaction Kernel as Velocity-Dependent Force

It was a second application of the method of resonating group structure to arrive at a new view of the elementary interaction between nucleon and nucleon. It had been a complication in the analysis of the interaction between cluster and cluster to encounter exchange, or velocity-dependent, terms as well as direct terms. The first are inescapable. The second are not. They too would become velocity-dependent terms if the elementary interaction between nucleon and nucleon were itself velocity-dependent. Or, as I wrote at the time, "A distinction must be made between possible velocity dependence of the forces between the fundamental particles themselves, and the variations of intergroup forces with velocity, of which we speak here." Who is to say that the elementary interaction between particles is not itself velocity-dependent? At least four kinds of coupling between nucleon and nucleon had already been proposed and were then currently the subject of investigation, differing one from another by their exchange properties and by their consequences for nuclear binding and the saturation of nuclear forces. If in the interaction between one cluster and another exchange of constituents automatically gave rise to a velocity-dependent component in the effective interaction, and if the nucleons themselves were also complex entities, would it not be natural for their own "elementary" coupling to have an exchange character? Katherine Way and I discovered[217] that a velocity-dependent elementary interaction of the mathematically simplest character one could possibly imagine, a Gaussian exponential function, gave a reasonable account of the saturation of nuclear forces as well as the binding of the canonical lightest nuclei. Therefore, we had two motivations—simplicity and reasonableness—for adopting, and did adopt, velocity-dependent elementary forces in our further applications of the method of resonating group structure to problems of nuclear binding and nuclear scattering.

Detail, Detail? Compute, Compute?

What next? It would have been conceivable to try to apply the method of resonating group structure in all detail from the lightest nuclei to the heaviest nuclei. However two difficulties stood in the way of such a program, one calculational, the other foundational. If such an approach were to be feasible at all, it ought to be feasible for the two-α-particle system which had led me into resonating group structure in the first place. In endeavoring to treat this problem I did weeks and weeks of number work which would take a negligible time with today's computers. If it took so much time to determine the three parts of the interaction starting with one choice for the original internal α-particle wave function, it would take hopelessly long to redo the calculations with alternative choices for the internal α-particle wave function. That meant that there would be no way to have proper assurance about the stability of the answer for calculated phase shifts. After all, the α-α coupling is the small residual left over after enormous internal coupling forces are taken into account. Being in effect the small difference between large terms, it could be evaluated reliably only by using calculational facilities not then available. Therefore, I dropped these calculations.

Second, it was already beginning to become apparent that it

would be a never-ending struggle to arrive by semiempirical means at an adequate description of the interaction between nucleon and nucleon. The observations had brought to light or were bringing to light such effects as the quadrupole moment of the deuteron, the difference in binding between the ^1S and ^3S states of the deuteron, the binding energies of H^3, He^3, and He^4; and soon Heinz Barschall and I were to find from the scattering of 2.5 MeV neutrons in helium the extraordinary strength of the spin-orbit coupling.[218] The more experimental detail one got, the more parameters one had to introduce in the "elementary interaction." It did not look as if this were destined to be the best way of finding out about atomic nuclei. There are "almost insuperable mathematical difficulties in the way of attempts to make nuclear spectroscopy alone yield any complete and unambiguous account of nuclear *forces*. Concerning the symmetry properties and certain other general features of the nuclear *wave function*, however, it appears possible to draw definite conclusions which depend on only a few very general assumptions about the exchange nature of nuclear forces—not, for example, on the number of particles involved in the interactions, nor particularly on the way the forces may vary with velocity."[219]

What recommended itself in preference to detailed calculations was a statistical or broad-brush treatment of a wide sweep of nuclear properties, supplemented by symmetry considerations, the principle of microscopic reversibility, and other dependable guides including the Breit-Wigner[220] theory of nuclear resonances.

If I had ever had any faith in the philosophy "give me the Hamiltonian and I will give you everything," that faith was now gone. If such a program could not be carried out for liquids, where one knew the interaction, how could it for nuclei, where one didn't? Moreover, I had lost all hope that one could ever determine the elementary interaction between nucleon and nucleon from phase shifts for scattering. I had had to recognize that the elementary interaction between objects with a specified relative angular momentum in principle depends on their relative velocity as well as on their separation; or, in mathematical terms, requires the "kernel of an integral operator," again a function of two variables. However, there is no way whatsoever of determining an arbitrary function of two variables from a function of one variable: here, the dependence of phase shift on velocity. Was it not time to conclude that the 1934 program was dead? By 1937–38 was it not appropriate to be guided by the catch-as-catch-can methods of molecular physics and solid state physics? Had not Einstein given us the definition of a scientist as "an unscrupulous opportunist?" Instead of unthinkingly "turning the crank of the standard Hamiltonian sausage grinder" was not one going to require a separate method to recognize each distinct aspect of nuclear behavior? What then was more natural than to consider the α-particle model for α-particle nuclei, the liquid drop model and the compound nucleus model for the reactions of more complex nuclei?

The α-Particle Model and Nuclear Rotation

The α-particle model had a simple philosophy. Abandon first principles in favor of a purely empirical estimate of the effective coupling between α-particle and α-particle. Such an estimate Hafstad and Teller[221] proposed and supplied, using for the purpose the known binding energies of such nuclei as Be^8, C^{12}, O^{16}, Ne^{20}, and Mg^{24}. Even simpler to treat than binding energies on such a semiempirical foundation, I had already found, were the lowest rotational and vibrational excitations of these α-particle nuclei. All that entered was the quantum mechanics of a simple "molecule," supplemented by appropriate symmetry considerations. One has since achieved greater accuracy in calculating the energies of the low-lying excitations of these nuclei but never greater simplicity.

What about the simplest system, the one composed of two α-particles? For the α-α scattering the differential cross section was to be measured anew and with greater precision in 1937 by Mohr and Pringle[222] and in 1939 by Samuel Devons.[223] Their results gave occasion to lay out in detail in 1941 the methods I had developed to get phase shifts from scattering and to state what one knows in consequence about the phase shift in its dependence on energy. One rock upon which the analysis is founded appears in this sentence from the summary paper[224] (p. 19): "An argument given elsewhere[225] shows that the interchange of neutrons and protons in close collisions between two alpha-particles does not impair the validity of Eq. (1) [the formula for cross section in terms of phase shifts]."

A second, immediately following paper[226] evidences the new outlook, gained in seven years of contact with the a-a problem. It interprets the phase shifts, not in terms of a "law of force" between one a-particle and another, but in terms of the energy levels of the compound nucleus Be^8. It uses for that purpose all relevant data from whatever source. It puts special emphasis on rates of decay or breakup of the states in question.

What about the rotation of heavier nuclei? Nothing was more obvious than to see what one could get out of the liquid drop model.[227] The usefulness of this model had been demonstrated in von Weizsäcker's semiempirical formula for nuclear masses.[228] Could one not go further and consider the implications of a surface tension for nuclear dynamics? And what was more natural than to go from statics to dynamics by the intermediate step of a steady rotation? Katherine Way[229] and I asked each other what would be the order of magnitude of the magnetic moments one would expect for atomic nuclei on the liquid drop model. One day she came in and reported a difficulty. The equations gave no solution for the case of a sufficiently highly charged nucleus turning at a sufficiently great angular velocity. It was clear that one had to do in this case with some kind of instability. It took only 1939 and the discovery of Hahn and Strassmann to recognize the nature of the instability: nuclear fission. Why did we not go to the analysis of higher order terms in the deformation energy and predict fission in advance of its discovery? It was not any difficulty in the mathematics. It was a difficulty in the model. It failed to give the right magnitudes and the right trends for nuclear magnetic moments. We had to recognize —and K. Way pointed out and emphasized[230]—that in magnetic moments we were dealing with a nuclear property especially sensitive to what we would today call "individual particle effects," effects evident in the data and already beginning to be taken into account in the "odd-particle model."[231] We were not in the domain in which a statistical or liquid drop model of rapidly rotating nuclei is possible.

If turning as a rigid body did not give an acceptable picture for nuclear rotation, what model would? Edward Teller had ideas on this subject and I had case examples, which we developed together in a paper on nuclear rotation.[232] The main point was simple. The very low-lying rotational energy levels that one would have expected on the basis of rigid rotation would of necessity be "promoted" in energy for an elementary reason: A rotation, followed by a minor displacement of the particles in the system, would restore the set of particles to a configuration, identical up to a permutation, with the original configuration. Therefore, it was all important to know what kind of potential energy barrier had to be overcome to perform this slight displacement. We examined several idealized models. They showed how sensitive is the position of low-lying states of high angular momentum to the energy of the deformation. We concluded that it is reasonable to believe that certain low-lying levels are indeed rotational states.

A Look Ahead: Individual-Particle Properties and the Collective Model

A few words may be permitted about the sequel as I saw it, although it lies outside the timeframe of our conference. While exploring gravitational action at a distance at Paris, September 1949-January 1950, I made several trips to Copenhagen. There Niels Bohr and I were developing jointly the so-called collective model of nuclear properties (Hill and Wheeler [1953]).[233] By that time the experimental evidence on nuclear spins and magnetic moments, so impressively correlated by Maria Mayer[234] and by Haxel, Jensen, and Suess,[235] had made it inescapably evident that nuclei show "independent particle" properties. Niels Bohr and I were having to come to terms with The Great Accident of nuclear physics—the circumstances that the mean free path of particles in the nucleus is neither extremely short compared with nuclear dimensions (as assumed in the liquid drop picture) nor extremely long (as assumed in the earliest days of nuclear physics), but of an intermediate value. When we took this fact into account we found that we could understand how a nucleus could at the same time show independent particle properties and yet behave in many ways as if it were a liquid drop.

Returning to Paris on the train from one of those stimulating and very busy week-long stays at Carlsberg, I suddenly realized that the collective model makes sense also for nuclear rotation. Thus the individual particle will circulate around fairly freely inside the nucleus and will deform the nuclear "wall," imparting a large quadrupole moment to the system. The great magnitude of the observed quadrupole moments could be accounted for neither through the individual particle model by itself nor through the liquid drop model by itself but only by the two working in combination, as in the collective model. It took only a few weeks to learn that this picture of nuclear quadrupole moments had been anticipated and was being published by James Rainwater.[236] In retrospect I can well believe that Niels Bohr, the soul of integrity, had had through Aage Bohr some sixth sense of what Rainwater was doing and, not meaning to, had nevertheless let slip some tiny phrase—recorded only in the subconscious—that set me off on the same line of thought. Thereafter Bohr and I took his insight as the best of all ways to explain at once the necessity and the workability of the collective model, and as a lead-in to its application to nuclear structure, nuclear excitations, and nuclear reactions. Still in the future lay the discovery by Aage Bohr[237] and by him and Ben Mottelson[238] of how many features of atomic nuclei can be understood by analyzing the coupling of the angular momenta of individual particles to the collective motion of the nucleus.

Further Development of the Compound Nucleus Model
and Its Specialization, the Liquid Drop Model, 1935-38

In 1935 the liquid drop model implied that one was idealizing the mean free path of a nucleon as short as compared with nuclear dimensions. On a visit to the United States in 1937, Niels Bohr and Fritz Kalckar described how much could be done with that model in accounting for nuclear excitations and nuclear reactions. In effect this model provided a workable specialization of the more general compound nucleus model. The central point of Bohr's compound nucleus model of nuclear reactions is simple: The fate of a nucleus is independent of the mechanism by which it is formed. In other words, a nuclear reaction occurs in two well separated stages. First, the particle enters the nucleus and imparts excitation to the system. Then the resulting compound nucleus, with no reference whatsoever to how it got the energy, uses it for radiation, neutron or a-particle emission, or any other competing process.

Some Men and Moments in Nuclear Physics 269

The central concept in nuclear physics, according to this idealization, is not the nucleus but the quantum state of the nucleus. That state is characterized not only by its energy but also by the level broadening, Γ, associated with its finite life. More specifically, this level broadening is the sum of contributions associated with each of the elementary processes which have the possibility to take the nucleus out of the state in question; thus

$$\Gamma = \Gamma_\alpha + \Gamma_n + \Gamma_\beta + \Gamma_{r_1} + \Gamma_{r_2} + \ldots$$

Here the subscripts refer respectively to α-decay, neutron emission, β-decay, a radiation process that leads to the first or second or other state of the nucleus, and so on. Each contribution to the line broadening is directly proportional to the elementary Rutherford-Soddy or Einstein decay constant or probability per second associated with the process in question; thus,

$$\Gamma_\alpha(\text{erg}) = \hbar(\text{erg sec}) A_\alpha(\text{sec}^{-1}).$$

Know these decay constants for each process and each level, and one will know the story of nuclear transformations: that is the ideal. Of course no one would expect the compound nucleus model to apply to a system as light as deuterium or tritium. Neither does it make sense when one deals with the breakup of a deuteron in the Coulomb field of a heavy nucleus, nor in other situations where it is evident from the start that no proper compound nucleus has been formed. Great, however, is the number of cases where a system approximating a compound nucleus is formed; and often good in order of magnitude is the account that this model gives for the cross section and distribution in energy of the final products of the various competing reactions. No more attractive means offered itself to map out in broad outline the whole of nuclear physics.

I found myself very much taken up in this new line of development. It followed logically from the general compound nucleus model and the more specific liquid drop model that Bohr had conceived—together—in that 1935 seminar and shortly afterward had published,[239] that Bohr and Kalckar had so promisingly developed in their early 1937 lectures in the United States, that Bethe and Bacher, and Bethe, had so promptly made available,[240] that Bohr and Kalckar had expounded in their final paper,[241] and that Weisskopf had so effectively forwarded.[242] This model provided a "poor man's way" of calculating or estimating the magnitude of the elements of the reaction matrix in the energy continuum and—when supplemented by the Breit-Wigner formula—even near and at a resonance. What a different coloration nuclear physics was beginning to assume in my eyes in 1938 than it had had in 1934 when I gave my ten-minute paper in Washington! It fell to me to present nuclear physics from this new point of view in lectures at the University of Chicago in the summer of 1941, at Princeton in 1940,[243] and —still earlier—at the University of Michigan at Ann Arbor in the summer of 1939. There Fermi expounded cosmic ray physics and Heisenberg also lectured until he had to go home for what he thought—and what he told us—was to be machine gun practice in the Bavarian Alps but turned out unexpectedly to be work on uranium.[244]

Princeton

When I went to Princeton in 1938 I did not foresee that I would have a home there for thirty-eight years. I did not anticipate that there would be four years (1942-45) of heavy wartime involvement in nuclear fission and nuclear engineering,[245] followed by four years developing a cosmic ray-elementary particle laboratory (1945-49) and three years "on special commission" (1950-52). Still less did I realize that I would give twenty years (1953-73) to intensive exploration of the unsuspected consequences of Einstein's geometrical theory of gravity and would then turn to still another area, the quantum theory of measurement (1973-   ). I regarded nuclear physics as my central subject.

" 'Cepts; get the 'cepts." This Princeton student motto, heritage from Woodrow Wilson's introduction of the preceptorial system of instruction in 1905, advice to formulate in handy phrases the handful of absolutely central concepts of the subject, I found as useful to adopt in nuclear science as in other areas of physics. Students could joke that "taking physics at Princeton is like trying to take a drink of water out of a firehose," but more than half the students chose to take physics at some point in their four undergraduate years, a record unsurpassed in any liberal arts institution I know.

Another recent arrival, Milton G. White, together with Eugene Wigner and Rudolf Ladenburg, were my most active associates in nuclear physics. Edward Condon, who would also make important contributions in nuclear physics, had just left. Among the many other colleagues in the Princeton community were Louis A. Turner, later to be involved in fission; Walker Bleakney, destined to move from mass spectroscopy to shock-wave physics; H. P. Robertson, a contributor to relativity physics and cosmology, with an all-around talent in mathematical physics; Gaylord P. Harnwell, soon to move and eventually to become president of the University of Pennsylvania; and finally a man of remarkable judgment and force of character, H. D. Smyth, chairman of our department.

My office was located in Fine Hall. In it were housed also the members of the School of Mathematics of the Institute for Advanced Study until their move in 1939 to the new first building of the Institute for Advanced Study about a mile away. They included Albert Einstein,[246] Oswald Veblen, John von Neumann, and Hermann Weyl. The community of mathematics and physics was small enough so that all—students and professors—could meet together at the daily tea. In later years Robert Oppenheimer characterized it as "the place where we explain to each other what we don't understand."

Milton White and the cyclotron were soon dedicated to measurements of the β-decay energies of mirror nuclei inspired by the recent work of Feenberg and Wigner[247] and Wigner[248] on this topic. The best opportunity to pursue experiments on nuclear reaction rates and cross sections of the kind in which I was interested lay in the much more modest equipment elsewhere in the laboratory. With it Malcolm MacPhail and J. Giarratana measured the scattering of neutrons by C, N, Na, Mg, and Al in its dependence on energy.[249] I found it fascinating to see how these results fell into order.[250] Important ingredients in the interpretation went back to Breit's treatment of wave functions near resonance,[251] the Breit-Wigner formula,[252] and the work of Kapur and Peierls.[253]

It was equally interesting to follow the experiments of Ladenburg, Kanner, Barschall, and van Voorhis in the attic of Palmer Laboratory. There with what today would be called primitive apparatus they used the reaction

$$D^2 + D^2 \rightarrow He^3 + n^1$$

to produce neutrons and to measure cross sections[254] — in preparation, neither they nor I realized, for a cross section measurement of great importance for the future.

Among the courses I gave was one on nuclear physics, wonderful opportunity to review the whole sweep of the subject and connect it with my own ongoing work. I also had a responsibility for the Monday night Journal Club: to seek out "hot topics" and find colleagues at any level—student or faculty—to report them, typically three to an evening. We thoughtfully arranged it so chairs would be in short supply: get there at 7:30 P.M. or bring your own chair! How else could we meet the demand to be finished in time for the nine o'clock movie at the Garden Theatre?

Fission

Four months after my arrival at Princeton I was able to arrange a special treat for the weekly Journal Club. Monday, January 16, 1939, I met the Swedish-American Line's MS Drottingholm at the pier in New York. Across the stormy North Atlantic, unbeknownst to any of us, it brought the fateful find of fission from the Old World to the New: a few words spoken by Otto Robert Frisch to Niels Bohr right before he boarded the ship; a few words spoken by Bohr to me after his arrival. Coming with me to meet Bohr and his son Erik and Léon Rosenfeld were Enrico and Laura Fermi. They took father and son off for an overnight visit in New York before Bohr's three-month stay in Princeton, where he was to lecture on the quantum theory of measurement. Rosenfeld, his collaborator in the preparation of these lectures, I took on the train to Princeton and induced him to speak at the evening Journal Club on Hahn and Strassmann's discovery,[255] still unknown in the U.S., on Meitner and Frisch's proposal[256] that this was the breakup of a liquid droplet, and on Bohr's conclusion, worked out in conjunction with Rosenfeld on the voyage over, and written up four days later,[257] that this new process fits in naturally with the compound nucleus theory of nuclear reactions. Rosenfeld's account created great excitement! It also greatly distressed Bohr when he heard about it. He felt obliged to protect Meitner and Frisch until their publication should have appeared (February 11, 1939); and to protect Frisch until his intended ionization chamber experiment at Copenhagen should have been done (January 16) and published[258] (February 25, 1939). For that reason he had said nothing about the discovery either at Columbia or to the Fermis.

Bohr once arrived in Princeton, we set to work to go from Frisch and Meitner's broad-brush picture to a detailed analysis of the mechanism along the lines of the compound nucleus model and liquid drop model that Bohr—and I—had already been expounding and applying. This work took not only the three months of Bohr's stay in Princeton but two additional months of finishing up until I could send it in for publication (June 28, 1939). The topics that had to be taken up are seen in this quotation from the paper:[259]

[We] estimate quantitatively in Section I by means of the available evidence the energy which can be released by the division of a heavy nucleus in various ways, and in particular examine not only the energy released in the fission process itself, but also the energy required for subsequent neutron escape from the fragments and the energy available for beta-ray emission from these fragments.

In Section II the problem of the nuclear deformation is studied more closely from the point of view of the comparison between the nucleus and a liquid droplet in order to make an estimate of the energy required for different nuclei to realize the critical deformation necessary for fission.

In Section III the statistical mechanics of the fission process is considered in more detail, and an approximate estimate made of the fission probability. This is compared with the probability of radiation and of neutron escape. A discussion is then given on the basis of the theory for the variation with energy of the fission cross section.

In Section IV the preceding considerations are applied to an analysis of the observations of the cross sections for the fission of uranium and thorium by neutrons of various velocities. In particular it is shown how the comparison with the theory developed in Section III leads to values for the critical energies of fission for thorium and the various isotopes of uranium which are in good accord with the considerations of Section II.

In Section V the problem of the statistical distribution in size of the nuclear fragments arising from fission is considered, and also the questions of the excitation of these fragments and the origin of the secondary neutrons.

Finally, we consider in Section VI the fission effects to be expected for other elements than thorium and uranium at sufficiently high neutron velocities as well as the effect to be anticipated in thorium and uranium under deuteron and proton impact and radiative excitation.

The whole enterprise was very much to Bohr's taste, liking as he did to see any part of physics with which he was concerned brought together in a comprehensive and harmonious whole. In addition, he had always loved the subject of capillarity. For one of his first pieces of student research he had experimented on the instability of a jet of water against breakup into smaller drops.

The Fission Barrier

A new feature of capillarity entered in the case of fission, the concept of fission barrier. The very idea was new and strange. More than one distinguished colleague objected that no such quantity could even make sense, let alone be defined. According to the liquid drop picture, is not an ideal fluid infinitely subdivisible? And therefore cannot the activation energy required to go from the original configuration to a pair of fragments be made as small as one pleases? We obtained guidance on this question from the theory of the calculus of variations in the large, maxima and minima, and critical points. This subject I had absorbed over the years by osmosis from the Princeton environment, so thoroughly charged by the ideas and results of Marston Morse.[260] It became clear that we could find a configuration space to describe the deformation of the nucleus. In this deformation space we could find a variety of paths leading from the normal, nearly spherical configuration over a barrier to a separated configuration. On each path the energy of deformation reaches a highest value. This peak value differs from one path to another. Among all these maxima the minimum measures the height of the saddle point or fission threshold or the activation energy for fission. The fission barrier *was* a well-defined quantity!

Bohr knew from earlier days that a work of Lord Rayleigh would have something to say about the capillary oscillations of a liquid drop. We rushed up to the library on the next floor of Fine Hall and looked it up[261] in the *Scientific Papers* of Rayleigh.[262] This work furnished a starting point for our analysis. However, we had to go to terms of higher order than Rayleigh's favorite second-order calculations to pass beyond the purely parabolic part of the nuclear potential, that is, the part of the potential that increases quadratically with deformation. We determined—as soon also did Feenberg,[263] von Weizsäcker,[264] Frenkel,[265] and others—the third-order terms to see the turning down of the potential. They enabled us to evaluate the height of the barrier, or at least the height of the barrier for a nucleus whose charge was sufficiently close to the critical limit for immediate breakup.

We found that we could reduce the whole problem to finding a function f of a single dimensionless variable x. This "fissility parameter" measures the ratio of the square of the charge to the nuclear mass. This parameter has the value 1 for a nucleus that is already unstable against fission in its spherical form. For values of x close to 1, by the power series development mentioned above, one could estimate the height of the barrier and actually give quite a detailed calculation of the first two terms in the power series for barrier height, or f, in powers of $(1 - x)$. The opposite limiting case also lent itself to analysis. In this limit the nucleus has such a small charge that the barrier is governed almost entirely by surface tension. The Coulomb forces give almost negligible assistance in pushing the material apart.

Between this case (the power series about $x = 0$) and the other case (the power series about $x = 1$) there was an enormous gap. We saw that it would take a great amount of work to calculate the properties of the fission barrier at points in between. Consequently we limited ourselves to interpolation between these points. In the twenty-eight years since that time many workers have done an enormous amount of computation on the topography of the deformation energy as depicted over configuration space as a "base" for the topographic plot. Work by Wladyslaw J. Swiatecki and his collaborators[266] at Berkeley, by Strutinski and his colleagues in the USSR,[267] and by Ray Nix and his colleages[268] at Los Alamos has taught us much more than we ever knew before about the structure of this fission barrier. It continues to reveal many previously unsuspected features for values of x that are remote from the two simple, original limits.

Two Highly Fissile Nuclei: Pu^{239} and U^{235}

For our immediate needs, however, our simple "poor man's" interpolation was adequate. With it, knowing—or estimating from observation—the fission barrier for one nucleus, we could estimate the fission barrier for all the other heavy nuclei, among them plutonium 239. Thanks to the questioning of Louis A. Turner, soon to write his great and timely review of nuclear fission,[269] we came to recognize that this substance, which up to then one had never seen except through its radioactivity (Abelson and McMillan [June 15, 1940]),[270] would be fissile. This conclusion was soon to lead to a preposterous dream: by means of a neutron reactor such as never before existed, manufacture kilograms of an element never before seen on earth. With the estimate of barrier height the first step had been taken, if we could only have realized it, toward the giant three hundred million dollar[271] wartime plutonium project.

The barrier height of a compound nucleus against fission was not the only factor relevant for fission. Equally important in governing the probability of this process was the excitation, or "heat of condensation," delivered up by the uptake of a neutron to form the compound nucleus in the first place. On this point an important development occurred on a snowy morning when I was occupied with classes and not with Bohr. He, having breakfast at the Nassau Club with Rosenfeld and with an arrival of the night before, George Placzek, faced Placzek's continuing skepticism about the very existence of fission. Placzek asked, how can it possibly make sense that slow neutrons and fast neutrons cause uranium to split but not neutrons of intermediate energy? Bohr stopped but said not a word, left with Rosenfeld, crossed the campus to Fine Hall still without a word and there, when Placzek and I joined them, explained the great idea[272] that had just come to him: that the slow neutron fission takes place in the rare isotope U^{235} and the fast neutron fission in the abundant isotope U^{238}. Thus an incoming neutron delivers up a high heat of condensation when it enters into a nucleus with 143 neutrons, because it can form a new neutron pair. This excitation puts the compound nucleus U^{236} over the barrier summit. Therefore, the U^{236} must split when it is formed by slow neutron capture. Moreover, the cross section for fission of the rare U^{235}, like the cross section for the "fission" of boron,

$$n^1 + B^{10} \to He^4 + Li^7,$$

must exceed by far the geometrical cross section of the nucleus for sufficiently slow neutrons. That circumstance makes it understandable why an isotope present to only one part in 139 imparts to natural uranium the observed substantial fission cross section. However, the cross section must fall off inversely as the velocity of the neutron for U^{235} as for B^{10}; hence the negligible fission cross section of natural uranium for neutrons of intermediate energy. In contrast, when a slow neutron enters U^{238} to form the compound nucleus U^{239}, no new neutron pair is formed. The heat of condensation delivered up is not enough to exceed the fission barrier. Only neutrons of a substantial kinetic energy striking U^{238} can produce U^{239} with enough energy to surmount the barrier. Hence, the existence of fast neutron fission in natural uranium.

Was it reasonable to expect so great a difference between U^{235} and U^{238} from the estimated odd-even difference in neutron binding? Could not the fission barrier differ equally drastically from the one nucleus to the other? Might not this difference be the dominant factor? How could one be sure that the proposed attribution of slow fission to U^{235} and fast fission to U^{238} really made sense until one was clear about these energies? Fortunately Bohr and I had just been through the systematics of nuclear energies in the course of calculating the release of energy in various actual and potential fission processes. Therefore, we could estimate the difference between the excitation developed by neutron capture in the two uranium isotopes as almost a million volts, in favor of fission of U^{235}. From our interpolation for fission barriers we estimated on the other hand a barrier almost 1 MeV lower for U^{235} than for U^{238}. Thus we concluded there was about a 2 MeV margin in favor of the fission of the rare isotope. In later years, after the development of the collective model[273] it became clear[274] that individual particle effects can modify significantly barrier heights and barrier shapes from the predictions of the simple liquid drop model. However, the qualitative conclusions are not affected; U^{235} is the fissile nucleus.

Placzek, wonderful person that he was, a man of the highest integrity, often a thoroughgoing skeptic about new ideas, said to me over and over in those early spring days of 1939 that he could not believe that the small amount of U^{235} could be the cause of the slow neutron effects in natural uranium. I therefore bet him a proton to an electron, $18.36 to a penny, that Bohr's diagnosis was correct. A year later Alfred Nier at Minnesota had separated enough U^{238} to make possible a test and sent it to John Dunning at Columbia to measure its fission cross section.[275] On April 16, 1940, I received a Western Union money order telegram for one cent with the one-word message "Congratulations!" signed Placzek.[276]

The "Width for Fission"

Bohr and I knew we would not understand the fission processs if we limited attention to energies; we had to know rates. How great is the partial width of a nuclear state with respect to fission? What is the probability per second that a nucleus endowed with a specified excitation will undergo fission?

"Over the barrier" fission was the process of most immediate interest. For it we could give a simple picture. With the help of deformation parameters such as had been introduced by Lord Rayleigh, we were able to represent the shape of the liquid drop model by a single point in a many dimensional configuration space. To change from one shape to another in a specified time meant a certain kinetic energy. The system had also for any configuration a definite potential energy of deformation. It could be imagined to be plotted "vertically" over a base space having for coordinates the deformation parameters. The dynamics of the charged droplet could be visualized in terms of the frictionless rolling of a marble over a white plaster model of this potential energy surface. The surface evidently has a minimum at the origin, at the point of zero deformation, corresponding to a spherical system. Therefore, the marble, given a sufficiently limited excitation, will roll around forever.

The situation changes when the energy given to the marble in the first place exceeds by even a little the critical energy for fission. This energy measures the height of a critical point in the potential energy surface, the "saddle point" or "mountain pass" or minimax that leads "over the hill." Once the representative point or the marble passes this summit, fission is guaranteed; but fission has not yet occurred. For that the marble must roll downhill to the point of actual tearing, or "scission." What is the chance the marble will get over the mountain pass in the first place? The smaller the energy available in excess of the critical energy, the better the aim must be to get over. But no one is doing any aiming. Chance alone is at work. This circumstance simplified the analysis rather than complicated it. That was a lesson familiar to all acquainted with the history of cosmic ray physics. How could I forget the many detailed calculations Størmer and his associates made on the orbits of cosmic ray particles in the earth's magnetic field?[277] Or the happy idea that G. Lemaitre and Manuel Sandoval Vallarta had, to spare themselves[278] and later workers[279] almost all these details? They had only to employ Liouville's theorem. It said that the density of systems in phase space remains constant in time. In this way they got simple results for the flux of cosmic rays onto the top of the atmosphere.

Past seminars in the Princeton community had brought familiarity with another problem even closer in spirit to fission: the theory of mononuclear reactions. With it I knew that both Henry Eyring and Eugene Wigner had been occupied in times past. Wigner was laid up in the hospital by jaundice but I took the liberty to visit him and ask for references on the subject. It turned out that he was one of the principal contributors.[280] From a study of the key papers[281] I soon realized that in the absence of special symmetries the probability per second of going over the saddle is given by the same kind of ratio whether the process is dissociation of a molecule or escape of a cosmic ray particle from a magnetic trap or fission of a nucleus. The denominator of this ratio is the volume of phase space accessible to the representative point of the system within the zone of trapping; or rather, the increase in this volume per unit increase in the energy available to the system. The numerator is the volume of a "transition state" phase space. It is defined by fixing at its critical value the coordinate leading over the saddle, stilling all velocity of movement in that direction and letting the remaining degrees of freedom of the system have any energy between the saddle-point energy and the actually available energy. Bohr and I soon found that we could give this result an extremely simple form,

$$\Gamma_f = \begin{pmatrix}\text{partial width with} \\ \text{respect to fission}\end{pmatrix} = \hbar \begin{pmatrix}\text{probability per} \\ \text{second of fission}\end{pmatrix} = \hbar A_f$$

$$= \frac{\text{(level spacing)}}{2\pi} \begin{pmatrix}\text{number of accessible} \\ \text{levels of the transition} \\ \text{state nucleus}\end{pmatrix}$$

$$= \frac{\text{(level spacing)}}{2\pi} \begin{pmatrix}\text{number of channels} \\ \text{for fission}\end{pmatrix};$$

or,

$$\Gamma_f = (D/2\pi) N_f.$$

This formula applied not only to fission but also to many other reactions. It was more general than any other that had previously been available in reaction rate theory. The simple new formula gave considerable insight into the rate of passage over the fission barrier.

Years later, reading the works of Marie Sklodowska Curie in preparation for a celebration of her centenary,[282] I was astonished to discover that she had thought about the breakup of a nucleus in just such terms. She referred to Rutherford and Soddy and their 1902 and still standard theory of radioactive decay. She recognized, as they did, that they were dealing with a process that transcended the understanding of the time. At the 1913 Solvay Congress[283] she called attention anew to the mystery of the exponential law. She stressed the experimental evidence that an atom, if it had not yet decayed, had not aged at all, no matter how long it had lived. She proposed "to look in the interior of the atom for the element of disorder necessary to explain the application of the law of chance." She brought forward the suggestion of Debierne, first, that in the center of the atom there may exist a temperature much higher than the external temperature and second, that the mechanism involved may be identical with that of a monomolecular chemical reaction. She asks us to imagine "a molecule which is moving about in the interior of a box endowed with a tiny hole." Marie Sklodowska Curie goes on to say, "When the molecule in the course of its motion meets the hole it leaves the box and the system is radically changed. If we have a great number of boxes each containing one molecule, and if the initial velocities and positions of the molecules are random, it may happen that the escape phenomenon is governed by the rule of chance, even though the constitution of the system itself is relatively simple." Mme. Curie was in advance of her age. She put forward the right idea to describe nuclear fission during an epoch when she had to deal with the leakage of a-particles through a potential barrier!

It is interesting to note the caution with which Bohr adopted our channel formula. He would come in every other day or so, and we would analyze its scope and validity for perhaps half a day, trying out first this approach and then that approach. But his supreme caution was most evident when we wanted to interpret the number of levels accessible in the transition state. Today that number is called "the number of channels," and we use it as the central quantity in the channel-analysis theory of the fission rate.[284] Also we apply similar channel analysis considerations to other nuclear reactions. But at that time the idea that each one of these individual channels has in principle a definite experimentally observable significance was, for us, an open question. Still less did we appreciate, until the later work of Aage Bohr,[285] and of Aage Bohr and Ben Mottelson,[286] the possibility that each individual channel would have its individual angular distribution from which one could determine the K values of that channel.[287] The cautious phrase that was used in reference to that channel number appears in the following quotation (p. 443): "It should be remarked that the specific quantum-mechanical effects which set in at and below the critical fission energy may even show their influence to a certain extent above this energy and produce slight oscillations in the beginning of the Γ_f/d [yield] curve, allowing possibly a direct determination of N* [the number of channels]." Of course we know how later on in the 1950s these variations were observed by Smith, Henkel, and Nobles[288] and by Lamphere and Greene[289] and others[290] and how they led to direct measurement of the channel number.

It was not hard to generalize this treatment of "over the barrier fission" to "under the barrier fission" and spontaneous fission. It was clear that there had to be and was a continuous transition from the one regime to the other. We were fortunate that right in the building measurements were being carried out by Ladenburg, Kanner, Barschall, and van Voorhis.[291] They found that the fission cross section for uranium was about 0.5×10^{-24} cm^2 and changed little with energy for neutron energies from 2 MeV to 3 MeV. For thorium the results were similar except that the cross section was five times lower. Combining these observations with the reaction rate theory that we had developed, we concluded (p. 444) that "the critical energy for U^{239} is not far from 3/4 Mev in excess of the neutron binding ($\sim$ 5.2 Mev . . .):

$$E_f(U^{239}) \sim 6 \text{ Mev}"$$

and for Th^{232} "we are led . . . to a fission barrier 1 3/4 Mev greater than the neutron binding . . . ,

$$E_f(Th^{232}) \sim 7 \text{ Mev."}$$

This is not the place to go into prompt neutrons, delayed neutrons, the physics of fission product decay, and many another topic that came up, nor to detail the many impressive experimental results that were obtained on these and other topics week by week. Enough has been said to show what nuclear physics looked like to one participant in the prewar period.

Coming Shadows

What would nuclear physics look like if war came? On this Leo Szilard had strong ideas, and with Fermi and Wigner he pushed strongly the idea of a nuclear chain reaction. We—Bohr, Rosenfeld, Teller, Wigner, and I—had a critical meeting with him in Wigner's office, March 16, two months to the day after Bohr's arrival in America. Szilard outlined the Columbia data and the preliminary indications from it that at least two secondary neutrons emerge from each neutron-induced fission. Did this not mean that a nuclear explosive was certainly possible? Bohr, however, countered, emphasizing the enormous difficulty of separating the necessary quantities of U^{235}, "It would take the entire efforts of a country to make a bomb." Little did he foresee that the efforts of thousands of workers drawn from three countries would be needed to achieve that goal.

As I look back on those days I feel a great sadness. How did it come about that I looked on fission first as a physicist and only secondarily as a citizen? Why did I not look at it first as a citizen and only secondarily as a physicist? A simple survey of the records[292] shows that between twenty and twenty-five million people perished in World War II and more of them in the later years than in the earlier years. Every month by which the war was shortened would have meant a saving of the order of half a million to a million lives. Among those granted life would have been my brother Joe, killed in October 1944 in the Battle for Italy. What a difference it would have made if the critical date had been not August 6, 1945, but August 6, 1943. If there was anywhere that the Allied nuclear project lost time it was at the beginning. How many months it took to get any money! How many more months it took to get a few tons of graphite! How many further months it took to get onto an engineering basis! The unhappily slow start of the project in the United States, like the happily slow start—and continuance—of the project in Germany,[293] came most of all from failure to establish rapport in the very beginning between the scientific work and one big man near the top with a big heart accustomed to do big things in a big way.

Why did I not have the wisdom and sense of mission to identify such a man and enlist him as protector and driving force of the enterprise in early 1939? Why did I not start working full-time on uranium February 1, 1939, rather than February 1, 1942? Position? Nobody in America had a more central position in fission physics. Too young? Men of twenty-seven and younger, history tells us, have had the courage, judgment, and force to speak the decisive word at the decisive time. Unaware that a great war was coming? No one followed more closely the dispatches of the foreign correspondents of the New York Times. Insufficient sense of involvement? That was it. I did not grasp that the vital interests of the United States were at stake. Would that I had understood history better. Would that I had been taught as a child that Gibraltar-firm rock of American foreign policy: never let all Europe fall under the dominion of a single power (Jefferson [January 1814] ;[294] Wilson [April 2, 1917];[295] Roosevelt [October 21, 1944]).[296] What a difference it would have made to have been pushed on by that sense of history. Would it have meant two months and one million lives? Or two years and ten million lives?

Having ended, and anticipating questions, I beg your indulgence if I ask the first myself: Is nuclear physics done, and this conference its obituary? Let me answer: How dismaying that anybody should have any such impression! If a few in the larger community believe that nuclear physics is 90 percent complete and only 10 percent remains to be done, my impression of the proportion is the direct opposite. There are at least five great fields of nuclear physics on which we have made only a beginning: rapidly rotating nuclei and isometric states,[297] superheavy nuclei,[298] nuclear matter in bulk,[299] nuclei far from the valley of stability, and exotic nuclei.

About those exotic systems in which a mu-meson or a pi-meson or a heavier meson spends a substantial fraction of its time within the nucleus we know something,[300] a little about nuclei containing a hyperon,[301] but practically nothing about "hexagonal nuclei," if I may use that term for nuclei made of the whole collection of baryons that fill out the familiar hexagon of baryons,

$$\begin{array}{ccc} & \Xi^- & \Xi^0 \\ \Sigma^- & \Sigma^0,\Lambda^0 & \Sigma^+ \\ & n \quad p & \end{array}$$

or these supplemented by the ultrastrange Ω^--particle. Instead of everything in nuclear physics being done, almost everything remains to be done.

REFERENCES

Not included here are the relevant portions of the nuclear physics community's classified reports. Some are now in the process of being declassified. Were abstracts or even titles to be listed, they might reveal in a single small but brilliantly lighted picture what one understood of scattering by, capture in, and fission of nuclei by 1942–45 and of thermonuclear processes by 1950–53. Most of this particular writer's unclassified materials before 1976, including research notebooks, are on deposit of the Library of the American Philosophical Society, Philadelphia. The references below in no way aim at completeness. The author thanks colleagues for help in locating difficult-to-find items, especially Felix Bloch, Julian Boyd, Salley Brown, Howard E. Carr, James Davidson, Ugo Fano, Paul Gleichauf, Beverly Kuhn, Arthur Link, Enrico Rodrigo, Wallace Sale, Daniel Sperber, Judi Taylor-Williams, and Kenneth Watson. Parts of the treatment of fission and cosmic rays are adapted and expanded from the author's Lauritsen Memorial Lecture of January 27, 1975, at the California Institute of Technology, "Great Men and Great Moments in Nuclear Physics" (unpublished).

1. J. A. Wheeler, "Interaction between Alpha-Particles" [Abstract], *Physical Review*, 45 (1934), 746.
2. J. A. Wheeler, "Theory of the Dispersion and Absorption of Helium," *Physical Review*, 43 (1933), 258–263.
3. N. Bohr and J. A. Wheeler, "Mechanism of Nuclear Fission" [Abstract], *Physical Review*, 55 (1939), 1124.

4. N. Bohr and J. A. Wheeler, "The Mechanism of Nuclear Fission," *Physical Review*, 56 (1939), 426-450.

5. R. W. Wood, *Physical Optics* (New York: Macmillan, 2nd ed., 1911; 3rd ed., 1934).

6. W. Seabrook, *Doctor Wood* (New York: Harcourt Brace, 1941).

7. R. W. Wood, *How to Tell the Birds from the Flowers and Other Woodcuts* (New York: Dodd, Mead, 1917) and many subsequent editions, in many languages.

8. R. W. Wood, "The n-Rays," *Nature*, 70 (1904), 530-531.

9. F. D. Jones, *Mechanisms and Mechanical Movements* (New York: Industrial Press, 1918); updated, 1930, as *Ingenious Mechanisms for Designers and Inventors*.

10. T. A. Janvier, *Silver Mining at Zacatecas*, with 7 illustrations by Frederic Remington, large folio, appendixes, unbound (New York: Harper, 1889), issued as a special supplement to *Harper's Weekly*, October 19, 1889.

11. J. A. Thomson, *The Outline of Science*, 4 vols. (New York: Putnam, 1922).

12. C. P. Steinmetz with E. J. Berg, *Theory and Calculation of Alternating Current Phenomena* (New York: Johnston, 1897).

13. C. P. Steinmetz, *Four Lectures on Relativity and Space* (New York: McGraw-Hill, 1923; reprinted, New York: Dover, 1967).

14. H. A. Lorentz, *Problems of Modern Physics. A Course of Lectures Delivered in the California Institute of Technology*, ed. H. Bateman (New York: Ginn, 1927).

15. P. R. Heyl, "A Redetermination of the Constant of Gravitation," [U.S.] *National Bureau of Standards Journal of Research*, 5 (1930), 1243-1290.

16. R. D. Rose, H. M. Parker, R. A. Lowry, A. R. Kuhlthau, and J. W. Beams, "Determination of the Gravitational Constant G," *Physical Review Letters*, 23 (1969), 655-658; J. W. Beams, "Finding a Better Value for G," *Physics Today*, 24 (May 1971), 34-40.

17. B. O. Peirce, *A Short Table of Integrals*, 2nd rev. ed. (Boston: Ginn, 1910), present of high school days from a cousin, Charles Archibald Blake, gives on pages 95-96 eight sample cases of Fourier series.

18. F. D. Murnaghan, *Vector Analysis and the Theory of Relativity* (Baltimore: Johns Hopkins Press, 1922).

19. H. L. Dryden, F. D. Murnaghan, and H. Bateman, *The Report of the Committee on Hydrodynamics of the Division of Physical Sciences of the National Research Council* (Washington, D.C.: National Research Council of the National Academy of Sciences, 1932); reprinted as *Hydrodynamics* (New York: Dover, 1956).

20. J. S. Ames and F. D. Murnaghan, *Theoretical Mechanics* (Boston: Ginn, 1929; reprinted, New York: Dover, 1958).

21. C. W. Misner, K. S. Thorne, and J. A. Wheeler, "Bianchi Identities and the Boundary of a Boundary," chap. 15 in *Gravitation* (San Francisco: Freeman, 1973).

22. E. J. B. Goursat, *Cours d'analyse mathématique* (Paris: Gauthier-Villars, 1929).

23. H. W. Turnbull, *The Theory of Determinants, Matrices, and Invariants* (London: Blackie, 1928).

24. A. Wintner, *Spektraltheorie der unendlichen Matrizen. Einführung in den analytischen Apparat der Quantenmechanik* (Leipzig: Hirzel, 1929).

25. J. A. Bearden and J. A. Wheeler, "The Determination of e or λ by the Dispersion of X-rays" [Abstract], *Physical Review*, 43 (1933), 1059.

26. J. A. Wheeler and J. A. Bearden, "The Variation of the K Resonating Strength with Atomic Number," *Physical Review*, 46 (1934), 755-758.

27. G. R. Harrison, "William F. Meggers Dies; Was World-Famous Spectroscopist," *Physics Today*, 20 (January 1967), 147-151.

28. F. Hund, *Linienspektren und Periodisches System der Elemente* (Berlin: Springer, 1927).

29. W. F. Meggers and J. A. Wheeler, "The Band Spectra of Scandium-, Yttrium-, and Lanthanum Monoxides" [Abstract] *Physical Review*, 37 (1931), 106; "The Band Spectra of Scandium-, Yttrium-, and Lanthanum Monoxides," [*U.S.*] *National Bureau of Standards Journal of Research*, 6 (1931), 239-275.

30. E. Rutherford, J. Chadwick, and C. D. Ellis, *Radiations from Radioactive Substances* (Cambridge: Cambridge University Press, 1930).

31. J. J. Thomson, *Conduction of Electricity through Gases* (Cambridge: Cambridge University Press, 1903); third edition in two volumes, by J. J. Thomson and G. P. Thomson, 1928 and 1933.

32. E. Segrè, ed. *Enrico Fermi: Collected Papers*, vol. 1 (Chicago: University of Chicago Press, 1962).

33. H. Weyl, *Guppentheorie und Quantenmechanik*, 2nd ed. (Leipzig: Hirzel, 1931).

34. G. N. Lewis and M. Randall, *Thermodynamics and the Free Energy of Chemical Substances* (New York: McGraw-Hill, 1923).

35. E. E. Slosson, *Creative Chemistry* (New York: Century, 1920).

36. A. H. Compton and S. K. Allison, *X-Rays in Theory and Experiment* (New York: D. Van Nostrand, 1935).

37. A. E. Ruark and H. C. Urey, *Atoms, Molecules and Quanta* (New York: McGraw-Hill, 1930).

38. J. L. Lowes, *The Road to Xanadu: A Study in the Ways of the Imagination* (Boston: Houghton Mifflin, 1927).

39. J. Hutton, "Theory of the Earth, or an Investigation of the Laws Observable in the Composition, Dissolution, and Restoration of Land upon the Globe," *Transactions of the Royal Society of Edinburgh*, 1 (1785), 209-304.

40. C. Lyell, *The Principles of Geology, Being an Attempt to Explain the Former Changes of the Earth's Surface, by Reference to Causes Now in Operation*, 3 vols. (New York: Appleton, 1830-33).

41. H. Ward, *Charles Darwin: The Man and His Warfare* (Indianapolis, Indiana: Bobbs-Merrill, 1927).

42. W. Bothe and H. Becker, "Künstliche Erregung von Kern-γ-Strahlen," *Zeitschrift für Physik*, 66 (1930), 289-306.

43. I. Curie and F. Joliot, "Émission de protons de grande vitesse par les substances hydrogénées sous l'influence des rayons γ tres penetrants," *Comptes rendus*, 194 (1932), 273-275.

44. J. Chadwick, "Possible Existence of a Neutron," *Nature*, 129 (1932), 312; "The Existence of a Neutron," *Proceedings of the Royal Society of London*, 136[A] (1932), 692-708.

45. I. M. Frank, "Neutron Optics and Ultracold Neutrons," Joint Institute for Nuclear Research Report JINR-D3-7991, Dubna, USSR (Translation CONF-740419), 1974.

46. K. F. Herzfeld, *Kinetische Theorie der Wärme*, vol. 3 of *Lehrbuch der Physik*, ed. J. H. J. Müller and C. S. M. Pouillet (Braunschweig: Vieweg, 1925).

47. K. F. Herzfeld, "Klassische Thermodynamik," *Handbuch der Physik*, vol. IX, ed. H. Geiger and K. Scheel (Berlin: Springer, 1926), pp. 1-140.

48. K. L. Wolf and K. F. Herzfeld, "Absorption und Dispersion," *Handbuch der Physik*, vol. XX, ed. H. Geiger and K. Scheel (Berlin: Springer, 1928), 480-634.

49. K. F. Herzfeld and T. A. Litovitz, *Absorption and Dispersion of Ultrasonic Waves* (New York: Academic, 1959).

50. Wheeler, "Dispersion and Absorption of Helium" (reference 2).

51. From the song "Depression Days."

52. R. C. Stauffer, ed., *Charles Darwin's Natural Selection Being the Second Part of His Big Species Book Written from 1856 to 1858* [and cannibalized by Darwin in his writing *On The Origin of Species*] (Cambridge: Cambridge University Press, 1975), pp. 535-565.

53. R. D. Sommer, ed., "Guide to Scientific Instruments," *Science*, 194, no. 4267A (1976), 3-170; see especially, p. 7.

54. Lord Kelvin [W. Thomson], *Baltimore Lectures* [of 1884] *on Molecular Dynamics and the Wave Theory of Light* (London: C. J. Clay and Sons, Cambridge University Press Warehouse, 1904).

55. F. Bloch, "Reminiscences of Heisenberg and the Early Days of Quantum Mechanics," *Physics Today*, 29 (December 1976), 23-27.

56. P. Ehrenfest and G. Breit, "A Remarkable Case of Quantization," *Proceedings of the Koninklijke Akademie van Wetenschappen, Amsterdam*, 25 (1923), 2-5.

57. P. Ehrenfest and J. R. Oppenheimer, "Note on the Statistics of Nuclei," *Physical Review*, 37 (1931), 333-338.

58. National Research Council Fellowships were small in number, great in obligation. First-time awards in physics in 1933 numbered fourteen, according to a kind October 20, 1978, communication from Beverly Kuhn of the Fellowship Office of the National Academy of Sciences.

59. F. Darwin, ed. *Charles Darwin, His Life Told in an Autobiographical Chapter and in a Selected Series of His Published Letters* (New York: Appleton, 1892; reprinted as *The Autobiography of Charles Darwin and Selected Letters* (New York: Dover, 1958); the quoted passage is on p. 58 in both editions.

60. G. Breit, "On the Possibility of Nuclear Disintegration by Artificial Sources" [Letter], *Physical Review*, 34 (1929), 817-818.

61. *Ibid.*

62. G. Breit, M. A. Tuve, and O. Dahl, "A Laboratory Method of Producing High Potentials," *Physical Review*, 35 (1930), 51-65; M. A. Tuve, G. Breit, and L. R. Hafstad, "The Application of High Potentials to Vacuum-Tubes," *Ibid.*, 35 (1930), 66-71.

63. R. J. Van de Graaff, K. T. Compton, and L. C. Van Atta, "The Electrostatic Production of High Voltage for Nuclear Investigations," *Physical Review*, 43 (1933), 149-157.

64. M. A. Tuve and L. R. Hafstad, "The Emission of Disintegration-Particles from Targets Bombarded by Protons and by Deuterium Ions at 1200 Kilovolts" [Letter], *Physical Review*, 45 (1934), 651-653.

65. *Ibid.*; G. Breit, "The Dependence of Radio Fading on Modulation" [Abstract], *Physical Review*, 25 (1925), 589; G. Breit and M. A. Tuve, "A Test of the Existence of the Conducting Layer," *Ibid.*, 28 (1926), 554-575.

66. G. Breit, "The Effect of Retardation on the Interaction of Two Electrons" [Letter], *Physical Review*, 34 (1929), 375; "The Fine Structure of He as a Test of the Spin Interactions of Two Electrons," *Ibid.*, 36 (1930), 383-397; "Dirac's Equation and the Spin-Spin Interactions of Two Electrons," *Ibid.*, 39 (1932), 616-624.

67. G. Breit and I. I. Rabi, "Measurement of Nuclear Spin" [Letter], *Physical Review*, 38 (1931), 2082-2083.

68. G. Breit, "Quantum Theory of Dispersion," *Reviews of Modern Physics*, 4 (1932), 504-576; "Quantum Theory of Dispersion (continued). Parts VI and VII," *Ibid.*, 5 1933), 91-140.

69. N. P. Heydenburg, "Paschen-Back Effect of Hyperfine Structure and Polarization of Resonance Radiation. Cadmium (6^1P_1-5^1S_0)," *Physical Review*, 43 (1933), 640-647;

"The Nuclear Magnetic Moment of Caesium from the Polarization of Resonance Radiation," *Ibid.*, 46 (1934), 802–805.

70. G. Breit and E. Wigner, "Capture of Slow Neutrons," *Physical Review*, 49 (1936), 519–531.

71. G. Breit and E. Wigner, "The Saturation Requirements for Nuclear Forces," *Physical Review*, 53 (1936), 998–1003.

72. G. Breit and G. E. Brown, "Effect of Nuclear Motion on the Fine Structure of Hydrogen," *Physical Review*, 74 (1948), 1278–1284; G. Breit, G. E. Brown, and G. B. Arfken, "The Effect of Nuclear Motion on the Hyperfine Structure of Hydrogen," *Ibid.*, 76 (1949), 1299–1304; G. Breit and G. E. Brown, "Perturbation Methods for Dirac Radial Equations," *Ibid.*, 76 (1949), 1307–1310.

73. R. T. Cox, C. G. McIlwraith, and B. Kurrelmeyer, "Apparent Evidence of Polarization in a Beam of β-Rays," *Proceedings of the National Academy of Science [Physics]*, 14 (1928), 544–548. For reference to subsequent work of Cox and collaborators on this effect and of its being truly a first indication of parity nonconservation see Lee Grodzins, "Measurement of Helicity," in *Progress in Nuclear Physics*, vol. 7, ed. O. R. Frisch (New York: Pergamon Press, 1959), pp. 163–241; and historical articles by R. T. Cox and L. Grodzins and others, pp. 138–162 in chap. 3, "Discovery of Parity Violation in Weak Interactions," *Advances in Experimental Physics*, ed. B. Maglich, γ volume, 1973; and see the articles of F. E. Myers and R. T. Cox, *Physical Review*, 34 (1929), 1067–1068; and of C. T. Chase, *Ibid.*, 34 (1929), 1069–1074; *Ibid.*, 36 (1930), 984–987; *Ibid.*, 36 (1930), 1060–1065. We are deeply indebted to Professors Gregory Breit, Richard Cox, and Lee Grodzins for assistance in finding this material.

74. T. D. Lee and C. N. Yang, "Question of Parity Conservation in Weak Interactions," *Physical Review*, 104 (1956), 254–258; C. S. Wu, E. Ambler, R. W. Hayward, D. D. Hoppes, and R. P. Hudson, "Experimental Test of Parity Conservation in Beta Decay" [Letter], *Ibid.*, 105 (1957), 1413–1415.

75. J. A. Wheeler, gatherer, "Mercer Street and Other Memories [of Albert Einstein]," University of Texas at Austin preprint, 1978; to be published in *Albert Einstein, 1879–1979*, ed. P. C. Aichelberger and R. Sexl (Wiesbaden: Vieweg; London: Heyden; and, in German translation, in a book of the same title, Wiesbaden: Vieweg).

76. H. Mildrum and B. Schmidt, "The Allison Method of Chemical Analysis," United States Air Force Aeropropulsion Laboratory Technical Report AFA PLTR–66–52, 1966; contains extensive bibliography.

77. F. L. Yost, G. Breit, and J. A. Wheeler, "Coulomb Wave-Functions," *Journal of Terrestrial Magnetism and Atmospheric Electricity*, 40 (1935), 443–447; "Coulomb Wave Functions in Repulsive Fields," *Physical Review*, 49 (1936), 174–189.

78. J. A. Wheeler, "Wave Functions for Large Arguments by the Amplitude-Phase Method," *Physical Review*, 52 (1937), 1123–1127.

79. J. A. Wheeler, "Semiclassical Analysis Illuminates the Connection between Potential and Bound States and Scattering," in *Studies in Mathematical Physics: Essays in Honor of Valentine Bargmann*, ed. E. H. Lieb, B. Simon, and A. S. Wightman (Princeton, New Jersey: Princeton University Press, 1976), pp. 351–422.

80. J. A. Wheeler and G. Breit, "Li$^+$ Fine Structure and Wave Functions near the Nucleus" [Letter], *Physical Review*, 44 (1933), 948.

81. Wheeler, "Interaction between Alpha-Particles (reference 1).

82. G. Breit and J. A. Wheeler, "Collision of Two Light Quanta," *Physical Review*, 46 (1934), 1087–1091; preliminary report in G. Breit and J. A. Wheeler, "Collision of Two Quanta" [Abstract], *Ibid.*, 45 (1934), 766.

83. H. C. Urey, F. G. Brickwedde, and G. M. Murphy, "A Hydrogen Isotope of Mass 2 and Its Concentration," *Physical Review*, 40 (1932), 1-15.

84. *International Conference on Physics. London 1934. A Joint Conference Organized by the International Union of Pure and Applied Physics and the Physical Society. Papers and Discussions in Two Volumes. Vol. I. Nuclear Physics* (Cambridge: Cambridge University Press and the Physical Society, 1935).

85. C. Weiner, ed., *History of Twentieth Century Physics. Proceedings of the International School of Physics "Enrico Fermi," Varenna on Lake Como, July-August 1972* (New York: Academic, 1977); "Institutional Settings for Scientific Change; Episodes from the History of Nuclear Physics," in *Science and Values*, ed. A. Thackray and E. Mendelsohn (Atlantic Highlands, New Jersey: Humanities Press, 1974), pp. 187-212.

86. W. Bleakney, "Additional Evidence for an Isotope of Hydrogen of Mass 2," *Physical Review*, 39 (1932), 536.

87. J. D. Cockcroft and E. T. S. Walton, "Experiments with High Velocity Positive Ions. II. The Disintegration of Elements by High Velocity Protons," *Proceedings of the Royal Society of London*, 137[A] (1932), 229-242; "Disintegration of Light Elements by Fast Protons," *Nature*, 131 (1933), 23.

88. Chadwick, "Possible Existence of a Neutron"; "Existence of a Neutron" (reference 44).

89. E. O. Lawrence, M. S. Livingston, and M. G. White, "The Disintegration of Lithium by Swiftly-Moving Protons," *Physical Review*, 42 (1932), 150-151.

90. C. D. Anderson, "The Apparent Existence of Easily Deflectable Positives," *Science*, 76 (1932) 238-239; "The Positive Electron," *Physical Review*, 43 (1933), 491-494.

91. *International Conference on Physics* (reference 84).

92. E. Fermi, E. Amaldi, B. Pontecorvo, F. Rasetti, and E. Segrè, "Azione di sostanze idrogenate sulla radioattività provocata da neutroni.—I.," *La ricerca scientifica*, 5 (2) (1934), 282-283.

93. J. Chadwick and M. Goldhaber, "A 'Nuclear Photo-Effect': Disintegration of the Diplon by γ-Rays," *Nature*, 134 (1934), 237-238.

94. H. A. Bethe and R. Peierls, "Photoelectric Disintegration of the Diplon," *International Conference on Physics*, pp. 93-94 (reference 84).

95. H. A. Bethe and R. F. Bacher, "Theory of the Deuteron," in "Nuclear Physics. A. Stationary States of Nuclei," *Reviews of Modern Physics*, 8 (1936), 105-114.

96. W. Pauli, Open letter to physics meeting in Tubingen suggesting "that in β-decay a neutron [later to be called 'neutrino'] is emitted together with the electron, in such a way that the sum of the energies of neutron and electron is constant"; original German in W. Pauli, *Aufsätze und Vorträge über Physik und Erkenntnistheorie* (Braunschweig: Vieweg, 1961), p. 159; *Collected Scientific Papers*, vol. 2 (New York: Interscience, 1964), p. 1313; proposal of the neutrino in the discussion of Heisenberg's paper on the proton-neutron model of the nucleus, *Structure et propriétés des noyaux atomiques. Rapports et discussions du Septième Conseil de Physique Solvay, Brussels, October 22-29, 1933*, Paris, 1934; this and the further development of Pauli's proposal are recounted in L. M. Brown, "The Idea of the Neutrino," *Physics Today*, 31 (September 1978), 23-28.

97. E. Fermi, "Tentativo di una teoria dell'emissione dei raggi 'beta'," *La ricerca scientifica*, 4 (2) (1933), 491-495 (published in this journal when the original manuscript was rejected by another journal); "Tentativo di una teoria dei raggi β," *Nuovo cimento*, 11 (1934), 1-19; for the German version see "Versuch einer Theorie der β-Strahlen. I.," *Zeitschrift für Physik*, 88 (1934), 161-177; F. Rasetti's history of ideas leading to Fermi's beta-decay theory in *Enrico Fermi: Collected Papers*, ed. E. Segrè (Chicago, University of Chicago Press, 1962), vol. 1, pp. 538-540.

98. F. Reines, "The Early Days of Experimental Neutrino Physics," 24 pages and figures, September 12, 1978, preprint intended for submission to *Science*.

99. C. D. Ellis and N. F. Mott, "Energy Relations in the β-Ray Type of Radioactive Disintegration," *Proceedings of the Royal Society of London*, 141[A] (1933), 502-511.

100. C. D. Ellis, "The β-Ray Type of Radioactive Disintegration," London conference advance paper reprinted in *International Conference on Physics*, pp. 43-59 (reference 84).

101. L. H. Gray and G. T. P. Tarrant, "The Nature of the Interaction between Gamma Radiation and the Atomic Nucleus," *Proceedings of the Royal Society of London*, 136 [A] (1932), 662-691; "Phenomena Associated with the Anomalous Absorption of High Energy Gamma Radiation.—II," *Ibid.*, 143[A] (1934), 681-706; "III," *Ibid.*, 706-724.

102. C. D. Anderson and S. H. Neddermeyer, "Fundamental Processes in the Absorption of Cosmic-Ray Electrons and Photons," *International Conference on Physics*, pp. 171-187 (reference 84).

103. B. Rossi, "Some Results Arising from the Study of Cosmic Rays," London conference advance paper reprinted with October 1934 supplement in *International Conference on Physics*, pp. 233-247 (reference 84).

104. H. A. Bethe, "The Influence of Screening on the Creation and Stopping of Electrons," *Proceedings of the Cambridge Philosophical Society*, 30 (1934), 524-539.

105. H. A. Bethe and W. Heitler, "On the Stopping of Fast Particles and on the Creation of Positive Electrons," *Proceedings of the Royal Society of London*, 146[A] (1934), 83-112.

106. J. J. Thomson, *Electricity and Matter* (New York: Scribner's Sons, 1904).

107. S. Rozental, ed., *Niels Bohr: His Life and Work as Seen by His Friends and Colleagues* (Amsterdam: North-Holland, 1967), translated from the Danish book of the corresponding title (Copenhagen: Schultz, 1964); R. Moore, *Niels Bohr: The Man, His Science, and the World They Changed* (New York: Knopf, 1966).

108. *Biographical Memoirs of the Fellows of the Royal Society* (London: Royal Society, London, 1951), vol. 5, p. 387.

109. Rozental, *Niels Bohr*; Moore, *Niels Bohr* (reference 107).

110. A. Einstein, "Autobiographical Notes," in *Albert Einstein: Philosopher-Scientist*, ed. P. A. Schilpp (Evanston, Illinois: Library of Living Philosophers, 1949). I am indebted to Professor Hans Küng for emphasizing to me the influences of Spinoza on Einstein (Tübingen, June 12, 1978).

111. J. A. Wheeler, "No Fugitive and Cloistered Virtue," *The Presentation of the First Atoms for Peace Award to Niels Henrik David Bohr, October 24, 1957*, 1959, by the Ford Motor Company, University of Chicago Press, pp. 9-17; reprinted in *Physics Today*, 16 (January 1963), 30-32.

112. A. Osborne, *Your Creative Power* (New York: Scribner's Sons, 1948).

113. J. R. Oppenheimer, "The Theory of the Electron and Positive," at the Thursday, December 28, 1933, meeting of the American Physical Society in Boston, as cited in *Physical Review*, 45 (1934), 284.

114. W. H. Furry and J. R. Oppenheimer, "On the Limitations of the Theory of the Positron," *Physical Review*, 45 (1934), 903-904; compare with the earlier paper cited in reference 115.

115. W. H. Furry and J. R. Oppenheimer, "On the Theory of the Electron and Positive," *Physical Review*, 45 (1934), 245-262.

116. H. A. Bethe, "Discussion," in *International Conference on Physics*, p. 250 (reference 84).

117. C. D. Anderson, "Early Work on the Positron and Muon," *American Journal of Physics*, 29 (1961), 825-830; quote on pp. 827-828.

Some Men and Moments in Nuclear Physics 291

118. E. J. Williams, "Correlation of Certain Collision Problems with Radiation Theory," *Det kgl. danske videnskabernes selskab. Mathematisk-fysiske meddelelser*, 13, no. 4 (1935), 1-50.

119. C. D. Anderson and S. H. Neddermeyer, "Cloud Chamber Observations of Cosmic Rays at 4300 Meters Elevation and Near Sea Level," *Physical Review*, 50 (1936), 263-271.

120. N. Bohr, "On the Theory of the Decrease of Velocity of Moving Electrified Particles on passing through Matter," *Philosophical Magazine*, 25 (1913), 10-31; "On the Decrease of Velocity of Swiftly Moving Electrified Particles in passing through Matter," *Ibid.*, 30 (1915), 581-612.

121. E. Fermi, "Über die Theorie des Stosses zwischen Atomen und elektrisch geladenen Teilchen," *Zeitschrift für Physik*, 29 (1924), 315-327.

122. N. Bohr, "Über die Wirkung von Atomen bei Stössen," *Zeitschrift für Physik*, 34 (1925), 142-157.

123. E. J. Williams, "Applications of the Method of Impact Parameter in Collisions," *Proceedings of the Royal Society of London*, 139[A] (1933), 163-186.

124. Williams, "Correlation" (reference 118).

125. C. F. von Weizsäcker, "Ausstrahlung bei Stössen sehr schneller Elektronen," *Zeitschrift für Physik*, 88 (1934), 612-625.

126. B. Rossi, *High-Energy Particles* (New York: Prentice-Hall, 1952), section on the Bohr-Fermi-Weizsäcker-Williams method of the equivalent radiation field.

127. O. Klein and Y. Nishina, "Über die Streuung von Strahlung durch freie Elektronen nach der neuen relativischen Quantendynamik von Dirac," *Zeitschrift für Physik*, 52 (1928), 853-868.

128. Fermi, Über die Theorie des Stosses zwischen Atomen" (reference 121).

129. Anderson and Neddermeyer, "Cloud Chamber Observations" (reference 119).

130. Breit and Wheeler, "Collision of Two Light Quanta" (reference 82).

131. D. R. Brill and J. A. Wheeler, "Interaction of Neutrinos and Gravitational Fields," *Reviews of Modern Physics*, 29 (1957), 465-479; summarized in J. Wheeler, *Geometrodynamics* (New York: Academic, 1962), pp. 109-110.

132. J. R. Oppenheimer, "On the Elementary Interpretation of Showers and Bursts" [Abstract], *Physical Review*, 50 (1936), 389.

133. H. J. Bhabha and W. Heitler, "The Passage of Fast Electrons and the Theory of Cosmic Showers," *Proceedings of the Royal Society of London*, 159[A] (1937), 432-458.

134. J. F. Carlson and J. R. Oppenheimer, "On Multiplicative Showers," *Physical Review*, 51 (1937), 220-231.

135. Gray and Tarrant, "Interaction between Gamma Radiation and the Atomic Nucleus"; "Anomalous Absorption of High Energy Gamma Radiation" (reference 101).

136. L. Meitner and H. H. Hupfeld, "Über das Absorptionsgesetz für kurzwellige γ-Strahlen," *Zeitschrift für Physik*, 67 (1931), 147-168; C. Y. Chao, "The Abnormal Absorption of Heavy Elements for Hard γ-Rays," *Proceedings of the Royal Society of London*, 135[A] (1932), 206-213; J. S. Rogers, "Photographic Measurement of Absorption Coefficients of γ-Rays from Ra(B + C)," *Proceedings of the Physical Society* [London], 44 (1932), 349-366, and 367; C. Y. Chao, "Absorption and Scattering of Hard γ-Rays," Tsing Hua University, *Science Reports*, 1 (1932) 159-176 (see *Science Abstracts*, 36 [1933], 45).

137. Gray and Tarrant, "Anomalous Absorption of High Energy Gamma Radiation" (reference 101); quote on pp. 690-691.

138. Chao, "Absorption and Scattering of Hard γ-Rays" (reference 136); L. Meitner

and H. H. Hupfeld, "Über die Streuung kurzwelliger γ-Strahlung an schweren Elementen," *Zeitschrift für Physik*, 75 (1932), 705–715; L. Meitner and H. Kösters, "Über die Streuung kurzwelliger γ-Strahlung," *Ibid.*, 84 (1933), 137–144.

139. C. C. Lauritsen and J. R. Oppenheimer, "On the Scattering of ThC'' γ-rays" [Letter], *Physical Review*, 46 (1934), 80–81.

140. M. Delbrück, "Possible Existence of Multiply Charged Particles of Mass One," *Nature*, 130 (1932), 626–627.

141. M. Delbrück, foregoing reference as abstracted by "J. E. R. C." in *Science Abstracts*, 36 (1933), 45.

142. J. C. Jacobsen, "Über Absorption und Streuung von γ-Strahlen," *Zeitschrift für Physik*, 70 (1931), 145–158.

143. J. Cairns, G. Stent, and J. D. Watson, eds., *Phage and the Origins of Molecular Biology* (a festschrift for Max Delbrück) (Cold Spring Harbor, New York: Cold Spring Harbor Press, 1966).

144. R. Watson-Watt, *The Pulse of Radar, an Autobiography of Sir Robert Watson-Watt* (New York: Dial, 1959), p. 27 and p. 40.

145. K. Gödel, "Über formal unentscheidbare Sätze der Prinicipia Mathematica und verwandter Systeme I," *Monatshefte für Mathematik und Physik*, 38 (1931), 173–198.

146. R. S. Avery, "New Position Honors Distinguished Professors," *Caltech News*, 10, no. 6 (September 1976), 1.

147. J. R. Oppenheimer and M. S. Plesset, "On the Production of the Positive Electron" [Letter], *Physical Review*, 44 (1933), 53–55.

148. M. S. Plesset and J. A. Wheeler, "Inelastic Scattering of Quanta with Production of Pairs," *Physical Review*, 48 (1935), 302–306.

149. R. de L. Kronig, "On the Theory of Dispersion of X-Rays," *Journal of the Optical Society of America*, 12 (1926), 547–557.

150. H. A. Kramers, "Estratto dagli Atti," *Atti del Congresso Internazionale de Fisici, Como, 11–20 Settembre 1927* (Bologna: Zanichelli, 1928). This work, and that of Kronig in reference 149, are applied in the joint paper by R. de L. Kronig and H. A. Kramers, "Zur Theorie der Absorption und Dispersion in den Röntgenspektren," *Zeitschrift für Physik*, 48 (1928), 174–179.

151. J. Toll and J. A. Wheeler, "Some Pair-Theoretic Applications of the Dispersion Relation" [Abstract], *Physical Review*, 81 (1951), 654–655.

152. J. S. Toll, "The Dispersion Relation for Light and Its Application to Problems Involving Electron Pairs" (Ph.D. diss., Princeton University, 1952); available from University Microfilms, Inc., 300 N. Zeeb Road, Ann Arbor, Michigan 48106.

153. S. Chandrasekhar, "The Density of White Dwarf Stars," *Philosophical Magazine*, 11 (1931), 592–596; "The Maximum Mass of Ideal White Dwarfs," *Astrophysical Journal*, 74 (1931), 81–82.

154. Chadwick, "Possible Existence of a Neutron"; "Existence of a Neutron" (reference 44).

155. W. Baade and F. Zwicky, "Supernovae and Cosmic rays," *Physical Review*, 45 (1934), 138, abstract of paper presented at the December 15–16, 1933, Stanford meeting of the American Physical Society.

156. J. R. Oppenheimer and G. M. Volkoff, "On Massive Neutron Cores," *Physical Review*, 55 (1939), 374–381.

157. B. K. Harrison and J. A. Wheeler, "The Harrison-Wheeler Equation of State," first presented in *Onzième Conseil de Physique Solvay: La structure et l'évolution de l' univers* (Brussels: Editions Stoops, 1958); presented in greater detail in B. K. Harrison,

K. S. Thorne, M. Wakano, and J. A. Wheeler, *Gravitation Theory and Gravitational Collapse* (Chicago: University of Chicago Press, 1965); M. Wakano and J. A. Wheeler, "Equilibrium Configurations," same citation as above.

158. J. B. Hartle and K. S. Thorne, "Slowly Rotating Relativistic Stars. II: Models for Neutron Stars and Supermassive Stars," *Astrophysical Journal*, 153 (1968), 807–834.

159. A. G. Petschek, "Recent Progress in the Theory of Nuclear Matter," *Annual Review of Nuclear Science*, 14 (1964), 29–50.

160. Wakano and Wheeler, "Equilibrium Configurations" (reference 157).

161. J. H. Oort and Th. Walraven, "Polarization and Composition of the Crab Nebula," *Bulletin of the Astronomical Institutes of the Netherlands*, 12 (1956), 285–308.

162. J. A. Wheeler, "Superdense Stars," in *Annual Review of Astronomy and Astrophysics*, ed. L. Goldberg (Palo Alto, California: George Banta, 1966), vol. 4, pp. 393–432.

163. A. Hewish, S. J. Bell, J. D. H. Pilkington, P. F. Scott, and R. A. Collins, "Observation of a Rapidly Pulsating Radio Source," *Nature*, 217 (1968), 709–713.

164. Fermi, et al, "Azione di sostanze idrogenate" (reference 92).

165. N. Bohr, "Neutron Capture and Nuclear Constitution," *Nature*, 137 (1936), 344–348; and a summary of the same on p. 351.

166. Breit and Wigner, "Capture of Slow Neutrons" (reference 70).

167. Ruark and Urey, *Atoms, Molecules and Quanta* (reference 37).

168. P. A. M. Dirac, "The Quantum Theory of the Electron," *Proceedings of the Royal Society of London*, 117[A] (1928), 610–624; "Part II," *Ibid.*, 118[A] (1928), 351–361.

169. P. A. M. Dirac, "Discussion of the Infinite Distribution of Electrons in the Theory of the Positron," *Proceedings of the Cambridge Philosophical Society*, 30 (1934), 150–163.

170. W. Heisenberg, "Bemerkungen zur Diracschen Theorie des Positrons," *Zeitschrift für Physik*, 90 (1934), 209–231 (with correction, *Ibid.*, 92 [1934], 692); W. Heisenberg and H. Euler, "Folgerungen aus der Diracschen Theorie des Positrons," *Ibid.*, 98 (1936), 714–732.

171. H. A. Lorentz, "Het verband tusschen hoeveelheid van beweging en energiestroom. Opmerkingen over den bouw van electronen en atomen," *Verslag Koninklijke Akademie van Wetenschappen, Amsterdam*, 26 (1917), 981–995.

172. H. Poincaré, 'Sur la dynamique de l'électron," *Rendiconti del Circolo Matematico di Palermo*, 21 (1906), 130–175; reviewed in reference 173.

173. A. Pais, *Developments in the Theory of the Electron* (Princeton, New Jersey: informal publication, 1948).

174. W. Heisenberg, to J. A. Wheeler on metroliner from Washington, D.C., to Trenton, N.J., Thursday, April 26, 1973.

175. Chadwick, "Possible Existence of a Neutron"; "Existence of a Neutron" (reference 44).

176. W. Heisenberg, "Über den Bau der Atomkerne. I.," *Zeitschrift für Physik*, 77 (1932), 1–11; "II.," *Ibid.*, 78 (1932), 156–164; "III.," *Ibid.*, 80 (1932), 587–596; E. Majorana, "Über die Kerntheorie," *Ibid.*, 82 (1933), 137–145.

177. D. Iwanenko, "Sur la constitution des noyaux atomiques," *Comptes rendus*, 195 (1932), 439–441.

178. H. Yukawa, "On the Interaction of Elementary Particles," *Proceedings of the Physico-Mathematical Society of Japan*, 17 (1935), 48–57.

179. C. Bridenbaugh, *Vexed and Troubled Englishmen, 1590-1642* (New York: Oxford University Press, 1968).

180. Fermi, "Teoria dell'emissione dei raggi 'beta'"; "Teoria dei raggi β" (reference 97).

181. M. Curie, *L'isotopie et les éléments isotopes*, edité par la Société "Journal de Physique," Recueil des Conférence-Rapports de Documentation (Paris: La Presses Universitaires de France, 1924), vol. 9, p. 142: "La présence d'électrons dans le noyau est tout à fait en accord avec ce que nous savons sur l'émission des rayons β par les corps radioactifs (voir p. 18). Nous avons vu, d'autre part, comment l'émission de noyaux d'helium (rayons α) par les radioactives fait naître la conviction que ces noyaux constituent un element important de la structure des atomes et y existent avec une certaine indépendence (voir p. 17). Cela n'est pas prouvé, car ainsi que le fait remarquer Aston dans son livre sur les isotopes, la fumée ne préexiste pas dans le pistolet qui la produit lors du tir. Toutefois, cela semble extrêment probable, et l'on renoncerait difficilement à développer cette idée du point de vue de la radioactivité."

182. F. W. Aston, *Isotopes*, 2nd ed. (London: Arnold, 1924).

183. N. Bohr, "Chemistry and the Quantum Theory of Atomic Constitution" [Faraday lecture], *Journal of the Chemical Society, London, 1932*, part I (1932), 349-384; on p. 377 Bohr stresses that the "scope of the quantum mechanical symbolism is essentially confined, however, to problems where the intrinsic stability of the elementary electrical particles can be left out of consideration. . . . [The] so-called 'electron diameter' . . . [defines] a lower limit for the extension of the region where the idealisation according to which the electron is considered as a charged material point is justifiable." (This lecture was delivered at the Salter's Hall on May 8, 1930.)

184. D. M. Dennison, "A Note on the Specific Heat of the Hydrogen Molecule." *Proceedings of the Royal Society of London*, 115[A] (1927), 483-486; R. S. Mulliken, "Interpretation of the Atmospheric Absorption Bands of Oxygen," *Physical Review*, 32 (1928), 880-887; F. W. Loomis and R. W. Wood, "The Rotational Structure of the Blue-Green Bands of Na_2," *Ibid.*, 32 (1928), 223-236; R. S. Mulliken, "Band Spectra and Atomic Nuclei," *Transactions of the Faraday Society*, 25 (1929), 634-645.

185. Heisenberg, "Über den Bau der Atomkerne" (reference 176); Ehrenfest and Oppenheimer, "Statistics of Nuclei" (reference 57); H. A. Bethe and R. F. Bacher, "Nuclear Physics. A. Stationary Status of Nuclei," *Reviews of Modern Physics*, 8 (1936), 82-229; "Nuclear Physics. B. Nuclear Dynamics, Theoretical," *Ibid.*, 9 (1937), 69-244.

186. N. F. Mott and H. S. W. Massey, *The Theory of Atomic Collisions*, 3rd ed., (Oxford: Clarendon Press, 1965); see p. 217 for the magnetic moment of an electron in a tightly bound state according to Dirac's relativistic theory of the electron.

187. B. Edlén, "Atomic Spectra," in *Handbuch der Physik*, ed. S. Flügge (Berlin: Springer, 1964), vol. 27, pp. 80-220. I owe this and the following reference to the kindness of Dr. William C. Martin, U.S. National Bureau of Standards, Washington, D.C. He refers in particular to the higher ^3P states of neutral helium, in which the distance—in energy—between the J = 2 and J = 1 levels is an order of magnitude smaller than the distance to the J = 0 level. R. D. Cowan and K. L. Andrew, "Coupling Considerations in Two-Electron Spectra," *Journal of the Optical Society of America*, 55 (1965), 502-516.

188. H. P. Dürr, "Approximate Symmetries in Atomic and Elementary Particle Physics," in *Properties of Matter under Unusual Conditions*, ed. H. Mark and S. Fernbach (New York: Interscience, 1969), pp. 301-326.

189. K. Schwarzschild, "Zur Elektrodynamik. II. Die elementare elektrodynamische Kraft," *Nachrichten von der Köngl. Gesellschaft der Wissenschaften zu Göttingen. Mathematisch-physikalische Klasse* (1903), 132-141.

190. H. Tetrode, "Über den Wirkungszusammenhang der Welt. Eine Erweiterung der Klassischen Dynamik," *Zeitschrift für Physik*, 10 (1922), 317-328.

191. J. Frenkel, "Zur Elektrodynamik punktförmiger Elektronen," *Zeitschrift für Physik*, 32 (1925), 518-534.

192. A. D. Fokker, "Ein invarianter Variationssatz für die Bewegung mehrerer elektrischer Massenteilchen," *Zeitschrift für Physik*, 58 (1929), 386-393; A. D. Fokker and C. J. Gorter, "Die Kraftwirkungen zwischen bewegten Ladungen." *Ibid.*, 77 (1932), 166-169; F. Hoyle and J. V. Narlikar, *Action at a Distance in Physics and Cosmology* (San Francisco: Freeman, 1974).

193. R. P. Feynman and J. A. Wheeler, "Reaction of the Absorber as the Mechanism of Radiative Damping" [Abstract], *Physical Review*, 59 (1941), 683.

194. J. A. Wheeler and R. P. Feynman, "Interaction with the Absorber as the Mechanism of Radiation," *Reviews of Modern Physics*, 17 (1945), 157-181; "Classical Electrodynamics in Terms of Direct Interparticle Action," *Ibid.*, 21 (1949), 425-433.

195. R. P. Feynman, "The Development of the Space-Time View of Quantum Electrodynamics" [Nobel Prize lecture of December 11, 1965], *Science*, 153 (1966), 699-708.

196. R. P. Feynman, "The Principle of Least Action in Quantum Mechanics" (Ph.D. diss., Princeton University, 1942); available from University Microfilms, Inc., 300 N. Zeeb Road, Ann Arbor, Michigan 48106; R. P. Feynman and A. R. Hibbs, *Quantum Mechanics and Path Integrals* (New York: McGraw-Hill, 1965).

197. R. P. Feynman, "The Theory of Positrons," *Physical Review*, 76 (1949), 749-759; "Space-Time Approach to Quantum Electrodynamics," *Ibid.*, 76 (1949), 769-789; "Mathematical Formulation of the Quantum Theory of Electromagnetic Interaction," *Ibid.*, 80 (1950), 440-457.

198. Feynman, "Space-Time View of Quantum Electrodynamics" (reference 195).

199. E. C. G. Stueckelberg, "Remarque à propos de la création de paires de particules en théorie de relativité," *Helvetica Physica Acta*, 14 (1941), 588-594.

200. J. A. Wheeler, "From Relativity to Mutability," in *The Physicist's Conception of Nature*, ed. J. Mehra (Dordrecht, Holland: Reidel, 1973), pp. 202-247.

201. J. A. Wheeler, "Genesis and Observership," in *Foundational Problems in the Special Sciences*, ed. R. E. Butts and K. J. Hintikka (Dordrecht, Holland: Reidel, 1977), pp. 3-33; "The 'Past' and the 'Delayed-Choice' Double-Slit Experiment," in *Mathematical Foundations of Quantum Theory*, ed. A. R. Marlow (New York: Academic, 1978); "Law without Law," in "Frontiers of Time," University of Texas at Austin preprint, 1978, pp. 1-46; scheduled to appear in *Problems in the Foundations of Physics, Rendiconti della Scuola Internazionale di Fisica "Enrico Fermi," LXXII Corso*, ed. N. Toraldo di Francia and B. van Fraassen (Amsterdam: North-Holland, 1979).

202. K. R. Popper, *Conjectures and Refutations: The Growth of Scientific Knowledge* (London: Routledge and Kegan Paul, 1963; 5th ed., 1974), epigraph on p. vi.

203. E. Rutherford and J. Chadwick, "The Scattering of α-particles by Helium," *Philosophical Magazine*, 4 (1927), 605-620.

204. Wheeler, "Interaction between Alpha-Particles" (reference 1); J. A. Wheeler, "The Scattering of Alpha-Particles in Helium," *Physical Review*, 59 (1941), 16-26.

205. G. Breit, "Separation of Angles in the Two-Electron Problem," *Physical Review*, 35 (1930), 569-578.

206. E. A. Hylleraas, "Neue Berechnung der Energie des Heliums im Grundzustande, sowie des tiefsten Terms von Ortho-Helium," *Zeitschrift für Physik*, 54 (1929) 347-366.

207. J. A. Wheeler, "Molecular Viewpoints in Nuclear Structure," *Physical Review*, 52 (1937), 1083-1106; "On the Mathematical Description of Light Nuclei by the Method of Resonating Group Structure," *Ibid.*, 52 (1937), 1107-1122.

208. P. A. M. Dirac, "Note on Exchange Phenomena in the Thomas Atom," *Proceedings of the Cambridge Philosophical Society*, 26 (1930), 376-385.

209. K. Way and J. A. Wheeler, "Comparison of Majorana-Heisenberg and Velocity-Dependent Forces," *Physical Review*, 50 (1936), 675; K. Way, "Photoelectric Cross Section of the Deuteron," *Ibid.*, 51 (1937), 552-556; "Energy Levels of H^3, He^3, and He^5" [Abstract], *Ibid.*, 55 (1939), 678; published as "A Simple Picture of the Binding Energies of H^3 and He^5," *Ibid.*, 56, (1939), 556-561.

210. H. Parker, "Stable States of Two Deuterons" [Abstract], *Physical Review*, 51 (1937), 682.

211. D. L. Hill and J. A. Wheeler, "Nuclear Constitution and the Interpretation of Fission Phenomena," *Physical Review*, 89 (1953), 1102-1145; preliminary report in J. A. Wheeler and D. L. Hill, "The Collective Model of the Nucleus" [Abstract], *Ibid.*, 90 (1952), 366; J. J. Griffin, "The Variational Approach to Collective Nuclear Behavior," *Bulletin of the American Physical Society*, 30, no. 3 (1955), 48, for the April 28, 1955, Washington meeting; R. E. Peierls and J. Yoccoz, "The Collective Model of Nuclear Motion," *Proceedings of the Physical Society* [*London*], 70[A] (1957), 381-387; J. J. Griffin and J. A. Wheeler, "Collective Motions in Nuclei by the Method of Generator Coordinates," *Physical Review*, 108 (1957), 311-327.

212. M. V. Mihailović, D. Ocepek, M. Osredkar, and M. Rosina, eds., "Generator Coordinate Method for Nuclear Bound States and Reactions," *Fizika*, 5 (1973), supplement; A. Lumbroso, "α-α Scattering in the Generator-Coordinate Formalism," *Physical Review*, 10C (1974), 1271-1281; M. V. Mihailović and M. A. Nagarajan, "Center-of-Mass Motion in Complex Nuclear Collisions in the Generator Co-ordinate Representation," preprint, 1975, referred to in M. V. Mihailović, L. J. B. Goldfarb, and M. A. Nagarajan, "A Two-Centre Generator Coordinate Method for Complex Nuclear Collisions," *Nuclear Physics*, 273[A] (1976), 207-220; P. Bonche and B. Giraud, "Une application du formalisme de la coordonnée generatrice aux reactions de transfert entre ions lourds," *Recent Developments in Theoretical Nuclear Physics, La Toussuire, France, 5-10 February, 1973* (Villeurbanne, France: Institute of Nuclear Physics, 1974); B. Giraud, J. LeTourneux, and E. Osnes, "A Generator Coordinate Theory of Nuclear Reactions," *Annals of Physics*, 89 (1975), 359-392.

213. K. Wildermuth and W. McClure, *Cluster Representations of Nuclei* (Berlin: Springer, 1966); K. Wildermuth and Y. C. Tang, *A Unified Theory of the Nucleus* (New York: Academic, 1977); B. Giraud and B. Grammaticos, "On the Interpretation of Negative Collective Masses," *Annals of Physics*, 101 (1976), 670-681; T. Sasakawa, T. Sawada, et. al., eds., "Few-Body Problems in Nuclear Structure," *Progress of Theoretical Physics*, no. 61 (1977), supplement; K. Ikeda, R. Tamagaki, et al., "Microscopic Methods for the Interactions between Complex Nuclei," *Ibid.*, no. 62 (1977), supplement, pp. 1-294.

214. T. C. Griffith and E. A. Power, eds., *Nuclear Forces and the Few-Nucleon Problem*, vols. I and II, from the International Conference on Nuclear Forces and the Few-Nucleon Problem, held in London, University College, July 8-11, 1959, organized by L. Cranberg and H. Massey (Oxford: Pergamon, 1960); J. de Boer and H. J. Mang, "Cluster Structure" and "Collective Aspects" in *Proceedings of the International Conference on Nuclear Physics*, vol. 1 (Amsterdam: North-Holland, 1973 and New York: American Elsevier, 1973), pp. 107-163; *Proceedings of the International Conference on Clustering Phenomena in Nuclei*, Bochum, Germany, 1969 (Vienna: International Atomic Energy Agency, 1969); *Proceedings of the Second International Conference on Clustering Phenomena in Nuclei*, College Park, Maryland, 1975, National Technical Information Service, U.S. Department of Commerce, Springfield, Virginia 22161, report ORO-4856-26; *Pro-*

ceedings of the INS-IPCR Symposium on Cluster Structure of Nuclei and Transfer Reactions Induced by Heavy-Ions, Tokyo, 1975, IPCR Cyclotron Progress Report, supplement 4; Proceedings of the VII International Conference on Few Body Problems in Nuclear and Particle Physics, Delhi, India, 1975-76; Proceedings of the International Symposium on Nuclear Collisions and Their Microscopic Description, Bled, Yugoslavia, 1977, Fizika, 3 and 4 (1977), supplement.

215. Wheeler, "Mathematical Description of Light Nuclei" (reference 207); quotes on p. 1117 and p. 1119.

216. W. Heisenberg, "Die 'beobachtbaren Grössen' in der Theorie der Elementarteilchen," Zeitschrift für Physik, 120 (1942), 513-538; part II, Ibid., 120 (1942), 673-702; part III, Ibid., 123 (1944), 92-112.

217. Way and Wheeler, "Majorana-Heisenberg and Velocity-Dependent Forces" (reference 209).

218. J. A. Wheeler and H. H. Barschall, "The Scattering of 2.5-Mev Neutrons in Helium," Physical Review, 58 (1940), 682-687.

219. Wheeler, "Molecular Viewpoints in Nuclear Structure" (reference 207); quote on p. 1103.

220. Breit and Wigner, "Capture of Slow Neutrons" (reference 70).

221. L. R. Hafstad and E. Teller, "The Alpha-Particle Model of the Nucleus," Physical Review, 54 (1938), 681-692.

222. C. B. O. Mohr and G. E. Pringle, "The Scattering of Alpha Particles in Helium, Hydrogen, and Deuterium," Proceedings of the Royal Society of London, 160[A] (1937), 190-206.

223. S. Devons, "The Scattering of α-Particles in Helium," Proceedings of the Royal Society of London, 172[A] (1939), 559-567.

224. Wheeler, "Scattering of Alpha-Particles" (reference 204).

225. Wheeler, "Mathematical Description of Light Nuclei" (reference 207).

226. J. A. Wheeler, "The Alpha-Particle Model and the Properties of the Nucleus Be^8," Physical Review, 59 (1941), 27-36.

227. Bohr, "Neutron Capture" (reference 165); N. Bohr and F. Kalckar, "On the Transmutation of Atomic Nuclei by Impact of Material Particles. I. General Theoretical Remarks," Det kgl. danske videnskabernes selskab. Mathematisk-fysiske meddelelser, 14, no. 10 (1937), 1-40.

228. C. F. von Weizsäcker, "Zur Theorie der Kernmassen," Zeitschrift für Physik, 96 (1935), 431-458; updated in H. A. Bethe, "An Attempt to Calculate the Number of Energy Levels of a Heavy Nucleus," Physical Review, 50 (1936), 332-341.

229. K. Way, "Nuclear Quadrupole and Magnetic Moments" [Abstract], Physical Review, 53 (1938), 685; published as "The Liquid-Drop Model and Nuclear Moments," Physical Review, 55 (1939), 963-965.

230. Ibid.

231. Th. Schmidt, "Über die magnetischen Momente der Atomkerne," Zeitschrift für Physik, 106 (1937), 358-361.

232. E. Teller and J. A. Wheeler, "On the Rotation of the Atomic Nucleus," Physical Review, 53 (1938), 778-789.

233. Hill and Wheeler, "Nuclear Constitution" (reference 211); J. A. Wheeler, "Collective Model for Nuclei," Proceedings of the International Conference of Theoretical Physics, Kyoto and Tokyo, 1953 (Tokyo: Science Council of Japan, 1954), pp. 311-320; "Survey of Nuclear Models," Proceedings of the 1954 Glasgow Conference on Nuclear and Meson Physics, ed. E. H. Bellamy and R. G. Moorhouse (New York: Pergamon, 1955),

pp. 38-45; "Nuclear Models and the Collective Nucleus," *Nuovo cimento*, series 10, vol. 2, no. 3 (1955), supplement, 909-918.

234. M. G. Mayer, "On Closed Shells in Nuclei," *Physical Review*, 74 (1948), 235-239; "II.," *Ibid.*, 75 (1949), 1969-1970; "Nuclear Configurations in the Spin-Orbit Coupling Model. I. Empirical Evidence," *Ibid.*, 78 (1950), 16-21; "II. Theoretical Considerations," *Ibid.*, 78 (1950), 22-23.

235. O. Haxel, J. H. D. Jensen, and H. E. Suess, "Zur Interpretation der ausgezeichneten Nukleonenzahlen im Bau der Atomkerne. I.," *Die Naturwissenschaften*, 35 (1948), 376; "On the 'Magic Numbers' in Nuclear Structure," *Physical Review*, 75 (1949), 1766; "Modellmässige Deutung der ausgezeichneten Nukleonenzahlen im Kernbau," *Zeitschrift für Physik*, 128 (1950), 295-311.

236. J. Rainwater, "Nuclear Energy Level Argument for a Spheroidal Nuclear Model," *Physical Review*, 79 (1950), 432-434.

237. A. Bohr, "On the Quantization of Angular Momenta in Heavy Nuclei," *Physical Review*, 81 (1951), 134-138; "The Coupling of Nuclear Surface Oscillations to the Motion of Individual Nucleons," *Det kgl. danske videnskabernes selskab. Mathematisk-fysiske meddelelser*, 26, no. 14 (1952), 1-40; *Rotational States of Atomic Nuclei* (Copenhagen: Munksgaard, 1954).

238. A. Bohr and B. R. Mottelson, "Collective and Individual-Particle Aspects of Nuclear Structure," *Det kgl. danske videnskabernes selskab. Mathematisk-fysiske meddelelser*, 27, no. 16 (1953), 1-174.

239. Bohr, "Neutron Capture" (reference 165).

240. Bethe and Bacher; Bethe, "Nuclear Physics" (reference 185).

241. Bohr and Kalckar, "Transmutation of Atomic Nuclei" (reference 227).

242. V. Weisskopf, "Statistics and Nuclear Reactions," *Physical Review*, 52 (1937), 295-303.

243. J. A. Wheeler, "Nuclear Physics," notes by R. P. Feynman, pp. 1-62, with E. Fermi, "Cosmic rays," April 7, 9, 11, and 15, 1940, lectures at Princeton, notes by J. A. Wheeler, pp. 63-76, mimeographed notes for internal use at Princeton University; copies on deposit at the Niels Bohr Library of the American Institute of Physics, New York, and at the Library of the American Philosophical Society, Philadelphia.

244. W. Heisenberg, *Physics and Beyond: Encounters and Conversations*, translated from the original German by A. J. Pomerans (New York: Harper and Row, 1971), p. 172.

245. J. A. Wheeler, "Principles of Nuclear Power," 1943-45, unpublished, classified notes on the theory of the nuclear reactor, with special reference to control, thermal stability, power distribution, delayed neutrons, and poisoning by xenon and other fission products, with contributions by K. Way and cutaway perspective drawings by Harvey Hoehn; used to produce the also classified Hanford Engineer Works technical manual. D. F. Babcock, "The Discovery of Xenon-135 as a Reactor Poison," *Nuclear News—American Nuclear Society* (September 1964), 38-42.

246. Wheeler, "Mercer Street" (reference 75).

247. E. Feenberg and E. Wigner, "On the Structure of Nuclei between Helium and Oxygen," *Physical Review*, 51 (1937), 95-106.

248. E. Wigner, "On the Consequences of the Symmetry of the Nuclear Hamiltonian on the Spectroscopy of Nuclei," *Physical Review*, 51 (1937), 106-119; "On the Structure of Nuclei Beyond Oxygen," *Ibid.*, 51 (1937), 947-958.

249. M. MacPhail and J. Giarratana, "Total Cross Sections of Carbon, Nitrogen, Sodium, Magnesium, and Aluminum" [Abstract], *Physical Review*, 56 (1939), 207.

250. M. R. MacPhail, "Anomalous Scattering of Fast Neutrons," *Physical Review*, 57 (1940), 669-676.

251. G. Breit, "A Remark on Gamow's Treatment of Radioactive Disintegration" [Abstract], *Physical Review*, 40 (1932), 127; G. Breit and F. L. Yost, "Radiative Capture of Protons by Carbon," *Ibid.*, 48 (1935), 203-210; see especially pp. 206-207 on "Relations between Wave Functions near Resonance."

252. Breit and Wigner, "Capture of Slow Neutrons" (reference 70).

253. P. L. Kapur and R. Peierls, "The Dispersion Formula for Nuclear Reactions," *Proceedings of the Royal Society of London*, 166[A] (1938), 277-295.

254. R. Ladenburg and M. H. Kanner, "On the Neutrons from the Deuteron-Deuteron Reaction," *Physical Review*, 52 (1937), 911-918.

255. O. Hahn and F. Strassmann, "Über den Nachweis und das Verhalten der bei der Bestrahlung des Urans mittels Neutronen entstehenden Erdalkalimetalle," *Die Naturwissenschaften*, 27 (1939), 11-15.

256. L. Meitner and O. R. Frisch, "Disintegration of Uranium by Neutrons: a New Type of Nuclear Reaction," *Nature*, 143 (1939), 239-240.

257. N. Bohr, "Disintegration of Heavy Nuclei," *Nature*, 143 (1939), 330.

258. O. R. Frisch, "Physical Evidence for the Division of Heavy Nuclei under Neutron Bombardment," *Nature*, 143 (1939), 276.

259. Bohr and Wheeler, "Mechanism of Nuclear Fission" (reference 4); quote on pp. 427-428.

260. M. Morse, *The Calculus of Variations in the Large* (New York: American Mathematical Society, 1934); *Functional Topology and Abstract Variational Theory* (Paris: Gauthier-Villars, 1938); M. Morse and S. S. Cairns, *Critical Point Theory in Global Analysis and Differential Topology: an Introduction* (New York: Academic, 1969).

261. Lord Rayleigh [J. W. Strutt], *The Theory of Sound* (2nd ed., London, 1894; reprinted, New York: Dover, 1945), vol. 2, pp. 371-375.

262. Lord Rayleigh [J. W. Strutt], *Scientific Papers*, 6 vols. (Cambridge: Cambridge University Press, I, 1899; II, 1900; III, 1902; IV, 1903; V, 1912, VI, 1920; reprinted, New York: Dover, 1964).

263. E. Feenberg, "On the Shape and Stability of Heavy Nuclei," [Letter], *Physical Review*, 55 (1939), 504-505.

264. C. F. von Weizsäcker, "Zum Wefelmeierschen Modell der Transurane," *Naturwissenschaften*, 27 (1939), 133.

265. J. Frenkel, "On the Splitting of Heavy Nuclei by Slow Neutrons," *Physical Review*, 55 (1939), 987.

266. S. Cohen and W. J. Swiatecki, "The Deformation Energy of a Charged Drop. IV. Evidence for a Discontinuity in the Conventional Family of Saddle Point Shapes," *Annals of Physics*, 19 (1962), 67-164; "Part V: Results of Electronic Computer Studies," *Ibid.*, 22 (1963), 406-437.

267. V. M. Strutinski, N. Ya. Lyaschenko, and N. A. Popov, "Symmetrical Shapes of Equilibrium for a Liquid Drop Model," *Nuclear Physics*, 46 (1963), 639-659.

268. J. R. Nix, "Calculation of Fission Barriers for Heavy and Superheavy Nuclei," *Annual Review of Nuclear Science*, 22 (1972), 65-120.

269. L. A. Turner, "Nuclear Fission," *Reviews of Modern Physics*, 12 (1940), 1-29.

270. E. McMillan and P. H. Abelson, "Radioactive Element 93" [Letter], *Physical Review*, 57 (June 15, 1940), 1185-1186; J. Schintlmeister and F. Hernegger, 1940, according to Irving, reference 293, identified neptunium and plutonium in June 1940, but reported their findings only at the end of 1940.

271. W. Sale, October 12, 1978, recollection of appraisal—in the making of which he took part—of the Hanford, Washington plutonium plant at the time of its September 1, 1946, transfer from E. I. du Pont de Nemours and Company to the General Electric Company, including three reactors, two bismuth phosphate plutonium separation plants, and fuel storage facilities, $346 million, as kindly reported orally to the writer.

272. N. Bohr, "Resonance in Uranium and Thorium Disintegrations and the Phenomenon of Nuclear Fission," [Letter], *Physical Review*, 55 (1939), 418-419.

273. Hill and Wheeler, "Nuclear Constitution" (reference 211); Wheeler, "Collective Model for Nuclei"; "Survey of Nuclear Models"; "Nuclear Models" (reference 233).

274. J. A. Wheeler, "Nuclear Fission and Nuclear Stability," in *Niels Bohr and the Development of Physics: Essays Dedicated to Niels Bohr on the Occasion of his Seventieth Birthday*, ed. W. Pauli (New York: McGraw-Hill, 1955), pp. 163-184; "Fission Physics and Nuclear Theory," in *Physics; Research Reactors*, vol. 2 of *Proceedings of the International Conference on the Peaceful Uses of Atomic Energy, Geneva, 1955* (New York: United Nations Organization, 1956), pp. 155-163; "Fission," in *Proceedings of the International Conference on Nuclear Reactions, Amsterdam, 2-7 July 1956* (Netherlands Physical Society, 1956); also in *Physica*, 22 (1956), 1103-1114; V. M. Strutinsky, "Shell Effects in Nuclear Masses and Deformation Energies," *Nuclear Physics*, 95[A] (1967), 420-442.

275. A. O. Nier, E. T. Booth, J. R. Dunning, and A. V. Grosse, "Nuclear Fission of Separated Uranium Isotopes" [Letter], *Physical Review*, 57 (1940), 546.

276. G. Placzek, April 16, 1940, telegram, on deposit at the National Museum of Science and Technology, Washington, D.C.

277. C. Störmer, "Periodische Elektronenbahnen im Felde eines Elementarmagneten und ihre Anwendung auf Brüches Modellversuche und auf Eschenhagens Elementarwellen des Erdmagnetismus," *Zeitschrift für Astrophysik*, 1 (1930), 237-274; this paper carries references to Störmer's earlier work.

278. G. Lemaitre and M. S. Vallarta, "On Compton's Latitude Effect of Cosmic Radiation," *Physical Review*, 43 (1933), 87-91.

279. M. S. Vallarta and R. P. Feynman, "The Scattering of Cosmic Rays by the Stars of a Galaxy," *Physical Review*, 55 (1939), 506-507.

280. E. Wigner, "The Transition State Method," *Transactions of the Faraday Society*, 34, part 1 (1938), 29-41.

281. M. Polanyi and E. Wigner, "Bildung und Zerfall von Molekülen," *Zeitschrift für Physik*, 33 (1925), 429-434; "Über die Interferenz von Eigenschwingungen als Ursache von Energieschwankungen und chemischer Umsetzungen," *Zeitschrift für Physikalische Chemie, Haber-Band*, 139 [Abteilung A], (1928), 439-452.

282. J. A. Wheeler, 1967, "Maria Sklodowska Curie: Copernicus of the World of the Small," in *Maria Sklodowska-Curie: Centenary Lectures; Proceedings of a Symposium, Warsaw, 17-20 October 1967* (Vienna: International Atomic Energy Agency, 1968), pp. 25-32; also in *Science*, 160 (1968), 1197-1200.

283. M. S. Curie, "Sur la loi fondamentale des transformations radioactives," in *La structure de matière. Rapports et discussions du conseil de physique tenu à Bruxelles du 27 au 31 Octobre 1913*, Institut International de Physique Solvay (Paris: Gauthier-Villars, 1921), pp. 66-74; quotes on p. 69 and p. 71.

284. J. A. Wheeler, "Anomalies and Regularities in Fission," *Conference on Neutron Physics by Time-of-Flight, Gatlinburg, Tennessee, 1956*, Oak Ridge National Laboratory Report ORNL-2309, 1956, pp. 157-166; "Regularities and Anomalies in the Competition between Neutron Emission and Fission," *International Conference on the Neutron*

Some Men and Moments in Nuclear Physics 301

Interactions with the Nucleus, Columbia University, 1957, Department of Commerce Technical Information Division Report, TID, 1957, pp. 146-152; "Channel Analysis of Fission," in *Fast Neutron Physics*, ed. J. B. Marion and J. L. Fowler (New York: Wiley, 1963), vol. II, pp. 2051-2184; "Fission," in *Handbook of Physics*, ed. E. U. Condon and H. Odishaw (New York: McGraw-Hill, 1958), pp. 9:177-9:200; revised version of this article, jointly with I. G. Schroeder, in the second edition of this handbook, 1967, pp. 9:255-9:271.

285. Bohr, "Quantization of Angular Momenta"; "Coupling of Nuclear Surface Oscillations"; *Rotational States* (reference 237); "On the Theory of Nuclear Fission," *Proceedings of the [First] International Conference on Peaceful Uses of Atomic Energy*, Geneva (1955), 151-154.

286. Bohr and Mottelson, "Collective and Individual-Particle Aspects of Nuclear Structure" (reference 238).

287. I. Halpern, "Nuclear Fission," *Annual Review of Nuclear Science*, 9 (1959), 245-342.

288. R. K. Smith, K. L. Henkel, and R. A. Nobles, steps in the fast-neutron fission cross section of U^{238} as reported by A. Hemmendinger, *Proceedings of the [Second] International Conference on the Peaceful Uses of Atomic Energy*, Geneva (1958), paper P/663.

289. R. W. Lamphere, "Fission Cross Section and Fragment Angular Distribution for Fast Neutron-Induced Fission of U^{234}," *Nuclear Physics*, 38 (1962), 561-589.

290. J. S. Fraser and J. C. D. Milton, "Nuclear Fission," *Annual Review of Nuclear Science*, 16 (1966), 379-444.

291. R. Ladenburg, M. H. Kanner, H. H. Barschall, and C. C. Van Voorhis, "Study of Uranium and Thorium Fission Produced by Fast Neutrons of Nearly Homogeneous Energy," *Physical Review*, 56 (1939), 168-175.

292. Encyclopaedia Brittanica, 1964, article "World War II," in vol. 23, supplementary pages 792R to 793R, gives these military force losses (here rounded off and expressed in millions): US 0.3, UK 0.5, USSR 7.5, Germany 2.9, Italy 0.3, China 2.2, Japan 1.5, totaling Allies 10.7, Axis, 4.7. Vol. 6, p. 253 gives for the deaths in Auschwitz > 2.5; and vol. 16, p. 960 gives for them ~4.0. A proper bookkeeping would include military losses by other participants and death of civilians by starvation and otherwise, including war-induced famine in Burma.

293. D. Irving, *The German Atomic Bomb: The History of Research in Nazi Germany* (New York: Simon and Schuster, 1967).

294. T. Jefferson, January 1, 1814, letter to Thomas Leiper, ". . . It cannot be to our interest that all Europe should be reduced to a single monarchy. . . . And were the consequences even to be the longer continuance of our war, I would rather meet them than see the whole force of Europe wielded by a single hand," in *The Works of Thomas Jefferson*, ed. P. L. Ford, vol. 11 (New York: Putnam, 1905).

295. T. W. Wilson, April 2, 1917, Address to Congress advising that Germany's course be declared war against the United States: "Our object now is to vindicate the principles of peace and justice in the life of the world as against selfish and autocratic power and to set up amongst the really free and self governed peoples of the world such a concert of purpose and action as will henceforth ensure the observance of those principles." In *President Wilson's State Papers and Addresses* (New York: Doran, 1918), pp. 372-383.

296. F. D. Roosevelt, October 21, 1944, Address to the Foreign Policy Association, broadcast by radio, "enduring peace in the world has not a chance unless this Nation—our America—is willing to cooperate in winning it and maintaining it."

297. M. Goldhaber and R. D. Hill, "Nuclear Isomerism and Shell Structure," *Reviews of Modern Physics*, 24 (1952), 179-239; A. Bohr and B. R. Mottelson, "Moments of Inertia of Rotating Nuclei," *Det kgl. danske videnskabernes selskab. Mathematisk-fysiske meddelelser*, 30, no. 1 (1955), 1-24; "Collective and Individual-Particle Aspects of Nuclear Structure," *Ibid.*, 27, no. 16 (1953) (reference 238); see also H. Feshbach, "The 1975 Nobel Prize for Physics," *Science*, 190 (1975), 868-870; J. Wing, "Isomeric Yield Ratios," Argonne National Laboratory report ANL 6598, 1962; D. Sperber, "Equilibrium Configuration and Fission Barrier for Liquid Drop Nuclei with High Angular Momentum," *Physical Review*, 130 (1963), 468-477; R. A. Sorensen, "Nuclear Moment of Inertia at High Spin," *Reviews of Modern Physics*, 45 (1973), 353-377; S. Cohen, F. Plasil, and W. J. Swiatecki, "Equilibrium Configurations of Rotating Charged or Gravitating Liquid Masses with Surface Tension. II," *Annals of Physics*, 82 (1974), 557-596; earlier version in A. Ghiorso, R. M. Diamond, and H. E. Conzett, eds., *Proceedings of the Third Conference on Reactions between Complex Nuclei* (Berkeley: University of California Press, 1963), p. 32; F. S. Stephens, "Coriolis Effects and Rotation Alignment in Nuclei," *Reviews of Modern Physics*, 47 (1975), 43-65; G. Scharff-Goldhaber, C. B. Dover, and A. L. Goodman, "The Variable Moment of Inertia (VMI) Model and Theories of Nuclear Collective Motion," *Annual Review of Nuclear Science*, 26 (1976), 239-317; W. U. Schröder and J. R. Huizenga, "Damped Heavy-Ion Collisions," *Ibid.*, 27 (1977), 465-547.

298. F. G. Werner and J. A. Wheeler, "Superheavy Nuclei," *Physical Review*, 109 (1958), 126-144; G. T. Seaborg, "Elements beyond 100, Present Status and Future Prospects," *Annual Review of Nuclear Science*, 18 (1968), 53-152; G. N. Flerov and A. S. Ilyinov, *Na puti k sverkhelementam* (Moscow: Pedagogika, 1977).

299. Harrison and Wheeler, "Harrison-Wheeler Equation of State" (reference 157); Harrison et al., *Gravitation Theory* (reference 157); Hartle and Thorne, "Relativistic Stars" (reference 158); Petschek, "Progress in the Theory of Nuclear Matter" (reference 159); Wheeler, "Superdense Stars" (reference 162); G. Baym and C. Pethick, "Neutron Stars," *Annual Review of Nuclear Science*, 25 (1975), 27-77.

300. J. A. Wheeler, "Mu Meson as Nuclear Probe Particle," *Physical Review*, 92 (1953), 812-816; J. Rainwater, "Mu-Meson Physics," *Annual Review of Nuclear Science*, 7 (1957), 1-30; C. S. Wu and L. Wilets, "Muonic Atoms and Nuclear Structure," *Ibid*, 19 (1969), 527-606; G. Backenstoss, "Pionic Atoms," *Annual Review of Nuclear Science*, 20 (1970), 467-508.

301. W. F. Fry, "Hyperfragments," *Annual Review of Nuclear Science*, 8 (1958), 105-126; R. Seki and C. E. Wiegand, "Kaonic and Other Exotic Atoms," *Ibid.*, 25 (1975), 241-281.

INDEX TO REFERENCE NUMBERS

Abelson, P. H., 270
Allison, S. K., 36
Amaldi, E., 92
Ambler, E., 74
Ames, J. S., 20
Anderson, C. D., 90, 102, 117, 119
Andrew, K. L., 187
Arfken, G. B., 72

Aston, F. W., 182
Avery, R. S., 146
Baade, W., 155
Babcock, D. F., 245
Bacher, R. F., 95, 185
Backenstoss, G., 300
Barschall, H. H., 218, 291
Bateman, H., 19

Some Men and Moments in Nuclear Physics 303

Baym, G., 299
Beams, J. W., 16
Bearden, J. A., 25, 26
Becker, H., 42
Bell, S. J., 163
Bellamy, E. H., 233
Berg, E. J., 12
Bethe, H. A., 94, 95, 104, 105, 116, 185
Bhabha, H. J., 133
Bleakney, W., 86
Bloch, F., 55
Boer, J. de, 214
Bohr, A., 237, 238, 285, 297
Bohr, N., 3, 4, 109, 120, 122, 165, 183, 227, 257, 272
Bonche, P., 212
Booth, E. T., 275
Bothe, W., 42
Breit, G., 56, 60, 62, 65, 66, 67, 68, 70, 71, 72, 73, 77, 80, 82, 205, 251
Brickwedde, F. G., 83
Bridenbaugh, C., 179
Brill, D. R., 131
Brown, G. E., 72
Brown, L. M., 96
Cairns, J., 143
Cairns, S. S., 260
Carlson, J. F., 134
Chadwick, J., 30, 44, 45, 93, 203
Chandrasekhar, S., 153
Chao, C. Y., 136
Chase, C. T., 73
Cockcroft, J. D., 87
Cohen, S., 266, 297
Collins, R. A., 163
Compton, A. H., 36
Compton, K. T., 63
Conferences (International), 84, 214
Cowan, R. D., 187
Cox, R. T., 73
Curie, I., 43
Curie, M., 181, 283
Dahl, O., 62
Darwin, F., 59
Delbrück, M., 140, 141
Dennison, D. M., 184
Devons, S., 223
Dirac, P. A. M., 168, 169, 208
Dover, C. B., 297

Dryden, H. L., 19
Dunning, J. R., 275
Dürr, H. P., 188
Edlén, B., 187
Ehrenfest, P., 56, 57
Einstein, A., 110
Ellis, C. D., 30, 99, 100
Encyclopaedia Brittanica, 292
Euler, H., 170
Feenberg, E., 247, 263
Fermi, E., 92, 97, 121
Feynman, R. P., 193, 194, 195, 196, 197, 279
Flerov, G. N., 298
Fokker, A. D., 192
Frank, I. M., 45
Fraser, J. S., 290
Frenkel, J., 191, 265
Frisch, O. R., 73, 256, 258
Fry, W. F., 301
Furry, W. H., 114, 115
Giarratana, J., 249
Giraud, B., 212, 213
Gödel, K., 145
Goldfarb, L. J. B., 212
Goldhaber, M., 93, 297
Goodman, A. L., 297
Gorter, C. J., 192
Goursat, E. J. B., 22
Grammaticos, B., 213
Gray, L. H., 101
Griffin, J. J., 211
Griffith, T. C., 214
Grodzins, L., 73
Grosse, A. V., 275
Hafstad, L. R., 62, 64, 221
Hahn, O., 255
Halpern, I., 287
Harrison, B. K., 157
Harrison, G. R., 27
Hartle, J. B., 158
Haxel, O., 235
Hayward, R. W., 74
Heisenberg, W., 170, 174, 176, 216, 244
Heitler, W., 105, 133
Henkel, K. L., 288
Hernegger, F., 270
Herzfeld, K. F., 46, 47, 48, 49
Hewish, A. 163

Heydenburg, N. P., 69
Heyl, P. R., 15
Hibbs, A. R., 196
Hill, D. L., 211
Hill, R. D., 297
Hoppes, D. D., 74
Hoyle, F., 192
Hudson, R. P., 74
Huizenga, J. R., 297
Hund, F., 28
Hupfeld, H. H., 136, 138
Hutton, J., 39
Hylleraas, E. A., 206
Ikeda, K., 213
Ilyinov, A. S., 298
Irving, D., 293
Iwanenko, D., 177
Jacobsen, J. C., 142
Janvier, T. A., 10
Jefferson, T., 294
Jensen, J. H. D., 235
Joliot, F., 43
Jones, F. D., 9
Kalckar, F., 227
Kanner, M. H., 254, 291
Kapur, P. L., 253
Kelvin, Lord [W. Thomson], 54
Klein, O., 127
Kösters, H., 138
Kramers, H. A., 150
Kronig, R. de L., 149
Kuhlthau, A. R., 16
Kurrelmeyer, B., 73
Ladenburg, R., 254, 291
Lamphere, R. W., 289
Lauritsen, C. C., 139
Lawrence, E. O., 89
Lee, T. D., 74
Lemaitre, G., 278
Lewis, G. N., 34
Litovitz, T. A., 49
Livingston, M. S., 89
Loomis, F. W., 184
Lorentz, H. A., 14, 171
Lowes, J. L., 38
Lowry, R. A., 16
Lumbroso, A., 212
Lyashchenko, N. Ya., 267
Lyell, C., 40

McClure, W., 213
McIlwraith, C. G., 73
McMillan, E., 270
MacPhail, M. R., 249, 250
Maglich, B., 73
Majorana, E., 176
Mang, H. J., 214
Massey, H. S. W., 186
Mayer, M. G., 234
Meggers, W. F., 29
Meitner, L., 136, 138, 256
Mendelsohn, E., 85
Mihailovič, M. V., 212
Mildrum, H., 76
Milton, J. C. D., 290
Misner, C. W., 21
Mohr, C. B. O., 222
Moore, R., 107
Moorhouse, R. G., 233
Morse, M., 260
Mott, N. F., 99, 186
Mottelson, B. R., 238, 297
Mulliken, R. S., 184
Murnaghan, F. C., 18, 19, 20
Murphy, G. M., 83
Myers, F. E., 73
Nagarajan, M. A., 212
Narlikar, J. V., 192
Neddermeyer, S. H., 102, 119
Nier, A. O., 275
Nishina, Y., 127
Nix, J. R., 268
Nobles, R. A., 288
Ocepec, D., 212
Oort, J. H., 161
Oppenheimer, J. R., 57, 113, 114, 115, 132, 134, 139, 147, 156
Osborne, A., 112
Osnes, E., 212
Osredkar, M., 212
Pais, A., 173
Parker, H., 16, 210
Pauli, W., 96, 274
Peierls, R. E., 94, 211, 253
Peirce, B. O., 17
Pethick, C., 299
Petschek, A. G., 159
Pilkington, J. D. H., 163
Placzek, G., 276

Plasil, F., 297
Plesset, M. S., 147, 148
Poincaré, H., 172
Polanyi, M., 281
Pontecorvo, B., 92
Popov, N. A., 267
Popper, K. R., 202
Power, E. A., 214
Pringle, G. E., 222
Rabi, I. I., 67
Rainwater, J., 236, 300
Randall, M., 34
Rasetti, F., 92, 97
Rayleigh, Lord [J. W. Strutt], 261, 262
Reines, F., 98
Rogers, J. S. 136
Roosevelt, F. D., 296
Rose, R. D., 16
Rosina, M., 212
Rossi, B., 103, 126
Royal Society, London, 108
Rozental, S., 107
Ruark, A. E., 37
Rutherford, E., 30, 203
Sale, W., 271
Sasakawa, T., 213
Sawada, T., 213
Scharff-Goldhaber, G., 297
Schintlmeister, J., 270
Schmidt, B., 76
Schmidt, Th., 231
Schröder, W. U., 297
Schwarzschild, K., 189
Scott, P. F., 163
Seaborg, G. T., 298
Seabrook, W., 6
Segrè, E., 32, 92
Seki, R., 301
Slosson, E. E., 35
Smith, R. K., 288
Sommer, R. D., 53
Sorensen, R. A., 297
Sperber, D., 297
Stauffer, R. C., 52
Steinmetz, C. P., 12, 13
Stent, G., 143
Stephens, F. S., 297
Störmer, C., 277
Strassmann, F., 255
Strutinski, V. M., 267, 274
Stueckelberg, E. C. G., 199
Suess, H. E., 235
Swiatecki, W. J., 266, 297
Tamagarki, R., 213
Tang, Y. C., 213
Tarrant, G. T. P., 101
Teller, E., 221, 232
Tetrode, H., 190
Thackray, A., 85
Thomson, G. P., 31
Thomson, J. A., 11
Thomson, J. J., 31, 106
Thorne, K. S., 21, 157, 158
Toll, J. S., 151, 152
Tourneaux, J. Le, 212
Turnbull, H. W., 23
Turner, L. A., 269
Tuve, M. A., 62, 64, 65
Urey, H. C., 37, 83
Vallarta, M. S., 278, 279
Van Atta, L. C., 63
Van de Graaff, R. J., 63
Volkoff, G. M., 156
Van Voorhis, C. C., 291
Wakano, M., 157
Walraven, Th., 161
Walton, E. T. S., 87
Ward, H., 41
Watson, J. D., 143
Watson-Watt, R., 144
Way, K., 209, 229
Weiner, C., 85
Weisskopf, V., 242
Weizsäcker, C. F. von, 125, 228, 264
Werner, F. G., 298
Weyl, H., 33
Wheeler, J. A., 1, 2, 3, 4, 21, 25, 26, 29, 75, 77, 78, 79, 80, 82, 111, 131, 148, 151, 157, 162, 193, 194, 200, 201, 204, 207, 209, 211, 218, 226, 232, 233, 243, 245, 274, 282, 284, 298, 300
White, M. G., 89
Wiegand, C. E., 301
Wigner, E., 70, 71, 247, 248, 280, 281
Wildermuth, K., 213
Wilets, L., 300
Williams, E. J., 118, 123

Wilson, T. W., 295
Wing, J., 297
Wintner, A., 24
Wolf, K. L., 48
Wood, R. W., 5, 7, 8, 184
Wu, C. S., 74, 300

Yang, C. N., 74
Yoccoz, J., 211
Yost, F. L., 77, 251
Yukawa, H., 178
Zwicky, F., 155

DISCUSSION

Wilson: Professor Wheeler as usual gave us a number of perceptive perspectives on physics. I should remark to this particular group of *illuminati* before me that Professor Wheeler is an *American-born* physicist, and I think in some ways he was implicitly talking about American physics—he spoke about the roots of American physics, about the "Enlightenment Period" when Americans went to Europe for that enlightenment and then imported it to America, about the flowering of American physics. Some Cassandras have suggested that we are coming to a Byzantine era in physics—he scotched the notion, though. Thus we do not yet need to take up the decline and fall of American physics. Perhaps we could start by discussing the roots of American physics a little more. Wheeler made some very perceptive remarks about this, and I wondered if somebody in the audience would care either to ask more questions on that subject or to make some comments?

Segrè: I would like to make some comments first on Weyl's book [*Gruppentheorie und Quantenmechanik*]. It arrived in Rome and was brought to us by a bookseller. Fermi looked at it and read it. He told me that this was the last book in physics he read—after that he read papers or got information by other means. Majorana looked at it, and then—most extraordinarily—said: "This is a very good book. You really should read it." So I took it, read, I think, to the permutation group, and then my velocity went slower and slower and slower and then came to a standstill. I just couldn't go further. But Racah was visiting, and I said: "Look, you should learn group theory"—and this was my great contribution to the Racah coefficients, even more so Weyl's contribution. I gave this book which I had bought to Racah, with a little written explanation of what he should do with it, and he gave me Grotrian's book on spectra in exchange. So you see this is how group theory . . .

Some Men and Moments in Nuclear Physics   307

Wheeler: Was this responsible for Racah's interest in group theory?

Segrè: Most certainly. Although I don't know Racah coefficients, I have an important part in their origin! Now I have another remark on Weiszäcker-Williams. As everybody probably knows, the Weiszäcker-Williams method was really discovered in 1923 by Fermi. He was just out of school and hadn't yet discovered Fermi statistics. It was his first substantial contribution; you can find it in Fermi's *Collected Papers*. He sent the paper to Bohr, and Bohr for some reason misunderstood it or didn't appreciate it or didn't understand it and wrote some rather sharp derogatory remarks against it. This was a most unfortunate circumstance, because it produced a sort of, not resentment, but great unhappiness in this isolated, struggling, young fellow to find out that the great authority had said that what he had done was no good. When he knew of course that it was. And this left a trace . . .

Wheeler: Was Fermi aware that Bohr had already done it still earlier?

Segrè: When?

Wheeler: In the *Philosophical Magazine*, in the 1910–20 period.

Segrè: This I didn't know. No. I don't know whether Fermi knew, but it is not quoted in his paper. Bohr had done the Weiszäcker-Williams . . .

Wheeler: Yes, that's what I mentioned in my talk very briefly, but I should have said it in more detail.

Segrè: The Fourier analysis of the electromagnetic wave?

Wheeler: Yes.

Segrè: I didn't know that. I don't know whether Fermi knew, but I certainly didn't.

Wigner: Could you please repeat that? I don't think many of us understood what you just said.

Segrè: The main idea of the Weiszäcker-Williams system is of course to Fourier analyze the electric field and then use the absorption amplitudes, and so on. Now, this is very clearly done in a paper by Fermi of 1922 or 1923, which was probably one of his earliest important papers. He had written one on relativity, but this other paper was about his best thing before Fermi statistics, I believe. He was very proud of it and sent it to Bohr, And Bohr wrote a rather offhandish remark . . . it is reported, quoted what he said. Anyway, he didn't think much of it, and this had an influence, a rather severe influence, on Fermi at that time.

Wilson: I must say, if I could speak to that point as an experimentalist, that William's method was extremely important to us experimentalists. To me, what Williams had done was not just to make a technical Fourier transformation of the field but to apply this method to problems that seemed to have to do with very high energy, beyond the then known validity of our present laws, to reduce the problems to low energy phenomena which we understood very well. He emphasized that point over and over in his paper. That was the principal spirit of the Williams-Weiszäcker method.

Segrè: This is all contained in it. All that is needed is the transition probabilities . . .

Wilson: But in that paper Williams gave so many arguments . . . it is full of so many examples in which he explicitly made the statement that the laws of physics as we know them applied to what appeared to be mysterious high energy phenomena, and consequently it was not necessary to look for a breakdown of quantum mechanics, or conservation of energy.

Segrè: Yes. The next thing I wanted to say is related to this. I was invited to a Copenhagen conference, and I traveled with Euler from Germany to Copenhagen, possibly with somebody else, but I remember Euler very clearly. He explained to me in great detail the question of the penetrating particles, and he left me completely convinced that there had to be a new kind of particle. I had no great doubt that it had to be so, that you couldn't make do with a breakdown of electrodynamics or anything.

Wheeler: When was this?

Segrè: That was in 1937.

Wheeler: Yes, well, this is the fall of 1934 or spring of 1935.

Segrè: Yes, there probably was a difference in time.

Wheeler: Yes.

Bethe: Here is one remark, which is somewhat incidental, on the cosmic radiation. At the same conference in London in 1934, Arthur Compton and I got together and concluded that the primary cosmic radiation must be protons. Our paper, in *Nature*, didn't say anything about the cosmic radiation that comes down to earth, but only about the primary radiation. Our paper apparently was not noted in Copenhagen.

I also have a question. Could you explain again what made you and Bohr conclude that U^{235} was the isotope responsible for fission rather than U^{238}? I remember your bet with Placzek.

Wheeler: Placzek was always so stimulating but also always so skeptical, and this made him a good conscience for us. As I mentioned the bet was $18.36 to a penny. I wanted to put the proton to electron mass ratio in. A year later Nier made the measurements and confirmed the result, and as I said I got a telegraphic money order for one penny with the word "Congratulations" from Placzek. That was a very important point, but I am not really responsible for that point. It came about midway through our intense period of working together on the theory of fission and deriving fission barriers, deriving the rate of fission, and so on. But that particular point of taking apart the U^{235} and U^{238} comes from Bohr when he was walking back from the Nassau Club one snowy morning. The argument had to do with the fact that the Ladenburg-Kanner-Barschall-van Voorhis result had shown that the U^{238} cross section only comes in—I've forgotten the exact energy—I think it is at about 1 million volts where the cross section first comes up. So the cross section falls off strongly below that point, and therefore one concludes that the U^{238} is a nucleus endowed with a big barrier to be climbed over to get to fission, and one can only get over the barrier by shooting in a fast enough neutron. Yet, one knew—experiments had shown—that slow neutrons cause fission, but neutrons of intermediate energy have negligible effect in producing fission. The absence of significant effect at intermediate energy meant that one was there well below the fission barrier of U^{238}. Moreover, the U^{235}, present to less than 1 percent abundance, was also not contributing, but for a different reason: low abundance, low chance to be hit, even if high chance to split *when* hit and still lower neutron energy, because of the l/v law that small chance of a hit raises to a high chance of a hit only by attributing the slow neutron fission to U^{235} could one make sense of the situation. That was the nature of the argument.

Wilson: Still you would agree that when the isotopes were really separated by Al Nier here at Minnesota and sent to Dunning, that

was a more direct and simple and persuasive argument which . . .
Wheeler: I felt relieved of financial obligation.
Wilson: Big deal!
Nier: There's an interesting story connected with this. I met Fermi at the Physical Society meeting in April 1939, shortly after the discovery of fission. I was introduced to him by John Dunning, who was then of course very much interested in this. Fermi was also at Columbia and they knew each other. Someone raised the question the other day about how is it that there was such a delay in people doing some experiments. I remember Dunning and Fermi tried to impress on me that with the mass spectrometers I had, which had earlier been used to measure the 139 to 1 ratio, I ought to be able to produce separated samples of uranium isotopes. The cross sections were so large that one should be able to separate enough material for measurement. But I was busy on other things and didn't do anything about this for several months. Then I got a letter from Fermi in October 1939; he said he wondered how I was coming along, or some words to that effect, and urged me again to look into this. He made the comment that to determine whether it was the 235 or the 238 isotope would be of considerable theoretical interest and possibly of practical importance—the understatement of the year at that time. As you know, we did go ahead with the separation that winter, and I think on February 29, 1940, we completed the first separation. You've heard the stories about the samples. It was truly on a Friday afternoon when I mailed the first separated samples to Dunning, Grosse, and Booth at Columbia University—we were colleagues on this. I dropped the letter containing the samples off at the post office (in those days at six o'clock in the afternoon there was still somebody at the desk at the post office) and sent it by registered air mail. The samples were pasted on the margin of a piece of paper written in long hand, because it was after hours when I had written the letter. They received it on Saturday, worked all night Sunday night, and called and woke me early Sunday morning, to tell me that the target which I had called U^{235} (you could see nothing because there was so little there) was the one that gave the slow neutron fission.

Wheeler: Wonderful.

Wigner: May I change the subject very drastically? You made some remark that an electron is not an elementary particle. Could you comment a little on that? Why isn't it an elementary particle?

Wheeler: Why is it not an elementary particle? Not everything can be made out of it.

Wilson: You meant, did you not, constituent particle rather than elementary particle?

Wheeler: Yes, the basic particle; it is not the basic particle.

Wigner: Not the only basic particle.

Wheeler: It is not the entire story.

Wigner: I see. Well, of course on that we agree. Thank you very much.

Wheeler: I didn't mean for you to take it for granted, because for me it was not at all clear that one could not make all particles out of electrons. One can think of the magnetic moment of the neutron or the proton. They are so very much smaller than the magnetic moment of an electron, so how in the world could one ever possibly hope . . . ? But if you will look up the book of Mott and Massey, you will find there a formula that shows how much the magnetic moment of an electron that's bound drops off when it gets bound with relativistic velocity.

Wigner: The proton has a positive charge and the nuclei have positive charges. Could they conceivably be created from electrons, or am I entirely reoriented?

Wheeler: How could one make everything out of positive and negative charges?

Wigner: The other possibility was that the proton is made of, I don't know, 900 electrons and 901 positrons. You see, it is not clear to me how one could imagine that the proton is made of electrons.

Wheeler: Well, one would simply have a group of positive and negative electrons inside going around.

Wigner: I see. The other possibility was that the proton consists of 900 electrons, say, and 901 positrons, and this is what was refuted.

Wheeler: No, not 900, because they would have to be moving with relativistic velocities. Only if the velocities are relativistic should the force exerted by 1 nucleon on another be so strong as to be 100 times Coulomb forces.

Wigner: Yes, well, good, it's not 900, but 1000 and 1001, or 2000 and 2001. But this was the idea that was to be refuted, that the proton consists of electrons and positrons.

Wheeler: Yes, or 20 electrons or positrons, Not 2000, but 20.

Wigner: But there must be some attraction to keep them together. A proton doesn't disintegrate very fast. Thank you; you answered the question.

Wilson: I suggest that you both publish a paper subsequently on this subject.

Wheeler: Saying why the proton is not made of an electron? We can collaborate wholeheartedly in slaying that idea!

Wilson: I think there is even some experimental help on that subject!

Sopka: I'd like to return to the line of discussion that Professor Wilson suggested, about looking at the background of physics in America, and see if we could get some commentary from the people who are here on what seems to me to be the shifting of the center of gravity of physics toward America that took place in the earlier decades of this century.

Wilson: Let me call on a few. Professor Fowler, could you make some comments on that?

Fowler: I think the major comment that I would make concerns the growth of the first theoretical school in the United States under Robert Oppenheimer. At least for us in Pasadena, the fact that once a year Robert Oppenheimer, who had a joint appointment at Berkeley and at Pasadena, descended on the theoretical desert at Caltech and brought with him Bob Serber, Leonard Schiff, George Volkoff, Willis Lamb, Julian Schwinger, and others (not all at the same time) —the fact that there was a very active school of students and of professors who were in contact not only with what was going on on the East Coast, but also with what was going on in Europe, made the great difference as far as we were concerned, without casting any aspersions on the theoretical physicists at Caltech at that time. There was Epstein, and I was very fond of Epstein, but he was a loner and he wasn't in what was going on—he had done one great thing in a sense, with the Stark effect. Of course, we did have Tolman, but Tolman was more in relativity and in thermodynamics, and he was in the Chemistry Division. This relates to one issue that we haven't touched on very much, the neutron star developments. John Wheeler, our speaker this morning, has played such a great role there. A great deal of credit is always given to Landau, and of course to Serber and Oppenheimer, who wrote, I think, the first paper moving us toward our understanding that neutron stars could have masses near that of the sun and which led to the famous Oppenheimer-Volkoff paper in the *Physical Review* of 1939. In addition, Richard Tolman wrote a number of papers almost simultaneously and published them in the very same volume; the difficulty was that he used models and polytropes, and they weren't nearly as clean as when finally Oppenheimer and Volkoff, following Serber and Oppenheimer, made a solution which, if you neglect nuclear forces, is still the correct one—it comes out .7 of a solar mass. And all this argument about whether it is one solar mass or two is very important. That was the thing that made it very exciting. But of course there was a great development, as you well know, in American physics then; however, in the 1930s it was the Oppenheimer school on the West Coast that made the big difference.

Wilson: In some sense though, would you not agree that Oppenheimer was essentially importing a European product to the West Coast, a product that he had acquired during the time he had spent abroad? I think as you [Wheeler] were implying, the indigenous American physics came out of the work of say, Wood, Rowland, Lyman, or Michelson.

Wheeler: We see it especially in the experimental physics. It was reviewed in an earlier conference on the history of experimental physics. While we are speaking of experimental physics, of course, I think of Tuve and the infectious enthusiasm he brought to everything, as well as Ernest Lawrence. I can recall Tuve's introducing me to Dahl and telling me Dahl was a man who liked every adventure—he had traveled to the Arctic, he had gone on mountain climbing expeditions, he had carried a caravan across the Sahara—but now he had gone into physics because that's where the excitement lay. I thought that really was a beautiful example of the kind of spirit that infected the people who went into experimental physics, and physics generally.

Fowler: Well, I would take exception, in the sense that the students were indigenous. Robert [Oppenheimer] had learned some physics in Europe, but he raised a group of indigenous students who could then contribute to physics.

Wilson: As an American chauvinist, I would agree with you. The truth of the matter is that having exported American physics to Europe by Benjamin Thompson, we were just getting it back again!

Sopka: Heisenberg wrote of his visit to the United States in 1929, when he had been surprised and delighted to find at a large number of American universities not only a warm reception but also the fact that there were in the audience, a surprising (to him) number of people who did know what he was talking about.

Wheeler: It would be interesting to trace the political background for that, because, after all, it surely came partly as a consequence of universities' changing their patterns.

Stuewer: Yes, I would remark too that I think it must have been a fairly widespread phenomenon. Of course, at Minnesota in the mid-1920s Van Vleck and Breit were here, for example, and I am certain that there were other distinguished people at many other universities. I am just wondering whether Professor Wheeler or others who were at East Coast universities could identify similar figures who were obviously part of the same widespread movement that Heisenberg found so sympathetic when he arrived.

Wheeler: Yes, I am sure there has been a study of this. If I started naming names, I would leave out so many I would be ashamed to even start.

Bethe: I think one should mention the summer schools in Ann Arbor in the 1930s, which were very influencial, and they did bring European theorists, Fermi very prominently, Bohr, Heisenberg, I don't know whom all, to America, and the audience as I noticed later on [in 1936] was extremely well prepared to absorb the information. I think Ann Arbor summer schools were extremely important to start or foster physics in America.

Badash: My perception of this is that during the first two decades of this century there was a fair amount of physics being pursued in this country, laboratories being built and improved, staffs being expanded; and yet if you look at the 1920s and 1930s the people who were doing the work that is recognized as important were not those trained by the American physicists of the previous generation. It would appear that the important physicists were those who either went to Europe as graduate students, postdocs on Rockefeller Foundation Fellowships, or participated in the summer schools at Michigan and other places. The question I would like to ask is, What would have been the nature of American physics had the European connection not been so strong?

Wheeler: It wouldn't have been.

Wilson: I beg to disagree, because I would point out as counterexamples Ernest Lawrence, Merle Tuve, coming from the Dakotas. We can find other names I am sure. Going to Yale, there was a very strong physics tradition going back to Silliman's journal. I think that there was a very strong tradition of physics in this country that would have come perhaps to a . . .

Wheeler: Benjamin Franklin is our most famous . . .

Wilson: Benjamin Franklin; Benjamin Thompson, if you please . . .

Livingston: I'd like to speak up for the indigenous experimental tradition in American physics which was very strong. There was R. W. Wood; there was, for example, Swann, who was the leader of an experimental group that bred Lawrence and Beams and who else? In other words, there were indigenous experimental groups all over the country—Millikan, the Comptons . . . I think we should recognize that there was a very strong local American tradition largely on the experimental side.

Wheeler: We ought to remember that the Michelson experiment originated in the days when Michelson was sitting in the row while Kelvin was giving his Baltimore Lectures; Kelvin suggested the importance of such an experiment, and Michelson took it up.

Wigner: I would like to point out how important a personal contact is. I remember how much I learned directly from Einstein, both in his lectures and privately. It isn't that you learned that 2 times 3 is 6, but you learned something of the spirit, of the devotion and the type of understanding, and you realized what was not yet known, what ideas were still in the air and unknown. And that is terribly important, and from this point of view it isn't surprising that there is a local development. Schrödinger was in Berlin, and he talked to us and influenced us strongly. I know how much of an influence Sommerfeld had in Germany. A great many of the most famous and most successful theoretical physicists came from Sommerfeld's school, so the personal influence should not be entirely disregarded.

Wheeler: Speaking of this personal influence, nothing is more famous than the series of nine chemists, one of whom influenced the other and the other and the other, all the way down to Hans Krebs, going back to Otto Warburg and Liebig, and still farther to Berthelot and Lavoisier in Paris. There were nine of them over a period of 160 years, about 20 years apart.

Wigner: Very good.

Badash: The comments that we've heard seem to indicate a strong feeling of internationality in science, of communication regardless of national boundaries. Would anyone have any comments about feelings of nationalism?

Wheeler: There's of course the famous comment of Kettering when he was director of research at General Motors: "He who locks the laboratory door, locks out more than he locks in."

Swenson: One correction on the Michelson experiment: Certainly Simon Newcomb, and Maxwell *in absentia*, and Helmholtz had a lot more to do with that initiation than the Baltimore Lectures, although Kelvin did help in the resuscitation of the classical experiment with Morley later on. To change the subject, however, I think it would be very helpful to those of us who are historians and on the spectator's side of this beautiful conference to hear more comments from the actor's side on the literature of the time. I asked Hans Bethe yesterday a little bit about his opinion of the Hewlett and Anderson study, *The New World*, which was the semi-official history of the Atomic Energy Commission before it was the Commission, that is, before the legislation, and about David Irving's book on the German atomic bomb project. On books of that sort, your opinions would be most valuable to us. Could we have some discussion in that vein?

Wilson: Would you care to comment, Emilio?

Segrè: Before we leave the argument of the International exchange unresolved, I think one should mention by all means the Rockefeller Foundation, which had a really tremendous effect on world physics. The Rockefeller Foundation financed the international fellowships; that's how the Americans came to Europe. But it was also through the Rockefeller Foundation that Rasetti went to Pasadena; Heisenberg was a Fellow, Fermi was a Fellow, and so on and so forth. The roster of the Fellows of the Rockefeller Foundation illustrates the Foundation's absolutely amazing performance in the late twenties in its selection of Fellows, especially because it caught all these people when they were age twenty-four or twenty-five—they became famous afterward, but at that time they were not famous. And how it selected them really is very interesting, and even a useful question to pursue. It is known in part.

Wilson: I would like to add one thing to what we have been saying. Not that this field would have developed by American efforts. I think one of the speakers [Fowler] used the term "Second Renaissance" in describing this particular chapter in history, and we Americans would surely have to acknowledge that it was a European-centered Renaissance . . . probably imported, though, from New Zealand!

Fowler: One comment on nationalism. There may have been some, but it was secondary to the love of travel.

Wilson: Would someone care to comment on the literature though?

Fowler: Well, three cheers for the *Physical Review*.

Wilson: I think though that you [Swenson] meant a different kind of literature. The Hewlett books, for example, really cover a later period; they refer to the exploitation of nuclear physics. I think we've been talking more about an earlier time than that, and I suspect that there is not a very good popular book in that sense that covers this period. Can somebody point out a book in which you can read about the kind of things we have been discussing these past few days?

Member of audience: I think I can: *Brighter than a Thousand Suns*.

Wilson: No, I think that covers the war period. Mr. Jungk does have a fine introduction dealing with this period, but I think that in the sense of what we have been discussing, I still don't know of a book focusing on this period. Perhaps you do.

Fowler: Yes, I think that the greatest single book in nuclear physics as far as my personal experience was concerned was written by Rutherford, Chadwick, and Ellis. When I first came to Pasadena as a graduate student in 1933 to work with Charlie Lauritsen, Rutherford, Chadwick, and Ellis, *Radiations from Radioactive Substances* had been published three years earlier, and in our seminar each week a graduate student read a chapter and discussed it with the rest of us. It was full of experimental techniques and ways of looking at things, and included the whole history of—it wasn't even called nuclear physics in those days—the whole history of radioactivity before 1930. It played, I think, a terrific role. It must have done the same elsewhere.

Wilson: Yes, I share that experience. I read that book from cover to cover as my first introduction to the field.

Hiebert: To talk about the history of the subject in this way prompts me to ask a much more general question. Traditionally, it has often been said that a historian is primarily a person who wasn't there. Now since we have here a prestigious group of physicists, I wonder if we might put a pointed question to them and ask what they think a symposium of this kind really can accomplish. History has so many different faces, and is written in so many different ways, that one cannot tell everything. Carl Sandburg in his book on Abraham Lincoln says in the preface, "If I were to tell everything it would take me longer to tell than it took to enact in history." So I'd really like to know what is worth telling. We've had a good bit of anecdotal history here; we've had some attempt, it seems, to really set the record straight once and for all; and I remember that Professor Wheeler, especially, in one of the previous symposia, talked about there being conceivable practical or pragmatic consequences from doing history of this kind—to stimulate young minds or something of that sort. What do you think is the real function of writing a history of nuclear physics or of conducting a conference of this kind? So that the historians have the source materials to do it? I think this is a difficult question for us as historians, because there are so many options open; we really need advice.

Wheeler: It seems to me that there are two very different parts to the enterprise: the one, getting the record straight; the other, telling the lessons. Telling the lessons is an enterprise which takes, I would think, different skills than getting the record straight, and to do the former I could imagine somebody with the right spirit producing a marvelous book but being really unwilling to undertake it until the underbrush, all this difficulty of dates, sources, priorities, is straightened out. So it seems to me that we are doing a very important thing here. I recall for comparison Slosson's book on creative chemistry after World War I, where chemistry set the scene and there were stories of individual people and inspiring stories of individual things that were done. That was done without benefit of a historical project like this, and it could be done again for something less war-oriented than that book, but I do feel that if we wait too long in getting going we will have lost a lot of the punch.

Bethe: In my opinion the best thing we can give to our historian friends is emphasis—which things in the historical development did we consider important? And I hope to some extent we have done so.

Peierls: I really got up to make sure that the commendation of Jungk's book, *Brighter than a Thousand Suns*, did not stay on the record uncontradicted. Those who have seen the faces of people around me will have no doubt, but the record doesn't show that. In my opinion, it's a book that has a few facts right but is completely unperceptive in its understanding and interpretation. That perhaps allows me also to make a remark to this last question. It is precisely not only, as Hans Bethe said rightly, the emphasis—what was important—but also the interpretation, the interrelationships of things, and, as was said yesterday, the spirit and the way things felt, which I think might be difficult for historians to pick up otherwise. It is certainly easier following such discussions. It is of course ultimately for the historians to say whether they find this useful and what use they want to make of it.

Wilson: I'd like to make a negative comment by repeating a cliché. Was it Santayana who said, "He who ignores the lessons of history is doomed to relive it."? Well, I would be willing to ignore the lessons of nuclear history if only I could relive it. It was a joyous, marvelous time.

Stuewer: I would just like to make one comment on this also. From the point of view of one of the organizers of this symposium, this was precisely the thing that I'd hoped would come out of it—that we would learn from the distinguished physicists here what the benchmarks are, what sorts of emphases a historian should be on the lookout for. And I think that the splendid response we have had indicates that you have felt that it was important to give us this kind of guidance as well, and we are certainly very grateful to all of you for doing it.

Wilson: Thank you very much for those comments.

Wheeler: Concerning this question of what a history might contain, I don't know how you feel, Bob [Wilson], but would you feel that the way that *élan* and morale is communicated is more important and more difficult to capture, more transient, than anything else in the whole story of how people get infected with enthusiasm?

Wilson: I guess what I would most hope about the history is that it would be written in such a manner that it would reflect those things, not just to us physicists, but rather to the nonspecialists—that these humanistic elements could help bridge Snow's two cultures.

Badash: I've heard it said that because of Rutherford's great success in experimental physics, the Cavendish Laboratory in particular, and England in general, produced a generation of experimental physicists, that students wished to become little Rutherfords. Would anyone care to comment on the development of theoretical physics in England, and whether it was so affected by Rutherford's stature?

Wheeler: I might again mention one small thing. E. J. Williams switched from experimental physics at the Cavendish Laboratory to theoretical physics (that was before he came to Copenhagen), because the laboratory was shut by Rutherford's orders in the evening, and he wanted to keep working.

Wilson: Could you comment, Rudi?

Peierls: I could try a few comments on theoretical physics in England. I think that for the place that had seen the work of Newton and Maxwell, it was interesting that theoretical physics was not recognized as a subject. It was regarded as a branch of mathematics, and in general theoretical physics students—I mean students doing what we now call theoretical physics—were brought up as mathematicians and never saw the inside of a laboratory. There were some exceptions, people who switched later, or people who were brought up as experimental physicists and then turned to theory. There were some outstanding people in most of Rutherford's time, but rather few compared with other places, and this was aggravated by the fact that the system in England at that time was to select the best people by a very efficient screening procedure, and get the most promising students to Cambridge or Oxford and the ones in science to Cambridge only. Now, this made a very narrow approach therefore. Fowler was the main teacher of theoretical physics in those days. He was a very good man but had a tendency of putting students on parts of a large field, of which he could see the whole perspective, but the students generally couldn't, and they then went off to their jobs in small universities and essentially continued —many of them for the rest of their careers—to extend their Ph.D. theses. This was not a very efficient system, so on the whole at the time when Rutherford had an outstanding school of experimental physics, there was not at all a corresponding school of theoretical physics. Dirac of course came to Cambridge at that time, but he wasn't the product of any school.

Goldhaber: Of the directors of the Cavendish Laboratory some of the most famous were what we would now call theoretical physicists—it started with Maxwell, Rayleigh, and in my opinion J. J. Thomson was a transition between a theoretician and experimentalist. Rutherford was probably a crypto-theoretician, because he always had a very clear model of what he was doing, but he thought it was simpler to do it than to calculate it. He even had a little bit of depreciation, at least when he talked in public, of theoreticians. Though he admired Dirac, he would still say in front of Dirac, "Oh, these theoreticians, I was always one nose ahead of them." He really lived at a time when it was easy to test most of his ideas; it was faster than to calculate, and perhaps it would have been premature to calculate in such detail. As we have learned from Johnny Wheeler's talk, even at a much later time a-a scattering was still a formidable theoretical problem. But an experiment, at least if it is done well, answers something and gives you some hints about the direction in which you can go. So you have to be careful at what period of time you look. The modern development, which has led to a division, at least among the younger people, between theory and experiment, I think ought to be deplored. We ought to somehow make more whole physicists who are not so easily put into one box or the other . . .

Wheeler: Like Fermi . . .

Goldhaber: Like Fermi—make more Fermis! At least, the way physics is taught (as I hear from my colleagues in modern universities) it gives the young and very impressionable physicist the idea that all of physics is theoretical physics, but sometimes one has to go down to some dirty details, and these are done by some people somewhere else. They do not get the impression from the way it is taught at most places (not of course by our distinguished colleagues here) that it often starts with an experimental surprise, which is then understood and further developed theoretically and is only really deeply understood if one of the theoretical consequences is again tested experimentally. I think Toynbee once said something about science and technology which we could probably modify and say about theoretical and experimental physics, that it's really like a dance with two partners, first one takes a step, then the other takes a step, and it's this endless dance which makes it interesting. I have sometimes said that it's like a game of tennis—one cannot serve all of the time.

Wilson: As chairman, I would like to bring the session to a close with a light story apropos of Professor Goldhaber's remarks. As the director of the Fermi National Accelerator Laboratory, I was entertaining Professor Bogolyubov, who is the director of the Dubna Laboratory. We began to converse, and at a slow moment in the conversation, to keep it going, I traced his career from that of a pure mathematician, to a theoretical physicist, to an administrator, and finally to director of Dubna. I asked him when his fall to a director occurred (I thought I could ask that), and he answered without any hesitation: "Once I had descended to theoretical physics from mathematics, I could do anything!"

Indexes

Name Index

Abélard, Pierre, 249
Abelson, P. H., 276
Ahrens, J., 92
Allen, L., Jr., 92
Allison, S. K., 225
Amaldi, E., 42, 51-52, 54-56, 70, 166, 169
Ames, Joseph Sweetman, 137, 219, 223
Anderson, Carl D., 198, 236, 241-42, 244, 248
Anderson, Oscar E., 316
Astin, Allen, 227
Aston, F. W., 46, 256
Auger, P., 76, 249
Avogadro, A., 35

Baade, W., 250-51
Bacher, Robert F., 10, 14, 16, 160, 168-69, 175, 236, 269
Badash, Lawrence, 61, 110, 314, 316, 320
Bainbridge, K. T., 103, 160
Baldwin, G. C., 95-96
Baldwin, Lida F., 221
Bang, Bernhard, 250
Barber, N.F., 150
Bardeen, John, 23
Barnes, R. Bowling, 222, 227
Barnett, S. J. and L. J. H., 139
Barschall, H. H., 24, 26n23, 29, 77, 112, 158, 177, 264, 271, 281, 309
Bartlett, James H., Jr., 16, 30
Barton, H. A., 123
Bateman, H., 223
Bauer, Louis A., 138-39
Bayman, B. F , 161
Beams, Jesse W., 125, 315
Bearden, Joyce A., 223, 225
Beck, Guido, 160
Becker, H., 225
Becquerel, [A.] H., 113-14
Beil, H., 96
Bell, S. J., 252
Beltrami, Eugenio, 35
Bergère, R. 96
Berkner, L. V., 139
Berman, B. L., 97, 99
Bernardini, G., 38-39, 42, 56, 62
Berthelot, Pierre E. M., 316
Bethe, Hans A., 10, 14, 21, 23, 27, 29-30, 48, 57, 59, 66, 69, 78, 89, 100, 110, 112, 148, 151, 154, 158, 160, 164, 167-69, 174-77, 180, 183, 188-89, 194-96, 203-8, 210, 236-37, 239, 241-44, 269, 308, 314, 316, 319

325

Name Index

Bhabha, H. J., 245
Bjerge, T., 73, 236
Bjerrum, Niels, 240, 250
Blackett, P. M. S., 67-68
Bleakney, Walker, 271
Bloch, Felix, 70, 284
Blondlot, René-Prosper, 220
Bogolyubov, N. N., 322
Bohr, Aage, 24, 102, 268, 281
Bohr, Christian, 238
Bohr, Erik, 272
Bohr, Margarethe, 250
Bohr, Niels, 10, 20, 22-23, 29-30, 42, 46, 68-69, 71-72, 77-78, 161, 163, 165-66, 169-70, 176-77, 194-95, 208, 217-18, 234-39, 241-44, 247, 249-50, 253, 267-69, 272-74, 276-80, 282, 307, 309, 314
Bolsterli, M., 100
Booth, E. T., 310
Born, Max, 40, 235
Bothe, Walther, 6, 46-47, 65, 95, 166, 225
Boyce, Joseph, 127
Boyd, Julian, 284
Brady, J. J., 130-32
Bramblett, R. L., 99
Brasch, A., 119-20, 128, 143, 153-54
Breit, Gregory, 20-22, 30, 89, 115-18, 130-34, 136-37, 140, 168-69, 175-76, 180, 190, 202-4, 208, 217, 229-34, 238, 247-48, 253, 256, 260, 264, 270-71, 314
Bretscher, E., 88
Brickwedde, F. G., 84, 162
Brill, D. R., 245
Broglie, Maurice de, 249
Bromberg, Joan, 78, 173
Brostrøm, K. J., 68, 73, 236
Brown, G. E., 94, 100, 205, 231
Brown, Salley, 284
Buechner, W., 151
Buon, J., 92
Burkhardt, James L., 99

Cannizzaro, Stanislao, 35
Cantoni, C., 37
Carlos, P., 96
Carlson, J. F., 245

Carr, Howard E., 284
Cassen, Bernard, 190, 204
Chadwick, James, 5-6, 12, 14, 47-48, 60, 76, 86-88, 90, 93, 95, 108-9, 153, 162, 164-66, 170, 184, 189, 217, 224-25, 236, 260, 317
Chalmers, T. A., 68, 94, 153
Chamberlain, Owen, 196
Champion, F. C., 150
Chandrasekhar, Subrahmanyan, 250-52
Childs, Herbert, 127, 133
Churchill, Winston, 237
Cockcroft, John D., 5, 86, 108, 118, 121, 127-28, 135, 140-42, 144-48, 150-51, 155
Cohen, Abraham, 223
Compton, Arthur H., 42, 46, 138, 140, 166, 225, 228, 244, 247, 308, 315
Compton, K. T., 123, 133, 315
Condon, Edward U., 168, 190, 204, 232, 261, 271
Conversi, M., 56
Coolidge, W. D., 115, 117-18, 132, 139
Corbino, Orso Mario, 38-40, 43, 46, 49, 55, 62, 79
Coster, D., 71, 235
Cox, Richard T., 231
Crane, H. R., 150, 165
Cremona, Luigi, 35
Crew, Henry, 231
Crew, William, 231
Critchfield, C., 197
Curie, I., 6, 47-48, 50, 57-58, 67, 147-48, 153, 162, 225, 249
Curie, Marie Sklodowska, 46, 58-59, 256, 280

D'Agostino, O., 52, 54, 166, 169
Dahl, Otto, 128, 139, 230, 313
Danos, M., 97
Darwin, Charles, 225, 229
Daudel, R., 103
Davidson, James, 284
Debierne, A., 280
Debye, Peter, 46, 65, 71, 228
Delbrück, Max, 7, 165, 237, 246-49
Delsasso, L. A., 165, 203-4
Devons, Samuel, 265
Dieke, Gerhard, 219-20, 223

Name Index 327

Dirac, P. A. M., 66, 110, 254-56, 261, 321
Dryden, H. L., 223
Dunning, John R., 54, 160, 278, 309-10

Eckert, J. P., Jr., 227
Eddington, Arthur, 250
Edlefson, N. E., 126, 155
Ehrenfest, Paul, 40, 138, 228-29
Einstein, Albert, 36-37, 42, 138, 165-66, 232, 240, 259, 264, 269-71, 315
Ellis, C. D., 46, 65, 86, 224, 236, 317-18
Elsasser, W. M., 16-17, 30, 160, 176
Enriques, F., 36, 38
Eppler, H. B., 92
Epstein, P. S., 312
Euler, H., 191-92, 194, 205, 308
Eyring, Henry, 279

Fano, Ugo, 284
Farkas, A., 66
Faxén, H., 218
Feather, Norman, 87-88, 94, 224
Feenberg, Eugene, 17, 19, 60, 155, 176-78, 189-90, 192, 203-4, 271, 274
Fermi, Enrico, 10, 14, 19, 21, 34, 37-44, 47-52, 54-57, 59-60, 62, 66, 68-70, 77, 93-94, 108, 164-67, 169-71, 224, 235-36, 243, 252-53, 256, 270, 272-73, 282, 306, 308-9, 317, 322
Fermi, Laura, 6, 77, 272-73, 322
Ferraris, Galileo, 61
Ferretti, B., 56
Feshbach, Herman, 24, 27, 82, 85, 88, 107, 150, 210
Feynman, Richard P., 258
Fisk, J. B., 15, 196
Fleming, J. A., 133-34
Fokker, A. D., 257
Fournier, Georges, 184
Fowler, R. H., 86-87, 110, 321
Fowler, William A., 34, 61-62, 64, 78, 106, 108, 115, 117, 119, 121, 149, 154, 165, 203-4, 209-10, 214, 312-13, 317
Franck, James, 42, 69, 237
Franklin, Benjamin, 315
Frayne, J. G., 134
Frenkel, J., 257, 274

Friedlander, G., 95
Frisch, Otto Robert, 64, 75-78, 94, 237, 272-73
Fröhlich, H., 198
Fuller, Everett G., 90, 97-98, 104
Fuller, Leonard, 126
Fultz, S. C., 99
Furry, Wendell H., 241

Galeotti, Gino, 40
Galilei, Galileo, 35, 38
Galvani, Luigi, 35
Gamow, George, 61, 142-44, 151, 170
Garbasso, Antonio, 39, 62
Geiger, Hans, 66, 166
Gell-Mann, M., 171, 209
Gerlach, W., 66
Gentner, W., 95
Giarratana, J., 271
Gilman, Daniel Coit, 219
Gimm, H., 92
Gleichauf, Paul, 284
Gödel, K., 248
Goeppert-Mayer, Maria, 18-19, 26n19, 26n23, 27, 101, 225, 254, 267
Goldberger, M. L., 221
Goldhaber, Maurice, 14, 29, 82, 90, 102, 104, 106-8, 110, 152-53, 165-66, 189, 236, 321-22
Goldschmidt, Viktor Moritz, 27, 29
Golgi, Camillo, 61
Gordy, Walter, 254
Goudsmit, S., 46
Goursat, E. J. B., 223
Gracco, V., 92
Gray, L. H., 236, 245, 249
Greene, R. E., 281
Greinacher, H., 140-41, 143
Grodzins, Lee, 82
Grosse, A. V., 310
Grotrian, W., 306
Guggenheimer, K., 16, 17, 30, 176, 189, 203
Guliano, Balbino, 55

Hadley, J., 196
Hafstad, Lawrence R., 128, 139, 167-68, 191, 227, 265
Hahn, Otto, 65, 71-72, 169, 174, 266, 272

328   Name Index

Halban, Hans, 70
Hall, Harvey, 231
Halpern, Otto, 231
Hamermesh, Morton, 70
Hardy, James D., 227
Harkins, William D., 19-20, 30, 161
Harnwell, Gaylord, P., 271
Harrison, B. K., 251
Harteck, P., 71, 85, 106, 153
Hartree, D. R., 110
Haxel, O., 28, 101, 267
Hayward, Evans, 97, 104
Hegner, Janette La Tournette, 252, 254
Heisenberg, W., 12-13, 30, 48-49, 60, 89, 162-63, 167, 175, 180, 184-85, 191, 203-4, 236, 239-40, 246, 254-55, 262, 270, 314, 317
Heitler, W., 161-62, 198, 236, 241, 243-45
Helmholtz, Hermann von, 316
Héloise, 249
Henderson, Malcolm, 148, 150
Henkel, K. L., 281
Hepner, W., 195
Herb, R. G., 123, 128, 139
Hertz, Heinrich, 39
Herzberg, G., 161-62
Herzfeld, Karl F., 217, 225-26, 228-29, 254
Hevesy, Georg von, 68-69, 234, 236
Hewish, Anthony, 252
Hewlett, Richard, 316-17
Heydenburg, Norman P., 167-68, 231-32
Heyl, Paul R., 221, 233
Hiebert, Erwin N., 318
Hill, D. L., 267
Himmler, H., 71
Hintz, Norton, 106
Hitler, A., 50, 55, 67, 170, 174, 236, 245
Høffding, Harald, 240
Hoisington, L. E., 190
Holtsmark, J., 218
Houtermans, F., 174
Hubbard, John C., 223
Huby, R., 191
Hulburt, E. O., 137
Hulthén, L., 92
Humphreys, W. J., 224
Hund, F., 191, 224

Hutton, J., 225
Hylleraas, E. A., 260

Inglis, D. R., 191
Irving, David, 316
Ising, Gustaf, 124, 142-43, 147
Iwanenko, D., 48, 184, 255

Jacobsen, J. C., 236, 240, 246, 250
Jastrow, Robert, 196
Jefferson, Thomas, 283
Jensen, J. H. D., 27-28, 99, 101, 267
Johnson, M. H., 197, 206, 231
Johnson, T. H., 125, 130, 131
Joliot, F., 6, 47-48, 50, 57-58, 67, 147-48, 153, 162, 225
Jungk, Robert, 317, 319

Kalckar, Fritz, 23, 237, 268-70
Kamerlingh Onnes, Heike, 138
Kanner, M. H., 271, 281, 309
Kaptiza, P., 145
Kapur, P. L., 271
Kayser, Heinrich J. G., 224
Kellogg, J. M. B., 195
Kelly, M. A., 99
Kelvin, Lord [W. Thomson], 228, 315-16
Kemmer, N., 198-99
Kempton, A. E., 106
Kerst, D. W., 123, 128, 139
Kettering, Charles F., 316
Kiess, C. C., 224
Kirsch, Gerhard, 139
Klaiber, G. S., 95-96
Klein, O., 149, 236, 244
Koch, H. W., 97
Koch, Jørgen, 70, 73
Koffler, Henry, 4-5
Kojoian, Gabriel, 107
Kolster, F. A., 137
Kopfermann, H., 160
Kramers, H. A., 166, 239, 249
Krebs, Hans, 316
Kronig, R. de L., 249
Kröning, M., 92
Kuhn, Beverly, 284

Ladenburg, Rudolf, 271, 281, 309
Lamb, Willis, 77, 312

Name Index 329

Lamphere, R. W., 281
Landau, L., 79, 313
Lange, F., 119-20, 143, 153-54
Laslett, Jackson, 68
Lattes, G. C., 56
Lauritsen, Charles C., 120-21, 128, 132, 149-50, 154, 165, 203-4, 246, 317
Lavoisier, Antoine-Laurent, 316
Lawrence, Ernest O., 5, 68, 106, 112, 117, 122, 124-28, 130-33, 135-36, 138-39, 147-53, 155, 313, 315
Lea, D. E., 93
Lee, T. D., 205
Lefrancois, J., 92
Lehmann, P., 92
Lamaitre, G., 279
Lepretre, A., 96
Levi, Hilda, 237
Levi-Civita, T., 35-36
Levinger, J. S., 100
Lewis, Gilbert N., 84, 147, 150-52, 225
Liebig, Justus von, 316
Lilienthal, David E., 154
Lincoln, Abraham, 318
Link, Arthur, 284
Liouville, Joseph, 279
Litovitz, T. A., 226
Livingston, M. Stanley, 5, 10, 29-30, 112, 120, 126, 131, 147-50, 154-55, 315
Lorentz, H. A., 42, 221, 244, 255
Lowen, Irving S., 231-32
Lowes, John Livingston, 225
Lyell, C., 225
Lyman, Theodore, 313

McCord, R. V., 173
McKeown, M., 96
McMillan, Edwin M., 112, 130, 132-33, 136-37, 140-41, 143-44, 146-49, 151, 153-55, 180, 276
MacPhail, Malcolm R., 271
Majorana, Ettore, 13, 37, 42, 47-49, 54, 57, 89, 185, 187, 306
Malley, Marjorie C., 58, 75-76, 107, 174
Marconi, G., 46, 55
Margenau, Henry, 16
Massey, H. S. W., 87, 236, 257, 311
Mateosian, E. ter, 96
Mauchly, John W., 227

Maxwell, James Clerk, 316, 320-21
May, John, 62
Mayer, Joseph, 225, 254
Mayer, Maria. *See* Goeppert-Mayer, Maria
Meggers, William F., 224
Meitner, Lise, 47, 65-66, 71-73, 75-76, 84, 107, 169, 246, 272-73
Mendeleeff, D., 17
Merkel, B., 92
Michelson, Albert, 228, 313, 315-16
Migdal, A. B., 100
Mihelich, J. W., 104
Miller, L. C., 170-71
Millikan, Robert A., 46, 78, 121, 149, 228, 315
Mitchell, Madeline, 214
Møller, C., 78, 199, 237, 253
Mohr, C. B. O., 265
Moon, P. B., 160
Morley, Edward W., 316
Morris, J., 123, 135
Morse, Marston, 274
Morse, Philip M., 15, 196
Moszkowski, S. A., 101
Mott, N. F., 86, 236, 257, 311
Mottelson, B. R., 24, 102, 268, 281
Motz, Lloyd, 231
Muehlhause, Carl, 96
Mueller, D., 123
Murnaghan, Francis D., 223
Murphy, G. M., 84, 162
Murphy, J. J., III, 92
Murray, Robert T. K., 222, 225
Mussolini, B., 40, 42, 55
Myers, W. D., 100

Nagle, B. C., 92
Neddermeyer, Seth H., 236, 241-42, 244
Ne'eman, Yuval, 171
Neumann, John von, 271
Newcomb, Simon, 316
Newton, Isaac, 64, 110, 320
Nielsen, Walter, 254
Nier, Alfred O., 6, 8, 14, 27, 277, 309
Nishina, Y., 149, 244
Nix, J. R., 275
Nobles, R. A., 281
Nordheim, L. W., 19, 254

Nørdlung, Niels Erik, 250
Nutt, H., 88

Occhialini, G., 39, 42, 55-56
Okamoto, K., 97
Oliphant, M., 90-91, 106, 122, 144, 153
Oort, Jan H., 252
Oppenheimer, J. R., 23, 120, 229, 241, 245-46, 248, 251, 271, 312-13
Orthmann, Wilhelm, 65

Pancini, E., 56
Parker, Herman, 254, 261
Parkinson, D. B., 123, 128
Parravano, N., 46
Partovi, F., 92
Paschen, F., 42
Pasteur, Louis, 250
Pauli, W., 49, 66, 160-62, 168, 190, 214, 236, 239
Peano, Giuseppe, 35
Peebles, J., 100
Pegram, George B., 54, 160
Peierls, Rudolf, 14, 27, 59-60, 79, 89, 94, 110, 151, 163, 167, 180-81, 189, 195, 202-3, 205-11, 236, 271, 319-20
Penney, W. G., 144
Perlman, M. L., 95
Perrin, Francis, 184
Persico, Enrico, 38-39
Pettersson, Hans, 139
Pfund, A. H., 219-20
Phenneger, M. C., 92
Piccioni, Oreste, 38, 56
Pierce, G. W., 138
Placzek, George, 69-70, 72, 75, 237, 276-78, 309
Plantin, Christophe, 234
Pleasonton, Frances, 173
Plesset, Milton S., 237, 248-49
Pochettino, Alfredo, 37
Poincaré, [Jules] Henri, 114, 255
Polanyi, M., 22
Pontecorvo, Bruno M., 51, 54, 166, 169
Porter, C. E., 24
Powell, C. F., 56
Present, R. D., 168, 190
Pringle, G. E., 265

Pringsheim, Peter, 66
Puccianti, Luigi, 37, 42

Rabi, I. I., 66, 195, 230, 232
Racah, Giulio, 62, 306-7
Radicati, L. A., 56
Rainwater, James, 102, 268
Raman, C. V., 44
Ramsey, N. F., Jr., 195
Randall, M., 225
Rarita, W., 27, 195
Rasetti, Franco, 40-42, 44-45, 47, 51-52, 54, 62, 166, 169, 312
Rasmussen, E. H., 236
Rayleigh, Lord [J. W. Strutt], 220, 274, 278, 321
Read, John, 149-50
Regge, T., 56
Reiche, Fritz, 40
Richardson, O. W., 46
Riehn, P., 92
Righi, Augusto, 37
Ringuet, Louis Leprince, 249
Riska, D. O., 94
Ritz, Walter, 260
Robertson, H. P., 232, 271
Robson, J. M., 171
Rodrigo, Enrico, 284
Röntgen, W. C., 114
Roosevelt, F. D., 283
Rosenfeld, Léon, 77-78, 199, 237-38, 240, 258, 272, 276, 282
Rosenthal, Jenny E., 231
Rossi, B., 39, 42, 46, 62, 236, 241
Rowland, Henry Augustus, 219, 313
Roy, P., 92
Rozental, S., 237
Ruark, A. E., 225, 254, 256
Rubens, H., 96
Rumford, Count [Benjamin Thompson], 314-15
Runge, Carl D. T., 224
Rutherford, Ernest, 42, 46-47, 51, 60-61, 64, 78, 85-86, 88, 90-91, 93, 106-7, 110, 113, 115, 118, 121, 128, 132-34, 136, 141-43, 145-46, 148, 151, 153, 161, 217, 224, 235, 237, 260, 269, 280, 317-18, 320-21

Name Index

Salam, Abdus, 208
Salant, Edward, 231
Sale, Wallace, 284
Sandburg, Carl, 318
Santayana, George, 319
Saxton, D., 173
Scharff, Gertrude, 82
Schenkel, M., 140-41
Scherrer, P., 150
Schiff, Leonard I., 15, 196, 312
Schmidt, T., 18-19, 28
Schrödinger, Erwin, 7, 82, 86, 166, 184, 228, 315
Schüler, H., 109
Schwarzschild, K., 257
Schwinger, Julian, 15, 27, 195, 312
Sciasia, Leonardo, 57
Scribner, B. F., 224
Segrè, Emilio, 13, 26n*19*, 34, 42, 44, 51-54, 57-62, 67, 78, 103, 108, 113, 166, 169, 208, 210, 306-8, 316
Serber, Robert, 23, 120, 180, 198, 203-4, 207, 312-13
Share, S., 190
Shin, Y. M., 92
Shrader, E. F., 173
Siegbahn, Manne, 71
Silliman, Benjamin, 315
Simon, A. W., 166
Skopik, D. M., 92
Slater, John C., 103, 166
Slepian, J., 143
Slosson, E. E., 225, 319
Smallwood, H. M., 222
Smith, C. G., 155
Smith, R. K., 281
Smyth, H. D., 134, 271
Snell, A. H., 170-71
Snow, C. P., 320
Soddy, Frederick, 269, 280
Sommerfeld, Arnold, 10, 39, 43, 46, 140, 315-16
Sopka, Katherine, 151, 312, 314
Sperber, Daniel, 284
Spinoza, Benedict, 240
Sproul, Robert G., 126-27
Stark, Johannes, 312
Staub, Hans, 70
Steckel, Rushworth, 221

Steinmetz, Charles P., 221
Steinwedel, Helmut, 99
Stern, Otto, 46, 66-67, 126
Størmer, C., 279
Strassmann, Fritz, 71-72, 169, 266, 272
Strauss, Lewis, 154
Strømgren, Bengt, 250, 252
Strømgren, Elis, 250
Strutinski, V. M., 275
Stueckelberg, E. C. G., 258
Stuewer, Roger H., 29, 314, 319
Suess, Hans E., 18, 27, 29, 100, 267
Sugawara, M., 92
Sunyar, A. W., 82, 101-2
Swann, W. F. G., 131, 133-35, 138-39, 315
Swenson, Loyd, 30, 209, 316
Swiatecki, Wladyslaw J., 100, 275
Szilard, Leo, 7, 68-70, 78, 94, 119, 153, 282

Tamm, I. E., 48
Tarrant, G. T. P., 236, 245, 249
Tate, John T., 3, 6, 133, 214
Taylor, H. M., 217-18
Taylor-Williams, Judi, 284
Teller, Edward, 14, 96, 197, 206, 235, 237, 260, 264, 266, 282
Teller, Mici, 235
Tetrode, H., 257
Thibaud, Jean, 67
Thomas, L. H., 188
Thomson, J. Arthur, 221
Thomson, J. J., 64, 142, 224, 235, 237, 321
Tillman, J. R., 160
Toll, John S., 249
Tolman, Richard, 312
Townsend, J. S., 135
Toynbee, Arnold J., 322
Trabacchi, Giulio C., 58
Trump, John, 151
Turner, Louis A., 169, 271, 276
Tuve, Merle A., 12, 115-20, 123, 125, 128, 130-32, 136-39, 148, 151-53, 167-68, 190-91, 230, 313, 315

Uhlig, R., 97
Urey, Harold C., 5, 84, 162, 225, 254

332 Name Index

Vallarta, Manuel Sandoval, 279
Van de Graaff, R. J., 118-19, 122-24, 128, 135, 139, 150, 230
Van Vleck, J. H., 6, 138, 314
Veblen, Oswald, 271
Veksler, V., 112
Veyssiere, A., 96
Volkoff, George M., 195-96, 251, 312-13
Volta, Alessandro, 35
Volterra, Vito, 35-36, 38-39
Voorhis, C. C. van, 271, 281, 309

Wäffler, H., 92
Wakano, M., 251
Walecka, J. D., 206-7
Waller, Ivar, 236
Walraven, T., 252
Walton, E. T. S., 86, 108, 118-19, 121-22, 128, 130, 135, 140-41, 143-44, 146-47, 150-51
Warburg, Otto, 316
Watson, G. N., 137
Watson, Kenneth, 284
Watson-Watt, Robert, 248
Way, Katherine, 254, 261, 263, 266
Weiner, Charles, 127
Weiss, M. S., 98
Weisskopf, V. F., 24, 101, 194, 270
Weizsäcker, C. F. von, 191, 236, 143, 266, 274, 307-8
Wertenstein, L., 68
Weyl, Hermann, 140, 225, 271, 306
Wheeler, John A., 26n23, 29, 60, 77, 114, 169-70, 202, 213-15, 217-18, 263-64, 267, 306-16, 318, 320-22
Wheeler, Joseph, 282
White, Milton G., 167, 190, 271

Whittenmore, L. E., 137
Wick, Gian Carlo, 56, 66, 196
Wideröe, Rolf, 124-25, 127, 130, 132, 142-43, 146-47
Wiegand, Clyde E., 103, 196
Wigner, Eugene, 13-15, 17-22, 24, 30, 89, 94, 158, 163, 169, 173-77, 180, 183, 188, 191, 203-4, 206-8, 214, 231-32, 253, 262, 264, 269, 271, 279, 282, 307, 310-12, 315-16
Wilkinson, D. H., 100
Williams, Dudley, 254
Williams, E. J., 237-38, 242-44, 307-8, 320
Williams, John H., 8
Williamson, John R., 223
Wills, Lawrence A., 231
Wilson, Elizabeth [Mrs. Charles Baker], 103-4
Wilson, Robert R., 57, 108, 170, 207-8, 214, 217, 306, 308-20, 322
Wilson, T. Woodrow, 270, 283
Wintner, Aurel, 223
Wolfe, Hugh, 231
Wood, Robert W., 119, 219-20, 224, 230, 313, 315
Wooster, W. A., 65
Wyckoff, J. M., 97

Yamanouchi, T., 195
Young, L. A., 178, 189, 191
Yukawa, H., 30, 190, 198, 255

Zacharias, J. R., 195
Zieger, A., 92
Zieger, B., 92, 97
Zunti, W., 150
Zwicky, F., 250-51

Subject Index

Aarhus, University of, 68
Academic Assistance Council of Great Britain, 67
Accelerators: centers of activity by *1936*, 128; importance of, 114-15, 180; linear, 116, 127, 146; pressurized, 123, 128, 151; Rutherford's encouragement of development, 115, 142, 143, 148, 151; as tools for nuclear research, 148-51
— work of: Brasch and Lange, 119-20, 143, 153-54; Breit and Tuve, 116-19, 132-40, 230-31; Cockcroft and Walton, 68, 118, 119, 121-22, 140-47; Herb, Parkinson, and Kerst, 123, 128, 139; Ising, 124, 142, 143, 147; Lauritsen, 120-21, 149-50; Lawrence and Livingston, 124-26, 130-32, 154-55, 235; Van de Graaff, 118-19, 121-24, 127, 132, 135, 139, 150-51, 230; Widerøe, 124, 125, 130, 132, 142, 143, 146, 147
Action at a distance: electromagnetic theory of, 257-59; gravitational, 259
Allison effect, 232
Alpha-decay: M. Curie's ideas on, 280; Gamow theory of, 23, 121, 142, 144, 151

Alpha-particles(s): binding energy of, 13, 192-94; Bohr theory of stopping of, 242-43; detection of, 224; Rutherford's scattering experiments with, 135, 217-18, 260; stability of, 13, 163, 175. *See also* Nuclear models
American Cyanamid Company, 227
American Institute of Physics, conferences on history of nuclear physics, 11, 12, 15, 20, 21, 22, 24, 25n*1*, 30, 113
American Physical Society: meetings, 226, 228, 229, 232, 233, 241, 310; mentioned, 231
Angular momentum. *See* Spin
Antiproton, 34
Argonne National Laboratory, 27, 96
Artificial disintegration: of lithium, 118, 119, 122, 145, 146-47, 235; of nitrogen, 135; mentioned, 122
Artificial radioactivity: discovery of, 50-51, 57, 147, 232; induced by α-particles, 68; induced by neutrons, 51, 68, 235; of potassium, 67
Atomic Energy Commission, 316
Auger effect, inverse, 22

333

334 Subject Index

Baltimore Lectures. *See* Johns Hopkins University
Barnes Engineering Company, 227
Bartol Research Foundation, 135
Berlin: Kaiser-Wilhelm Institut für Chemie, 65; Physikalisch-Technische Reichsanstalt, 58, 65; University of, 6, 66, 82, 84, 85, 112, 119, 128, 227, 315
Beryllium, a-particle bombardment of, 47, 162, 225
Beta-decay: conservation of energy in, 161, 163, 165, 236; continuous spectrum, 65, 75-76; discrete spectrum, 76; early ideas on, 164; Fermi theory of, 49-50, 164, 256. *See also* Neutrino; Parity
Betatron, 116, 136, 139. *See also* Accelerators
"Bethe Bible." *See Reviews of Modern Physics*
Bethe-Goldstone equation, 205
Bethe-Heitler theory, 236, 242-44 *passim*
Big Bang theory, 100, 240
Binding energy: of light nuclei, 264, 265; of nuclei, 163
— work of: Feenberg and Wigner, 17; Weizsäcker, 191, 266. *See also* Alphaparticle(s); Deuteron(s); Nuclear matter
Birmingham, University of, 181
Black hole, 251
Bohr Institute. *See* Copenhagen
Bohr-Kramers-Slater theory, 166
Bose statistics, 161
Bothe-Geiger experiment, 166
Breit-Wigner theory. *See* Neutron(s), resonances
Bremsstrahlung, 243-44, 247
Breslau, University of, 227
Brookhaven National Laboratory, 100, 103
Brown-Boveri Company, 124
Bureau of Standards. *See* U. S. National Bureau of Standards

California, University of, at Berkeley: Radiation Laboratory, 112, 126-28, 147-48; mentioned, 34, 54, 106, 107, 117, 119, 120, 124, 125, 126, 127, 131, 133, 135, 152, 196, 312
California Institute of Technology: High Voltage Laboratory, 115, 119, 120, 128, 132, 149-50; rivalry with Berkeley, 120, 150, 152; mentioned, 95, 106, 117, 152, 153, 154, 248, 312, 317
Cambridge, University of: Cavendish Laboratory, 6, 46, 58, 64, 85, 86, 88, 91, 100, 106, 108, 109, 110, 112, 121, 122, 128, 134, 135, 139, 141, 142, 144, 164, 189, 224, 237, 320, 321; Churchill College, 145; Mond Laboratory, 145; mentioned, 14, 29, 51, 146, 250
Capillarity, 274
Carlsberg Foundation, 250
Carnegie Institution: Department of Terrestrial Magnetism, 12, 116-19 *passim*; 128, 132-36, 138-40 *passim*; 142, 227, 230; mentioned, 95, 115
Cartan's principle, 223
Causality, principle of, 249
Cavendish Laboratory. *See* Cambridge, University of
Chain reaction: conceived by Szilard, 153; first, 170; German attempts to achieve, 174
Channel number. *See* Nuclear fission; Nuclear reactions
Charge independence. *See* Nuclear forces
Charge symmetry. *See* Nuclear forces
Chicago, University of, 27, 133, 135, 138, 139, 161, 174, 270
Chicago World's Fair of *1933*, 237
Circulation theorem, 220
Clustering theory. *See* Scattering theory
Cockcroft-Walton accelerator. *See* accelerators
Columbia University, 14, 21, 54, 174, 232, 235, 273, 278, 282, 310
Como conference of *1927*, 42
Complementarity, principle of, 239-40
Compound nucleus: Bethe's application, 23; Bohr theory, 20, 22-23, 26n28, 69, 176, 194, 253-54; Breit-Wigner theory, 20-22, 175-76, 253-54; Harkins's speculations, 19-20, 30; mentioned, 265, 266, 268-70. *See also* Nuclear models, liquid drop
Compton effect, 42, 149-50, 166, 243, 245, 247

Subject Index 335

Compton-Simon experiment, 166
Computers: development of, 227; early lack of, 263
Conferences, historical, role of, 318-20
Conferences, international. *See* International conferences
Congrès International de l'Electricité, Paris, *1932*, 48-49
Copenhagen: Amalienborg Palace, 234; Bohr Institute, 20, 68, 69, 71, 72, 112, 186, 234-54 *passim*, 267, 273, 308, 309
Cornell University, 10, 29, 123
Correlations, nucleon-nucleon, 192-95
Cosmic ray(s): Compton-Millikan debate, 228; Fermi lectures on, 270; mini-showers, 247-48; orbits in earth's field, 279; penetrating radiation in, 236, 241-45, 308; postwar research in Italy, 56; primary, nature of, 308-9; showers, 234, 236, 244-45. *See also* Meson
Coulomb force, 12, 13, 185, 189, 204
Crab Nebula, 252
Cross section: deuteron absorption, 70; deuteron, photodisintegration of, 14, 89, 94, 189; gamma-neutron reactions, 95-99; neutron absorption, 54, 167, 169, 253, 309; neutron capture, 189, 253; neutron-proton scattering, 14; neutrons, thermal, 167; scattering, 166-67, 248-49, 284. *See also* Nuclear fission
Cyclotron: name coined, 127; near invention of, 155; mentioned, 68, 107, 108, 116, 180. *See also* Accelerators

Delbrück effect, 246, 249
Department of Terrestrial Magnetism. *See* Carnegie Institution
Depression, 226
Determinism, Einstein's belief in, 240
Deuterium, discovery of, 84, 162, 235; mentioned, 90
Deuteron(s): Bethe and Peierls theory of, 14, 89; binding energy of, 13, 106, 163-64, 188, 211, 236; composition of, 84, 90; instability of, presumed, 106-7, 151-53; photodisintegration of, 14, 84-94, 165, 189, 236; photodisintegration rate, 100; as projectiles, 106,

120, 147, 150, 151, 155, 165; quadrupole moment of, 27, 177, 195, 264; singlet and triplet states of, 14, 168, 174, 264; stability of, 163, 175
Diatomic molecules, band spectra of, 224
Dirac-Gordon equation, 149
Dirac Theory of electron, 66, 254-56
Direct interaction. *See* Scattering theory
Direct interparticle interaction, Wheeler-Feynman theory of, 258-59
Discovery, coincidence in, 114, 122, 126, 145
Disintegration, artificial. *See* Artificial disintegration
Dispersion theory: applied to nuclear scattering, 248-49; applied to pair theory, 249; mentioned, 226
DNA, 7
Doppler effect, 70
Dubna Laboratory, 322
Duke University, 254

Edinburgh, University of, 94
Effective-range theory: applied to photodisintegration of deuteron, 90, 189; mentioned, 14, 190, 195
Electron(s): as basic particle, 254-55, 311; in cosmic-rays, 241-45; Dirac theory of, 66, 254-56; dispersion of, 223-24; interaction between, 230; as nuclear particle, 254-59, 311-12; self-energy of, 186; theory of matter, 255. *See also* Nuclear models; Pair theory; Positron(s)
Elementary particle, concept of, 171, 180, 207-10
Energy, conservation of: Bohr questions, 161, 163, 165; Einstein's belief in, 165-66; experimental support for, 165; in nucleus, 185
England, physics in, 320-21
Enoch Pratt Library, Baltimore, 222
Equivalent radiation field method. *See* Williams-Weizsäcker method
Ethiopian war, 54
Exchange forces. *See* Nuclear forces
Exhibition Scholarship, *1851*, 144

Fascism, 40, 55-56
Federal Telegraph Company, 126

336 Subject Index

Fermi National Accelerator Laboratory, 322
Fermi statistics, 23, 42, 161, 187, 192, 307
Field theory, gravitational, 259
Fine structure: constant, 186; of lithium ion, 233
Fission. *See* Nuclear fission
Florence, University of, 39, 40, 42, 62
Forces, nuclear. *See* Nuclear forces
Fourier analysis. *See* Tide analyzer; William-Weizsäcker method
Franck-Hertz experiment, 42
Fusion. *See* Nuclear fusion

Gamma-rays: anomalous scattering of, 236, 245-48; as nuclear radiations, 76, 103-4; from radiothorium, 86; very high energy, 95, 128
Gamow conference of *1939*, 170
Geiger-Müller counters, 47, 148
General Electric Company, 115, 117, 139
General Motors Corporation, 227, 316
Generator, impulse, 122
Generator coordinates, method of, 261
George Washington University, 235
Germany, political situation in, 7, 11, 50, 67, 71, 86, 120, 153, 170, 174
Gödel's theorem, 248
Goldhaber-Teller model. *See* Resonances, giant dipole
Göttingen, University of, 40, 235
Gravitation, constant of, 221

Hafnium, discovery of, 235
Hamburg, University of, 27, 66, 67, 152
Hartree-Fock method, 17, 191, 194
Harvard University: Cruft Laboratory, 138; mentioned, 103, 154
Harwell, 64
Heat exchanger, design of, 223
Heaviside layer, 230
Heisenberg uncertainty principle, 160
High Voltage Engineering Corporation, 124, 139
Hyperfine structure, 44

Illinois, University of, 100
Institut du Radium. *See* Paris

Institute for Advanced Study, Princeton, 261, 271
Interaction: many-particle, 21; proton-neutron, 168; proton-proton, 168
Interaction kernel. *See* Scattering theory, velocity dependent force
Internal conversion, 76, 103-4
International Conference of Physics, London and Cambridge, *1934*, 235-36, 308
International conferences, 261, 296n214. *See also* Como conference; Congrès International de l'Electricité; Gamow conference; International Conference on Physics; Rome conference; Solvay conference
Invariance principles, Einstein's belief in, 166
Isotopic spin: conservation of, 202; independence, 168; notation, 186, 204; mentioned, 158, 162, 186-87, 199
Italian Academy, 55
Italian National Research Council, 55
Italy: career of Fermi in, 37-56; political situation in, 7, 11, 40, 54-56; science in, 35-40, 61-62

Johns Hopkins University: Baltimore Lectures, Lord Kelvin's, 228, 315, 316; Hospital, 227; New Plan, 222-23; research, teaching, studying at, 219-29; mentioned, 134, 137, 138, 139, 254

KDKA broadcasting station, 221
Kensington Museum of Science, 155
Klein-Nishina formula, 149-50, 244
Kolster decremeter, 137
Krypton, 158, 159

Laboratoire de Physique des Rayons X. *See* Paris
Lawrence Museum of Science, 155
Lecher wires, 134
Leipzig, University of, 246
Lenard rays, 139
Level density; observations on, 24; problem of, 22, 23
Leyden, University of, 40, 138
Library of Congress, 136

Liouville's theorem, 279
Liquid drop model. See Nuclear models
London, University of: Birkbeck College, 67; mentioned, 235
London International Conference of *1934*. See International Conference on Physics
Los Alamos Scientific Laboratory, 10, 24, 34, 64, 181

Magic nuclei, 16, 29
Magic numbers, 27-29
Magnetic moments. See Neutron(s); Nuclear magnetic moments; Proton(s)
Manchester, University of, 14, 237
Marx circuit, 115, 119
Mass spectrometer, separation of uranium isotopes with, 309-10
Massachuesetts Institute of Technology: Radiation Laboratory, 10, 150; mentioned, 101, 123
Mathematics, Fermi's use of, 59-60; in Italy, 35-36, 61
Medicine: infrared radiation in, use of, 227; Lawrence's interest in, 117; nuclear physics in, 54-55, 95, 108, 117, 149
Meson: discovery of, 198, 241-44, 308; lifetime of, 89; resonance, 91. See also Nuclear forces
Michelson experiment, 228, 315, 316
Michigan, University of, 54, 203, 270, 314, 315
Migration, intellectual, 6-7, 11, 71
Minnesota, University of, 3, 6, 7, 8, 34, 133, 134, 135, 138, 139, 214, 277, 309, 314
Mirror nuclei: β-decay energies of, 271; mentioned, 203, 204
Models, nuclear. See Nuclear models
Moderator: graphite as, 70; heavy water as, 70; solid carbon dioxide as, 71
Molecular beams, 232
Monte Carlo method, 52
Morse potential. See Nuclear forces, repulsive core in

National Bureau of Standards. See U.S. National Bureau of Standards
National Research Council Fellowships, 134, 138, 152, 217, 227, 229, 231, 248, 287n58
Neutrino: helicity of, 82; Pauli hypothesis of, 49, 66, 160, 161, 162, 168, 236; mentioned, 49, 164, 165. See also Beta-decay; Parity
Neutron(s): as composite particle, 175; decay of, 88-89, 164, 170-71; delayed, 86; discovery of, 12, 47-48, 84, 89, 162, 164, 170, 184, 186, 225, 235, 255; as elementary particle, 88, 184, 185, 207-8; existence of, speculations on, 60-61, 93, 161; magnetic moment of, 70; mass of, 86, 88-89, 106, 153, 164, 165, 205; polarized, 70; production by d-d reaction, 271; resonances, 22, 24, 70, 158, 169, 175, 231, 264, 270, 271; slow, 51-54, 69-70, 92-94, 109, 166-67, 189, 236, 252-53; source, 52-53, 68. See also Cross section; Nuclear fission; Nuclear forces; Nuclear models; Resonances, nuclear
Neutron stars, 206, 250-52, 313
New York, State University of, at Buffalo, 230
New York University, 139, 230-34 *passim*, 247, 260
North Carolina, University of, at Chapel Hill, 233, 252, 254-70 *passim*
Nuclear binding. See Binding energy; Scattering theory
Nuclear decay constants, 269
Nuclear energy, speculations on, 78-79
Nuclear fission: applications of, 174, 282-83; barrier, 274-81 *passim*, 309; Bohr-Wheeler theory of, 169, 218, 273-74, 309; channel formula, 279-81; cross section, 273, 276-78, 281, 309; discovery of, 71, 77-78, 169-70, 266, 272-73; fragments, 72-73, 169, 273; Frisch-Meitner explanation of, 71-72, 272; "heat of condensation" in, 276-77; origin of term, 72; possibility of predicting, 266; spontaneous, 281; width, 278-81
Nuclear forces: charge independence of, 18, 178, 189-91, 198-99, 202, 203-4; charge symmetry of, 162, 167, 191, 203-5; difficulty in determining, 264;

Subject Index

exchange, 13, 162, 185-86, 187, 188, 196, 262-64; field concept of, 197; as fundamental concepts, 209-11; many-body effects in, 186, 187, 197, 205-6; meson theory of, 197-200, 202-3, 206, 255; neutron-proton, 94, 189; range of, 15, 188, 192, 255; repulsive core in, 15, 186, 187, 196-197; saturation of, 13, 15, 162, 173, 185-86, 187, 188, 192, 195-96, 197, 231, 263; sigma, 205; spin dependence of, 188; static, 197; Strong-Force Credo, 255-58; tensor, 27, 177, 195-96, 199; two-body, 197-98, 199; velocity-dependent, 186, 187; Wigner, 15, 164, 188, 196; Yukawa, 191, 255; zero-range, 188. *See also* Scattering theory
Nuclear fusion, 78-79, 284
Nuclear isomers, 82, 100-104, 283
Nuclear level width, 269
Nuclear magnetic moments, 66, 230, 256-57, 266, 311
Nuclear masses, Weizsäcker semiempirical formula for, 191, 266
Nuclear matter: binding energy of, 191; as ideal neutron gas, 251
— calculations of: Euler, 191-93, 205; Walecka, 206-7
Nuclear models: α-particle, 184, 260, 265-66; cloudy crystal ball, 24; collective, 267-68, 277; electron-proton, 12, 44-45, 84, 135, 160, 184-85, 186; liquid drop, 23, 71, 194, 208-9, 253, 265, 268-70; neutron-proton, 12-15, 31, 48-49, 175, 184-86, 255; shell, 15-19, 27-30, 101, 176-77, 194, 197, 267; single-particle, 21, 176, 266, 267; statistical, need for, 264-65. *See also* Compound nucleus; Electron(s); Nuclear fission
Nuclear physics: beginning of, 11, 113-14; influence of other fields on, 173; influential books on, 317-18; nomenclature used in, 209; theory and experiment in, 208; transition of Rome group to, 43-46
Nuclear reactions, channel analysis of, 279-81
Nuclear reactor, design of, 223

Nuclear rotations, 260, 265-68
Nuclear statistics: of nitrogen, 44-45, 161, 162, 184, 256; of oxygen, 44-45; mentioned, 185, 257. *See also* Bose statistics; Fermi statistics
Nuclei: exotic, 283-84; heavy, structure of, 163, 164, 185, 189; light, bound excited states of, 101; rotating, 283; stability of, 167; superheavy, 283; very unstable, 283
Nucleus: as condensed matter, 218; quadrupole moment of, 268

Oak Ridge National Laboratory, 170
Optics, at Johns Hopkins University, 219-24
Oxford, University of, 123, 135, 181, 321

Padua, University of, 39, 62
Pair theory, 233, 244-49 *passim*, 256. *See also* Electron(s); Photon(s); Positron(s)
Palermo, University of, 54
Para-hydrogen, 66
Paris: Institut du Radium, 6, 58, 67; Laboratoire de Physique des Rayons X, 67
Parity, nonconservation of, 231-32, 288n73. *See also* Beta-decay; Spin
Paschen-Back effect, 42
Paschen series, 42
Patents, importance of, 116
Pauli principle, 28, 190, 194
Perturbation theory, use of, 192, 194, 198, 205
Phase shift analysis. *See* Scattering
Photoelectric effect: nuclear, 89-90, 95, 100; ordinary, 247. *See also* Deuteron(s)
Photons, collision of, 233, 244-45. *See also* Pair theory
Physics: change in character of, 107-8; completeness of, 186; instrumentation in, 228; relationship of theoretical to experimental, 322; spirit of, 159, 160, 183, 319
Physikalisch-Technische Reichsanstalt. *See* Berlin
Pion, discovery of, 56
Pisa: Scuola Normale Superiore, 37-38; University of, 39, 42, 62

Subject Index 339

Pittsburgh Veta Grande Mining Co., 221
Plutonium, fission of, 169-70, 275-76
Positron(s): annihilation radiation from, 67; discovery of, 235, 242
Poulsen arc, 126
Princeton University: Journal Club, 272; mentioned, 77, 118, 119, 122, 123, 124, 133, 135, 138, 139, 165, 169, 170, 173, 177, 214, 221, 231, 232, 258, 270-84 *passim*
Project Matterhorn, 249
Proton(s), magnetic moment of, 66
Pulsars, discovery of, 252

Quantum electrodynamics: challenges to, 234, 241-48; Feynman's contributions to, 258; renormalization technique, 246
Quantum mechanics, reception in the United States, 222, 225; in Italy, 36
Quantum theory: influence on Meitner, 75-76; and nuclear physics, 89; of observation, 259
Quark hypothesis, 180, 210

Rabi resonance method, 67
Racah coefficients, 306-7
Radioactivity: course at Johns Hopkins, 224; discovery of, 113-14; theory of, 280; mentioned, 318. *See also* Artificial radioactivity
Raman spectra, of oxygen and nitrogen, 44
"Rashomon effect," 83
Rask-Ørsted Foundation, 68
Rayen High School, 221
Raytheon Company, 155
Reactions, nuclear. *See* Nuclear reactions
Relativity, reception in Italy, 36
Repulsive core. *See* Nuclear forces
Research and discovery, compared, 6
Resonance(s): giant dipole, 82, 96-100; nuclear, 232-33; radiation, 219-20, 230. *See also* Neutron(s), resonances
Resonating group structure. *See* Scattering theory
Reviews of Modern Physics: "Bethe Bible," 10, 14, 15, 16, 22, 23, 29, 160, 168, 169, 175, 203, 269; Breit article on dispersion and absorption, 230; founded by Tate, 3, 6, 214

Rockefeller Foundation Fellowships, 315-17
Rome, University of: Physics Institute, 40-41; mentioned 14, 34, 35, 38-43, 51-54, 61-62, 69, 93, 112, 165, 167
Rome conference of *1931*, 46-47
Rotations, nuclear. *See* Nuclear rotations
Royal Society of London, 235
Russell-Saunders coupling, 17

Scattering: α-α, 217-18, 233, 252, 260, 265-66, 321; inverse, 233; length, 190; neutron-proton, 14, 188; of neutrons by light elements, 271; of neutrons by para- and ortho-hydrogen, 14; nucleon-nucleon, 231; proton-proton, 12, 167-68, 190; shape elastic, 22. *See also* Cross section
Scattering theory: direct interaction, 261-62; exchange interaction, 261-62; phase shift analysis, 196-97, 202, 218, 233, 260-67 *passim*; scattering-and-reaction matrix, 261-62, 270; velocity-dependent force, 261-63
Schenkel circuit, 140, 141
Scuola Normale Superiore. *See* Pisa
Self-energy, of point particle, 257-58
Shell Model. *See* Nuclear models
Signum Romanum, 53-54
Slepian circuit, 143, 144
Slow neutrons. *See* Neutron(s)
Smithsonian Institution, 145
Solid state physics, 235
Solvay conference: of *1913*, 280; of *1930*, 49; of *1933*, 49-50, 59, 152
Southern California Edison Company, 149
Space quantization, 229
Spectrograph, bismuth-crystal, 47, 57
Spectroscopy: at Johns Hopkins University, 219-24; in Rome, 42-43
Sperry Rand Corporation, 227
Spin: of N^{14}, 256; nuclear, 67, 86, 108-9, 229, 230, 256-57. *See also* Beta-decay; Isotopic spin; Nuclear forces; Parity
Spin-orbit coupling, 19, 26n23, 28, 29, 177, 197, 264
Stark effect, 312
Stationary states, nuclear, 163, 183

340 Subject Index

Statistics. *See* Bose statistics; Fermi statistics; Nuclear statistics
Stellar evolution, 251
Stern-Gerlach method, 66
Stockholm, Siegbahn's laboratory [Forskningsinstitutet för fysik], 71, 72
Super-multiplets, 177
Supernovae, 250-52
Symmetry: Einstein's belief in, 166; time-reversal, 262; Wigner emphasizes, 18, 24
Szilard-Chalmers method, 68, 94-95, 153

Technetium, 54-55, 103-4, 108
Tensor forces. *See* Nuclear forces
Tesla coil, 115, 117-18, 120, 122, 127, 131, 132, 135
Thomas-Fermi approximation, 191
Three-body problem, 260
Tide analyzer, 221, 233
Transformer, cascade, 120-21, 136, 149
Transuranic elements, 51, 73, 169

U.S. National Bureau of Standards, 90, 97, 104, 137, 221, 224, 227
United States: foreign policy of, 238; physics in, 214, 306, 312-17
Uranium: Fermi's experiments on, 51; fissionable isotope, 276-78, 309-10
Ursinus College, 227

Vacuum, properties of, 254
Van de Graaff generator. *See* Accelerators
Variational method, 260
Very Large Array, 34
Vienna: [Institut für Radiumforschung], 58, 139; University of, 65

Washington, University of, 181
Wave equation, separation of variables in, 229
Wave functions: amplitude-phase method for determining, 233; Coulomb, 233; nuclear 264
Weisskopf-Porter-Feshbach theory, 24
Weizsäcker semiempirical mass formula. *See* Binding energy; Nuclear masses
Wentzel-Kramers-Brillouin [WKB] method, 233
White dwarfs, mass of, 250-51
Widths. *See* Nuclear level width
Wigner forces. *See* Nuclear forces
Williams-Weizsäcker method, 242-44, 248, 307-8
Wisconsin, University of, 139, 158, 230, 262
Works Progress Administration, 233

X-rays, discovery of, 114

Yale University, 125, 135, 227, 230, 231, 315

79585

QC
773
S95
1977

SYMPOSIUM ON THE HISTORY OF NUCLEAR PHYSICS

DATE DUE

GAYLORD　　　　　　　　　　PRINTED IN U.S.A.